BOGATIN'S PRACTICAL GUIDE to TRANSMISSION LINE DESIGN and CHARACTERIZATION for SIGNAL INTEGRITY APPLICATIONS

ERIC BOGATIN

E-ISBN: 978-1-63081-692-6
Cover design by John Gomes

© 2020
Artech House
685 Canton Street
Norwood, MA 02062

ARTECH HOUSE

BOSTON | LONDON
artechhouse.com

Table of Contents

Chapter 1 What Are Transmission Lines and Why You Should Care

Every interconnect, from the 1 mm long leads in an IC package, to the 10 km long coax cables buried underground distributing cable TV, and everything in between, are transmission lines.

In most modern products, the electrical properties of interconnects are important and can make or break a product. Success will depend on the design and final verification (characterization) of the interconnects as transmission lines.

For a quick introduction to the purpose of this book, watch this video.

1.1 Do We Really Need Another Transmission Line Book?

Every traditional electrical engineering undergraduate program teaches electrical circuit analysis with resistors, inductors, and capacitors (RLC), referred to as *lumped circuit elements*. A typical RLC circuit, easily understood by all electrical engineers, is shown in **Figure 1.1**.

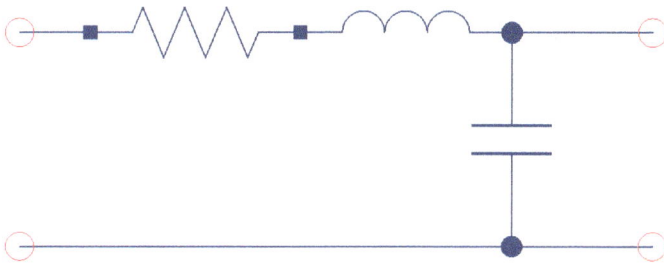

Figure 1.1 Typical RLC circuit with lumped circuit elements.

An engineer trained in electrical engineering will usually be armed with the ability to analyze simple lumped circuits. But this is inadequate to design, characterize, and analyze interconnects that behave as transmission lines.

When the topic of transmission lines is introduced in undergraduate or graduate electrical engineering classes, they are described in mathematical detail which, while absolutely correct, provides little physical insight.

Many classes that introduce transmission lines usually include them in the context of radio frequency (RF) designs. This means the analysis is often in the frequency domain. Any insight gained in the frequency domain behavior is difficult to translate to high-speed digital applications where performance is often analyzed in the time domain.

After taking a typical class, if an electrical engineering student ever thinks again about transmission lines, often with a bit of fear; all they can think about is the following equation, often referred to as "the transmission line equation" describing the input impedance of a uniform, lossless transmission line in the frequency domain:

$$Z_{in} = Z_0 \frac{Z_L + jZ_0 \tan(\beta l)}{Z_0 + jZ_L \tan(\beta l)} \tag{1.1}$$

where

Z_{in} = the input impedance of the transmission line

Z_0 = the characteristic impedance of the transmission line

$j = sqrt(-1)$

Z_L = the impedance of the load at the far end

β = the propagation constant of the transmission line = $(2\pi f)/v$, where f is the frequency and v is the speed of a signal on the transmission line

l = the length of the transmission line

While this relationship is absolutely correct, it provides little insight into what a signal sees in the time domain as it propagates down a transmission line.

Nowhere do we learn in traditional electrical engineering education any physical insight into how a signal interacts with a transmission line in the time domain, how the physical design of the interconnect affects its electrical properties, or what features we should adjust to optimize a design.

Using what you learn in an undergraduate or graduate course on transmission lines, you are hard pressed to answer questions like:

1. What is the initial input impedance you would measure in a transmission line open at the far end or shorted at the far end?

2. What is the initial voltage launched into a 50-ohm transmission line driven by a 10-ohm output impedance driver?

3. What does it mean to have a 50-ohm transmission line or a 75-ohm transmission line?

4. What would you measure, using an ohmmeter, at the front of a 50-ohm transmission line between the signal and return path?

5. If you launch a 1V signal into a 50-ohm transmission line, 20 mA will flow into the transmission line. When will 20 mA flow out the return path?

6. If you double the length of the transmission line, what happens to its characteristic impedance?

7. When a signal encounters a branch from one to two 50-ohm transmission lines, what impedance will it see?

Using what you learn in a course on transmission lines, you are not armed with the skill to decide between the trade-offs of using a microstrip or stripline geometry, as shown in **Figure 1.2**.

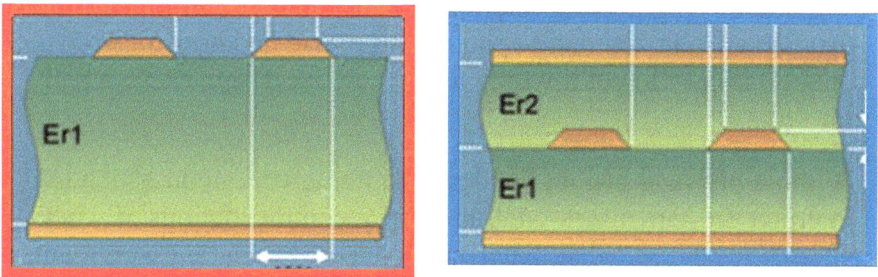

Figure 1.2 An example of a microstrip cross section and a stripline cross section, as shown in the Polar Instruments SI9000 tool.

It is *insight* or *engineering intuition* that we use in the design process. It is our intuition that we use to base our expectations and evaluate whether results are reasonable or unreasonable. It is our engineering judgment that we use to rapidly evaluate and select different design approaches.

Many otherwise experienced engineers will argue who needs intuition when we have a numerical simulation tool or an instrument where we can push a button and get an answer?

Especially in this era of machine learning where a software tool learns what input patterns create what output patterns, does understanding at a fundamental level really matter?

Using a numerical simulation tool is easy. Getting a result is almost guaranteed. But the result is more likely to be wrong than correct. There are far more paths to a wrong answer using a numerical simulation tool than paths to get an answer that is meaningful and useful.

> *Without the physical insight of what to expect in a simulation, we have no immediate way of checking how reasonable is the answer.*

It is just as easy to screw up a measurement as a simulation. Without the engineering intuition of what to expect, of the artifacts that can arise in simulation and measurement and how to avoid them, it is easy to get it wrong and send your design down a path for failure.

Without physical insight you have no framework upon which to interpret the sometimes-contradictory recommendations of what geometries to use, what materials to select, or what measurement approach is the right one.

The most common answer to all engineering questions is "it depends." It is by applying our engineering judgment and the full spectrum of analysis tools such as rules of thumb, analytical approximations, and numerical simulation tools and measurement characterization instruments that we answer "it depends" questions.

This video ebook is the missing manual on how to think about transmission lines from a physical and engineering perspective. There are enough books that provide the mathematical foundations. This is not to imply there is no math in this book and just hand-waving concepts.

The physical world seems to behave as described by Maxwell's equations. Every experiment ever done, every measurement ever done, is fully consistent with the predictions of Maxwell's equations, with no exceptions.

To as high a level of accuracy as anyone has measured, the real world is consistent with Maxwell's equations. If we add quantum mechanics to the principles with Maxwell's equations, which is called quantum electro dynamics (QED), we end up with a model of the real world that matches predictions to measurements to an astounding nine digits.

Signals interact with interconnects based on their electromagnetic properties. This is illustrated in **Figure 1.3**.

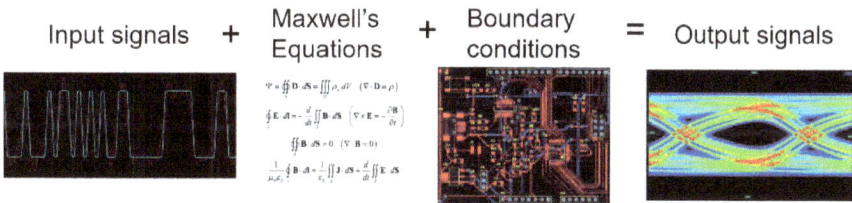

Figure 1.3 Signals interact with interconnects based on Maxwell's equations. But to analyze real problems we have to use selective simplification to apply our engineering intuition to solve real problems.

Maxwell's equations are ultimately the basis of how electromagnetic fields interact with the boundary conditions of the geometry and material properties of the interconnects. But we are not going to get to an answer quickly by attacking every problem starting out by solving Maxwell's equations.

In this video ebook, we use *strategic simplification* to distill these mathematical properties into a few *essential principles* and simplified *rules of thumb* and *approximations*. Where appropriate, we verify the approximations and the *strategic simplification* using numerical simulations and measurements, done correctly.

> *Our physical intuition is based on these essential principles.*

Many of the operations we perform when dealing with transmission lines, whether it is solving design problems or performing or analyzing measurements, are dynamic. They are difficult to convey just in words on a flat, static page.

As the missing manual, this book also contains links to videos that show dynamically many of the processes that are part of the design, characterization, or analysis process.

We are very selective in using a few commercial tools popular in the industry that will help us get to the answer faster. Where suitable, we also introduce free or open source tools that anyone can use. While it is generally true, you get what you pay for, some applications can be adequately tackled with free tools.

If you touch on interconnect design, fabrication, or characterization and have to make decisions about interconnects, this is the missing manual you can use to finally understand, in a *strategically simplified* way, how signals propagate and interact with interconnects and how the physical design of transmission line structures will impact performance.

1.2   All Interconnects Are Transmission Lines

All interconnects are transmission lines, always, no exceptions. A transmission line is composed of two conductors, a signal and a return path. Always, no exceptions. The properties of a transmission line depend as much on the return path as the signal path. **Figure 1.4** illustrates these two elements of every transmission line.

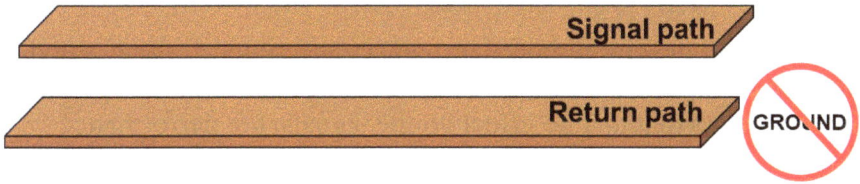

Figure 1.4 The two elements of every transmission line: a signal path and a return path.

One conductor is called the signal path. While it is common to refer to the second conductor as ground, it is important to get in the habit of calling the second conductor the *return path*. The return path does not care what is its DC voltage. The return path conductor could be at 0 V, 3.3 V, 5 V or even 12 V.

When we look at a circuit board, for example, we see a bunch of traces on the top and bottom surfaces. After all, we generally can't see the buried traces, only the surface traces. **Figure 1.5** shows an example of a close-up of a circuit board with surface traces.

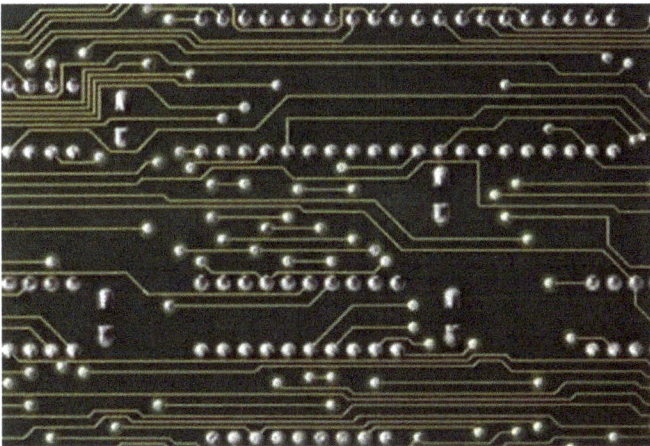

Figure 1.5 An example of the surface traces we see on the outer layers of any circuit board.

The return path for each transmission line is the plane on the adjacent layer in the board. In principle, it doesn't matter what the DC voltage of the plane is, just that it is an adjacent conductor.

19

The wires in the circuit board provide connectivity. If the ICs and passive components could be connected together with transparent interconnects that had no effect on the signals, the performance and the noise of the system would be determined only by the active devices.

Once connectivity is established by the interconnects, the real-world physical interconnects of copper traces, copper planes, dielectrics, and via structures can only screw up the signals and introduce more noise in addition to the intrinsic noise from the ICs.

The fundamental purpose of interconnect design is to limit to an acceptable level how much the interconnects screw up signals.

In addition to providing connectivity, the *only thing* interconnects can do is introduce noise. This will be in the form of *self-aggression* noise, when the interconnect a signal travels on distorts the signal, or *mutual-aggression* noise, when one signal propagating on an interconnect generates noise on another interconnect. This is also referred to as *cross talk*.

Interconnect design is about controlling the self-aggression and mutual-aggression noise to an acceptable level.

Understanding how the physical design of transmission lines affects the electrical properties will enable us to control the noise generated and keep it to an acceptable level, balancing the trade-offs of cost, schedule, and risk. The better we understand the root cause of the problems and how the physical design affects it, the faster we will come to an acceptable design that meets the cost and schedule targets.

1.3   The Importance of Measurement or Characterization

Designing the interconnects to reduce self and mutual aggression noise problems is the first step. It is always a balance between the performance you gain and the cost in terms of *dollars, schedule,* and *risk*. This is sometimes referred to as the *bang for the buck*.

One source of risk is confidence in your ability to accurately predict the performance of the product before you build it. How well does your analysis really include all the effects as predicted by Maxwell's equations, how well do you know all the geometry features in the final product, and how well do you really know the actual material properties?

Finally, how well does the final product actually match the design specification? Between the design sign-off and holding the final product in your hand are all the manufacturing implementation issues. Just because the design looks good on paper (or on a screen) does not mean the as-fabricated product matches the design.

This is where a measurement of the final product or of the manufacturing process is important to verify the product meets the performance spec. When it comes to the interconnects, one of the most important metrics that affects the final performance is the instantaneous impedance profile of the interconnect the signal would see. This is the primary information obtained in a time domain reflectometer (TDR) measurement.

For example, **Figure 1.6** shows an example of the measured impedance profile of an HDMI cable differential transmission line. From this measurement, the characteristic impedance of the transmission line and its uniformity can be extracted. In this example, the cable's differential impedance is 117 ± 2 ohms.

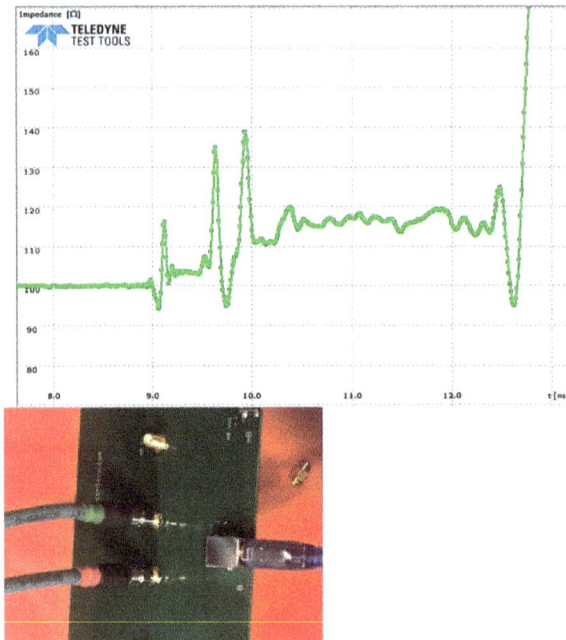

Figure 1.6 The measured differential impedance profile of an HDMI cable connected to a PCB break out board, measured by a TDR with a 50 psec rise time. Inset is the launch to the cable assembly that was measured.

An important element of this book is interpreting this sort of measurement to verify the performance of a transmission line and identify how well it matches a performance spec or where its weak links might be.

This video ebook is about establishing the foundations for designing and characterizing the electrical properties of interconnects in terms of their transmission line properties.

1.4   When Are Interconnects Not Transparent?

In some applications, the electrical properties of the interconnects are not important. In these cases, we say the interconnects are *transparent* to the signals and any design will probably work.

If your product can be built using a solderless breadboard with wires going every which way, such as shown in **Figure 1.7**, and it works just fine, the electrical properties of the interconnects are not important.

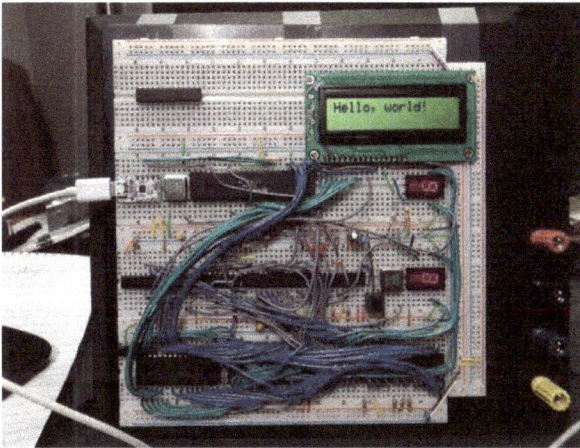

Figure 1.7 An example of a solderless breadboard digital circuit prototype.

The impact on electrical performance of a product from the interconnects and whether the interconnects are transparent depends on the combination of three parameters:

1.   The rise time of the signals

2.   The length of the interconnects

3.   The amount of noise acceptable

With a long enough rise time or short enough interconnects, all interconnects are transparent to signals. Generally, as rise times decrease, the first problem that breaks a product is too much cross talk between an aggressor and victim line. This is due to the mutual inductance between the signal-return paths and the di/dt in the aggressor.

The mutual inductance noise between an aggressor and victim line is roughly:

$$V_{noise} = M \frac{dI}{dt} = k \times 10 \frac{nH}{in} \times \text{Len[in]} \times \frac{V_{sig}}{50\,\Omega \times \text{RT[n sec]}} \quad (1.2)$$

where:

V_{noise} = the inductively coupled voltage noise in V

M = the mutual inductance between the aggress and victim signal-return paths

dI = the change in current ramp

dt = how long it takes for the current to increase

k = the coupling coefficient between the self-inductance and mutual inductance of a signal-return path pair, which might be on the order of 0.75 for a tightly coupled pair of loops

10 nH/inch = rough estimate of worst-case loop inductance of a signal-return path interconnect, per inch of path over each conductor

Len = the length of the interconnect path, in inches

V_{sig} = the signal voltage, in V

50 ohms = a rough estimate of the load impedance in the circuit

RT = the 10-90 rise time of a signal, in nsec

In most digital applications, an amount of cross talk less than 15% of the signal swing might be acceptable. This means the condition for the interconnects being transparent is that:

$$\frac{V_{noise}}{V_{sig}} < 0.15 \qquad (1.3)$$

If we use Len[in] and RT[nsec], this relationship for when interconnects are transparent simplifies to:

$$RT[n\sec] > Len[inches] \qquad (1.4)$$

This is the condition for interconnects to be transparent.

In a system with signals having a rise time of 10 nsec, if the longest interconnect length is shorter than 10 inches, the inductive noise will be low enough to not affect performance and all the interconnects will be transparent.

If the rise time is 2 nsec, and interconnects are longer than 2 inches, their electric properties may affect the performance of the product and must be considered early in the design phase.

Of course, this is just a simple rule of thumb, but it provides some guidance when considering when interconnects are no longer transparent. And, it points out the important principle that with a long enough rise time, or short enough interconnect, all interconnects are transparent.

A similar condition can be established for when self-aggression reflection noise would not be a problem. Of course, the reflection noise depends on the difference in impedance between the interconnect and the driver's output impedance, the rise time of the driver, and the time delay of the transmission line.

If we do not pay attention to the reflection noise, it can be a disaster. An example of the measured reflection noise in a particularly poorly mismatched system, with a very low source impedance and a high impedance at the receiver, is shown in **Figure 1.8**.

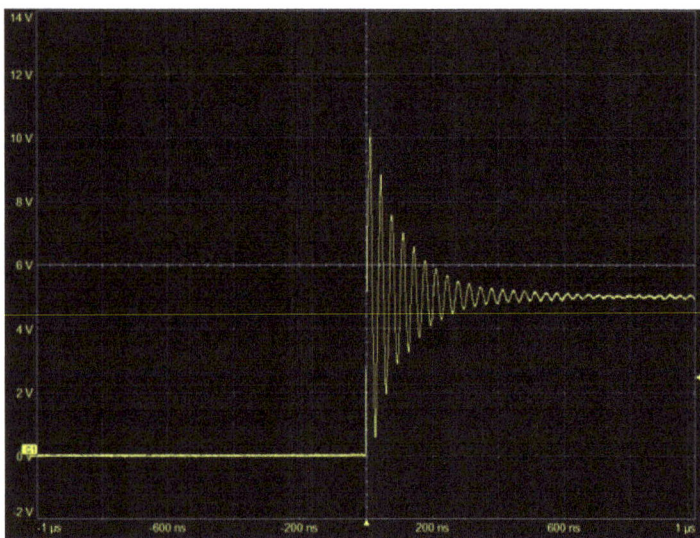

Figure 1.8 Example of the measured reflection noise at a high impedance receiver from a low impedance source connected by a well-designed transmission line.

Reflections in transmission line circuits with poor terminations will always occur independent of the rise time and interconnect lengths. It's just that when the rise time of the signal is long compared to the time between the peaks and valleys from the reflections, related to the time delay of the transmission line, the reflections will smear out during the rising edge. In such a case, the reflection noise may not be a problem.

Generally, in a typical case, with a driver impedance of 10 ohms driving a 50-ohm line, the impact on the signal at the receiver from reflections are dramatically reduced when the rise time of the signal is at least 4x the time delay of the interconnect. **Figure 1.9** shows examples of the received signal in the case of three rise times: RT = TD, RT = 2 x TD, and RT = 4 x TD.

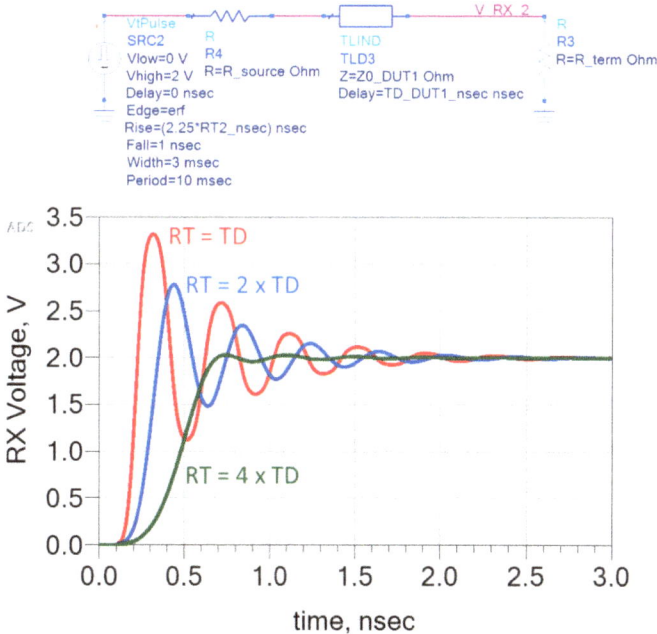

Figure 1.9 Top: example of a simple circuit with a signal source, a source impedance, driving a transmission line with a high impedance receiver. Bottom: the simulated voltage at the receiver as the rise time of the signal increases compared to the time delay of the transmission line. Simulated with Keysight's ADS.

From this specific example, the condition for an interconnect to be transparent and the reflection noise to be small enough to not be an issue is

$$RT > 4 \times TD \qquad (1.5)$$

where

RT = the 10-90 rise time of a signal

TD = the time delay of a transmission line

The connection between the TD of an interconnect and its physical length, if routed in FR4 with a dielectric constant, Dk ~ 4, is

$$TD[n\,sec] = \frac{Len}{v} = \frac{Len[in]}{6\frac{in}{n\,sec}} \qquad (1.6)$$

Combining these two simple relationships results in a condition for the interconnects to be transparent for reflection noise:

$$RT[n\,sec] > \frac{4}{6}Len[inches] = 0.67 \times Len[inches] \sim Len[inches] \quad (1.7)$$

Remarkably, the condition for reflection noise (self-aggression noise) to not be a problem due to transmission line reflection noise is the same condition for mutual inductance cross talk noise (mutual-aggression noise) to not be a problem.

Again, this points out that interconnects are transparent if they are shorter, in inches, than the rise time of the signal in nsec. If the rise time of signals is 5 nsec, all interconnects shorter than 5 inches are transparent.

Of course, this is a simple rule of thumb and should not be used as a strict design rule for design sign-off. It is meant to offer insight into the principles and offer a rough estimate of when an interconnect will not be transparent.

It is an example of one approach to answer the very important questions, which answer is often, "It depends." A rule of thumb offers a quick estimate to the answer of what it depends on, as well as what is its approximate magnitude.

This rule of thumb also identifies an important principle:

As rise times get shorter, typical length interconnects are less and less transparent and if we don't worry about their electrical properties at the beginning of the design cycle, the product may not work.

This is why we have to worry about the electrical properties of interconnects and the most important electrical properties are their transmission line properties.

1.5 Why the RF World is Different from the High-Speed Digital World

Interconnects are interconnects. Whether your application is for RF or high-speed digital, isn't the physics that govern the properties of the interconnects the same? Why should it matter if you are an RF or high-speed digital engineer?

While the properties of interconnects are the same for an RF signal as for a high-speed digital signal, the engineering design principles for designing optimal interconnects are different in these two applications.

The primary difference between RF design and high-speed digital design is that RF engineering typically involves signals with a specific frequency, the carrier frequency, with a narrow bandwidth about the carrier frequency. We care about the electrical properties of the interconnects primarily at the carrier frequency.

In RF applications the signal is narrow-band and it is the properties of the interconnects at a specific frequency, in a narrow bandwidth, which are important.

For high-speed digital applications, the frequency components of the signals are wideband. In a pseudorandom bit sequence (PRBS) signal with no encoding, a typical serial data pattern, the highest frequency component in the signal depends on the rise time of the signal.

This highest frequency might be on the order of 5x the Nyquist frequency, which is 2.5x the data rate. The lowest frequency component might extend to DC. This means the bandwidth of the signal extends all the way from DC up to the highest frequency component of the signal.

In high-speed digital applications, the signal bandwidths are wideband and it is the widebandwidth properties of the transmission line that are important.

This difference in the spectrum of signal components from these two world views is illustrated in **Figure 1.10**.

Digital Designers Live and Think in the Time Domain

Microwave Designers Live and Think in the Frequency Domain

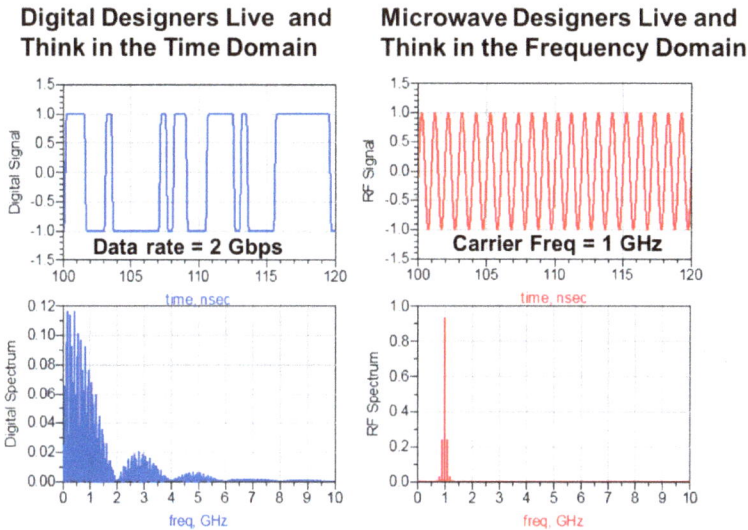

Figure 1.10 The primary difference between the RF designer and the high-speed digital designer is that signals in the high-speed digital world are typically wide bandwidth, while signals in the RF world are narrow bandwidth.

The wide bandwidth nature of signals in high-speed digital applications means the designer must worry about the wide bandwidth electrical properties of transmission lines.

Signals are typically analyzed in the time domain where performance is measured. This means thinking of how signals interact with the interconnects in the time domain will often get to an acceptable answer faster than thinking in the frequency domain.

For example, achieving a target impedance for a transmission line at one frequency can be achieved by using almost any characteristic impedance and tuning its length for the application frequency.

When designing a transmission line for a target impedance in a high-speed digital application over a wide frequency range, length has nothing to do with it.

31

Terminating a transmission line for an RF application means matching the impedance at one frequency. This can often be done with an LC circuit or another transmission line. Terminating a signal for a high-speed digital application means matching the impedance over a wide frequency range, which can often only be done with a resistor whose impedance is flat over a wide frequency range.

Even the concept of the impedance of a transmission line is radically different when viewed in the time domain rather than in the frequency domain.

This book primarily focuses on the time domain perspective of transmission lines, which is not often covered in most books that introduce transmission line properties. This makes it particularly useful for high-speed digital applications.

1.6 Review Questions

1. What is the difference between a microstrip and stripline transmission line geometry?

2. Why is engineering intuition an important skill?

3. How do you answer "it depends" questions?

4. What do Maxwell's equations describe and why are they so important?

5. Why don't we tackle every SI problem by grabbing a pencil and paper and solving Maxwell's equations?

6. What is a transmission line?

7. After connectivity, what is the only thing interconnects can introduce in an electronic product?

8. What is the primary tool we will use to measure or characterize a transmission line?

9. What are the two most common problems interconnects introduce as rise times get shorter?

10. If an interconnect is 12 inches long, like on a PC motherboard, what is the rise time below which the interconnects may not be transparent?

11. If an interconnect is 3 inches long, like on a small microcontroller board, what is the rise time below which the interconnects may not be transparent?

12. What is the biggest difference between signals in RF applications and high-speed digital applications?

13. Why can't we just design interconnects for high-speed digital applications with the same guidelines as they are designed for RF applications?

Chapter 2   Essential Principles of Signals on Interconnects

With just a few essential principles based on Maxwell's equations, we can gain the insight and intuition needed to understand how signals interact with interconnects. This will enable us to understand how physical design affects the important electrical properties of the interconnects that influence how signals are treated.

After all, the design process is a creative process. While we will always leverage analytical methods such as approximations and numerical simulations, we start the design process using our engineering judgement and intuition.

The better calibrated our engineering intuition, the farther along a successful design strategy we can start each project.

2.1   All Interconnects are Transmission Lines

In Chapter 1 we introduced the concept of a transmission line. Without exception, all interconnects used in any electronic system to connect a signal between two nodes or terminals is a transmission line.

Every transmission line always has two conductors. We label one of them a signal path and the other the return path.

Don't get in the habit of calling the return path ground. The ground path or a ground plane may be used as the return path, but this is a special case. Any conductor with any DC voltage may be used as a return path.

Often, the return path is called the *reference conductor*. This is very misleading and confusing. What does this mean? Can I identify any conductor as the reference? How do I tell the signal which conductor I selected as the reference?

The term *reference conductor* is also ambiguous. The words we use matter. They affect our intuition and how we conceptualize the conductors.

> *Get in the habit of calling the other conductor in a transmission line the return path and you will train your intuition to think about the return currents being in this conductor.*

As we will show in this chapter, the selection and design of the return path affects the signal propagation just as much as the signal path. If you only engineer the signal path, and do not engineer the return path, the signal will absolutely find a return path on its own.

Chances are, you will not like the return path the signal finds for itself. The signal propagation will not be as clean and low noise as if you engineered the return path on purpose.

2.2   Signals Are Dynamic

The purpose of an interconnect is to transport a signal from a transmitter to a receiver with acceptable distortion. While we focus on the design and characterization of the interconnects in this book, it is the quality of the signal that is transported by the interconnect that motivates our design choices.

When we launch a voltage between the signal and return path at the beginning of the transmission line, we are changing the electric field between these two conductors at the front end. A changing electric field generates a changing magnetic field, which then generates a changing electric field.

Once begun, these fields are *self-propagating*. They will propagate down the transmission line between the signal and return path conductors passing through the dielectric between them. This property of self-propagating changing electric fields is literally built into the fabric of space-time.

It was one of the profound contributions of James Clerk Maxwell to realize that changing electric and magnetic fields is really the nature of light. The propagating signal on a transmission line is literally light, though typically at a frequency in the radio spectrum of 1 MHz to 50 GHz, rather than the 500,000 GHz of red light.

It is this fundamental property of electromagnetic fields that creates the signal that propagates down the transmission line.

The signal that propagates is the changing electric and magnetic fields associating with the changing voltage between the signal and return path. The speed of the signal is really about how quickly the changing electric field between the signal and return path can create the changing magnetic field, which then generates the changing electric field in the dielectric material and self-propagates down the transmission line. This is illustrated in **Figure 2.1**.

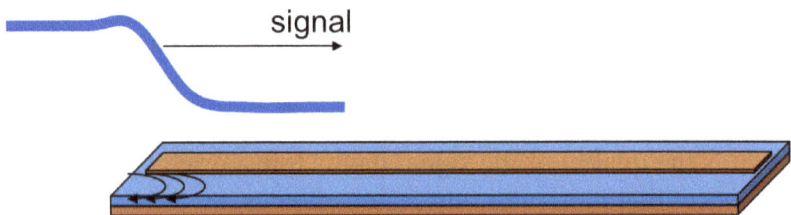

Figure 2.1 The changing and propagating electric fields between the signal and return path and the resulting instantaneous voltage profile that would be measured between the signal and return conductors.

As soon as the voltage is applied and ramps up between the signal and return conductors at one end of the transmission line, this changing voltage will create a changing electric field that will propagate down the line. This property is built into the very fabric of space-time.

There is absolutely nothing you can do to prevent this applied, changing voltage from propagating down the transmission line between the signal and return path at the speed of light in the material.

It is often incorrectly believed that the signal is traveling in the copper traces and is related to the resistance of the traces. Using this model, you would be inclined to believe that the speed of a signal is related to the resistance of the traces. A lower resistance would result in a faster speed. This is totally incorrect.

The signal is really the propagating, changing electric field between the signal and return path. This appears as a voltage between the signal and return path. This is why it's the dielectric material between the signal and return path that affects the speed of the signal.

In air, the speed of a changing, self-propagating electric field is 3x 10^8 m/sec or 30 cm/nsec or 11.8 inches/nsec. If the electric field is in a dielectric material, the dielectric constant of the material will slow the speed of propagation of the changing electric by the square root of the dielectric constant.

The speed of the signal (light) in a dielectric is:

$$V = \frac{11.8 \, ^{in}/_{n \, sec}}{\sqrt{Dk}} = \frac{30 \, ^{cm}/_{n \, sec}}{\sqrt{Dk}} \tag{2.1}$$

where

11.8 in/nsec is the speed of light in air

Dk = the dielectric constant of the laminate material.

With a typical dielectric constant for most laminates around 4, the typical speed of a signal is about 6 inches/nsec or 15 cm/nsec.

This is a very important rule of thumb to remember. The speed of a signal propagating down a transmission line made from an FR4 laminate is about 6 inches per nsec.

A signal launched into a 12-inch-long transmission line on a circuit board would take about 12 inches / 6 inches/nsec = 2 nsec to enter one end and reach the other end.

After the signal is launched into the transmission line, this voltage difference will propagate down the interconnect at the speed of light in the material.

If we could freeze time, we would see the voltage distribution across the transmission line such as shown in **Figure 2.2**.

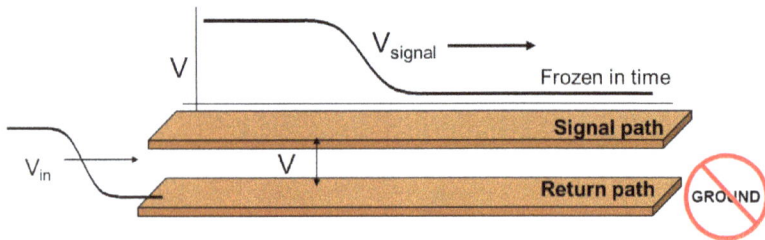

Figure 2.2 The voltage distribution on a transmission line frozen in time. The signal is this changing voltage that is propagating.

The signal is the voltage transition that is propagating down the transmission line with its associated changing electric field at the leading edge. The signal is, by its nature, dynamic, constantly in motion, propagating down the transmission line.

It is important to make the distinction between the voltage that would be measured on the transmission line between the two conductors, with an oscilloscope, for example, and the signal that propagates.

At any instant, there is a voltage difference between every two adjacent points on the signal and return path. If we were to place a scope probe between the signal and return path, the scope would measure the actual, instantaneous voltage between these two points.

However, the measurement of the voltage between two adjacent points alone tells us nothing about the *direction of propagation* of the signal. We sometimes refer to this voltage measured by a scope as a *scalar* voltage in that is it just one number, with no direction. This is distinct from the description of a *vector*, which would have direction consisting of propagation information.

We call the voltage pattern that is propagating on the interconnect, with information about its direction, the signal. It is the signal that is the propagating voltage. The voltage itself is just the scalar value as would be measured with a scope.

When there is just one signal propagating on the line, the voltage would be the same magnitude as the signal. But, when multiple signals propagate on the line. However, in opposite directions, each signal is clear and distinct, but a voltage measurement cannot distinguish between them. The voltage measured by the scope would be the sum of the propagating signals.

A scope measurement of the voltage on a transmission line tells us nothing about the direction of propagation of the signal. The propagating nature of the signal is an important feature of the signal and how it interacts with the interconnect, yet a scope measurement tells us nothing about this property.

If we learn all of our signal integrity just from scope measurements, we will have missed one of the most important properties of signals and our intuition may be completely wrong.

2.3 A Simple Free Tool to Illustrate the Propagation of a Signal

Signals are dynamic. It is hard to visualize this behavior on a flat, static page. To help calibrate your intuition, you can download this free simulation tool to show the dynamic nature of a signal.

Watch this video and I will walk you through using this simulation tool.

This simple simulation tool, based on flash and written by Yoshi Tsuji of Teledyne LeCroy, will be used throughout this book to illustrate some of the dynamic properties of signals propagating on transmission lines. While it has a simulation engine at its core, it is not a substitute for a circuit simulator like a SPICE-based simulator. It is really designed to illustrate the general principles of signals propagating on a transmission line and interacting with the instantaneous impedance it encounters.

In this first example we will use it simply to show the voltage on a transmission line as it propagates. **Figure 2.3** is a screenshot showing two instances of the same signal on a transmission line.

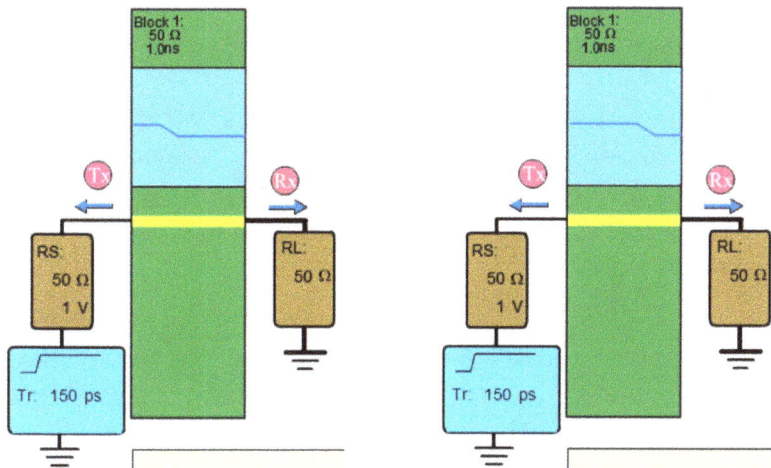

Figure 2.3 Two snapshots of a signal propagating down a transmission line using the free downloaded simulator.

The green region represents the return path of the transmission line. The yellow strip is the top view of the microstrip transmission line. It is driven with an adjustable ramp source with a source impedance. It is connected to an adjustable termination impedance.

Once the signal is launched from the source, the dark blue trace in the light blue field is a plot of the voltage profile along the microstrip transmission line as short time intervals as the signal propagates.

This tool allows you to visualize the rising edge of the signal as it propagates down this transmission line. Later in this book we will use this tool to visualize the reflections of the signal as it encounters changes in the instantaneous impedance.

2.4   Time Delay and Wiring Delay

An important property of a transmission line is its *time delay*. This is how long it takes a signal to enter the line and come out the other end. The time delay, also referred to as the *propagation delay*, of a transmission line is related to the length of the line and the speed of the signal by:

$$TD[n \sec] = \frac{Len[in]}{v[in / n \sec]} \qquad (2.2)$$

where

Len[in] = the physical length of the transmission line in inches

v[in/nsec] = the speed of the signal in the dielectric of the transmission line, roughly 6 inches/nsec in FR4 materials.

We sometimes refer to the *wiring delay* of an interconnect as 1/v = roughly 1/6[in/nsec] = 0.17 nsec/inch, in FR4. This is the delay per inch of the interconnect. It is not a speed, but a delay per inch, the inverse of the speed.

The time delay of a transmission line in FR4, using the wiring delay, is

$$TD[n \sec] = Len[in] \times 0.17[n \sec / inch]$$

For example, an interconnect 2 inches long will have a delay of 2-inches x 0.17 nsec/inch = 0.34 nsec.

Both the speed of a signal as 6 inches/nsec and the wiring delay of 0.17 nsec/inch are important rules of thumb to remember. They will help us quickly estimate the time delay of an interconnect.

2.5   Signals See an Instantaneous Impedance

The most important electrical property of an interconnect that determines the relationship between the voltage and current on it is its impedance. What makes this term so confusing is that there are multiple types of impedance, each with a different definition and each with a different application.

Regardless of the type of impedance, all definitions have their origin in the simple definition,

$$Z = \frac{V}{I} \qquad (2.3)$$

As the signal propagates on a transmission line, it sees an *instantaneous* impedance each step along the way. We can easily estimate this instantaneous impedance using a simple *physics model* for the transmission line and this definition of impedance.

The V term in the definition is just the signal voltage on the transmission line, 1V, for example. To determine the impedance the signal sees, we need to find the current associated with the signal as it travels down the transmission line.

The simplest path to determine the instantaneous impedance the signal sees is to take a zen approach and *be the signal* to see what it sees as it propagates.

Watch this video and I walk you through the analysis of the instantaneous impedance and the characteristic impedance of a transmission line.

Imagine, as in **Figure 2.4**, that we are the signal walking on the transmission line. Ahead of us, there is no voltage on the transmission line. Behind us, there is a 1V voltage we left in our wake between the signal and return path.

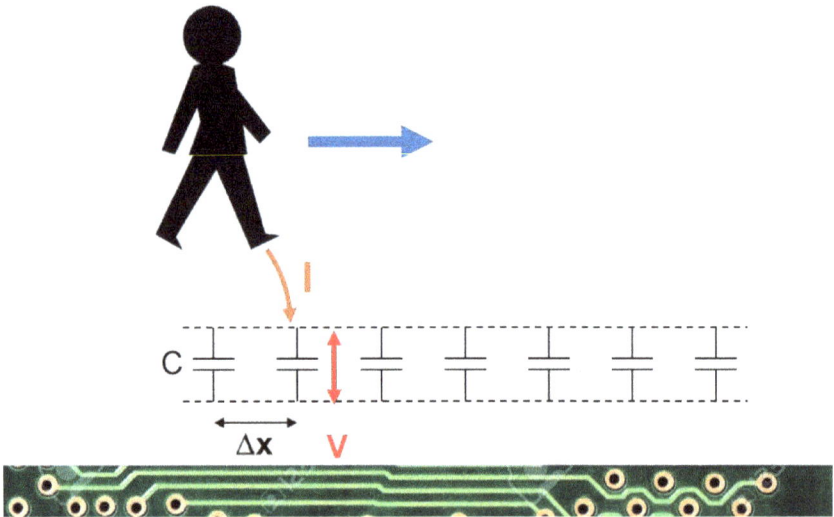

Figure 2.4 Imagine you are the signal, walking down the transmission line charging up the line as you go. Each step has a capacitance between the signal and return path.

In each footstep, there is some capacitance between the signal and return path. If the interconnect is uniform, and we travel down the interconnect at the same speed, the length of our footstep is the same and the capacitance in each footstep is the same. The capacitance in each footstep is

44

$$C = C_{Len} \times \Delta x = C_{Len} \times \Delta t \times v \qquad (2.4)$$

where:

C = the capacitance in each footstep

C_{Len} = the capacitance per length of the transmission line

Δx = the length of each footstep

Δt = the time it takes to step from footstep to footstep

v = the speed of the signal- the speed of light in the dielectric

With each step we take, we are charging up this capacitance in each footstep from 0V to 1V. There is some amount of charge, Q, we need to dump into the capacitor to raise it to 1V.

$$\Delta Q = C \times V \qquad (2.5)$$

As we step along the line, depositing the same charge in each footstep to leave the 1V in our wake, taking the same time between footsteps, there is a current coming out of our foot equal to:

$$I = \frac{\Delta Q}{\Delta t} = \frac{C \times V}{\Delta t} = \frac{C_{Len} \times \Delta t \times v \times V}{\Delta t} = C_{Len} \times v \times V \qquad (2.6)$$

As long as the transmission line is uniform, this current coming out of our foot to charge up the transmission line is constant.

This is a profound observation. As the signal travels down the uniform transmission line, there is a constant current associated with the signal that successively charges up each section of the transmission line in order to leave a 1V signal in its wake. This makes the instantaneous impedance the signal sees constant.

The ratio of the constant voltage of the signal to the constant current associated with it is the instantaneous impedance the signal sees. This instantaneous impedance is

$$Z = \frac{V}{I} = \frac{V}{C_{Len} \times v \times V} = \frac{1}{C_{Len} \times v} \qquad (2.7)$$

As we walk down the line, if the cross section is uniform, the capacitance per length and the speed of the signal is constant so the instantaneous impedance is constant. The signal sees one value of instantaneous impedance as it propagates.

If we double the length of the transmission line, the instantaneous impedance will not change. It will take us longer to travel down the transmission line, but each step will see the same instantaneous impedance.

If the line width were to increase in one region of the transmission line, the capacitance of that footstep would increase. The charge required to increase this capacitance to the same 1V would increase. If we take the same time to walk from footstep to footstep, and the charge has increased, the current required to charge up each footstep will increase. If the current increases for the same voltage signal, the instantaneous impedance will have decreased.

As the signal propagates down the uniform transmission line, if the capacitance per length increases, the instantaneous impedance decreases. If the capacitance per length decreases, the instantaneous impedance increases.

We all have intuition about how the physical design of an interconnect affects the capacitance per length. A larger width means a higher capacitance per length. The thicker the dielectric spacing between the signal and return, the lower the capacitance per length.

We can develop the same physical intuition about the impact from geometry and material properties on the instantaneous impedance by just inverting the behavior. Whatever feature increases the capacitance per length will decrease the instantaneous impedance. Whatever feature decreases the capacitance per length increases the instantaneous impedance.

2.6 Instantaneous Impedance and Characteristic Impedance

A signal sees an instantaneous impedance of a transmission line as it propagates down a transmission line. If the transmission line is uniform, the instantaneous impedance is constant down the length. There is one value of instantaneous impedance that characterizes a uniform transmission line.

We call this one value of instantaneous impedance the *characteristic impedance* of the uniform transmission line.

There are two figures of merit that describe any uniform transmission line:

- The time delay
- The characteristic impedance

These are the two figures of merit of an ideal, lossless, uniform transmission line and is the basis of the T-element model common to all versions of SPICE.

If we know these two properties of a transmission line, we know everything important about it and can predict how a signal will propagate on the transmission line.

The characteristic impedance, like the instantaneous impedance, does not depend on the length of the transmission line. It is only about the cross section and material properties.

The time delay of the transmission line will depend on the length of the transmission line.

We translate this cross-section information and material properties into the characteristic impedance using analysis tools. The most important analysis tool is a 2-D field solver.

This sort of tool will take the cross-section geometry information and calculate the electric field distribution given the boundary conditions of the conductors and dielectrics. It literally solves Maxwell's equations to calculate the electric field distribution and from this, the instantaneous impedance that is the characteristic impedance.

If length is provided, it will also calculate the time delay of the transmission line.

Figure 2.5 is an example of the input and output of the Polar Instruments SI9000, a popular 2-D field solver, for the special geometry of an uncoated microstrip transmission line. With a line width of 10 mils and a dielectric thickness of 5 mils, a 2:1 aspect ratio, the characteristic impedance of this microstrip is 47 ohms.

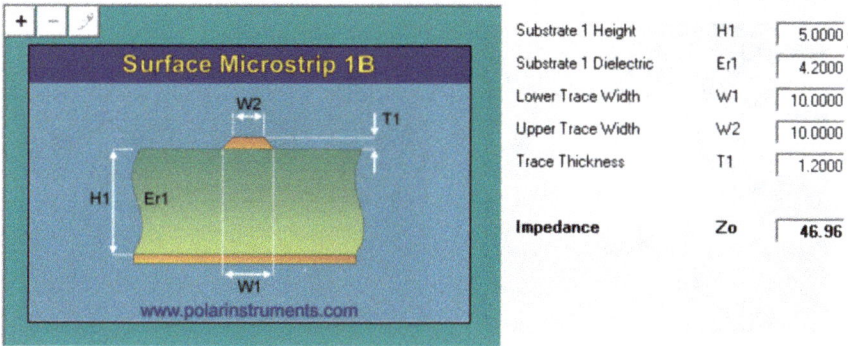

Figure 2.5 An example of the calculated characteristic impedance from geometry input information from Polar Instruments SI9000.

Every uniform transmission line has a characteristic impedance and time delay. Whether the transmission line is a 10-km long coax cable used to carry cable TV signals, or a 1-inch trace on a circuit board, each uniform interconnect has one value of characteristic impedance. This is illustrated in **Figure 2.6**.

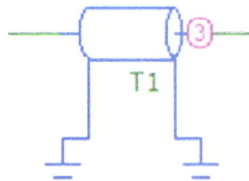

An ideal
transmission line
model: Z0, TD

Figure 2.6 A uniform transmission line can be an inch long or a km long and be described with the same two figures of merit, characteristic impedance, and time delay.

If the interconnect is not uniform, the instantaneous impedance will vary down its length and there is no one value of instantaneous impedance that characterizes it. It has no characteristic impedance.

By definition, a characteristic impedance ONLY applies to a uniform transmission line. If the interconnect is not uniform, we may separate it into uniform sections and define the characteristic impedance of each uniform section.

If the instantaneous impedance varies a little, we can still approximate the transmission line as a uniform transmission line with a characteristic impedance within some limit. **Figure 2.7** shows an example of the measured TDR response of five different RG174 coax cables. This is a direct measure of the instantaneous impedance of the transmission lines, showing the instantaneous impedance varying between the cables and constant to about ± 1ohm down their length. Each cable can be described by a characteristic impedance and time delay.

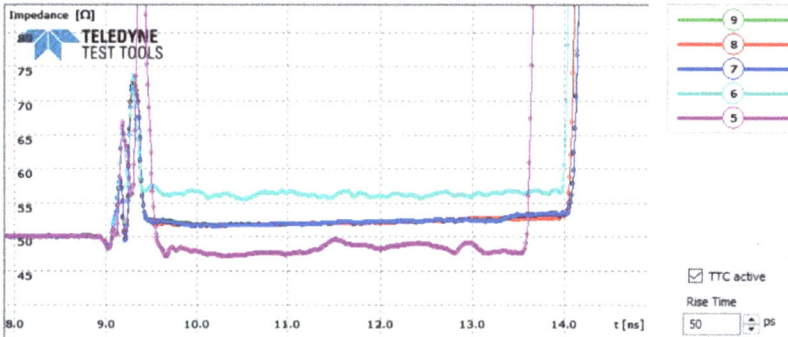

Figure 2.7 Example of the measured instantaneous impedance of five different RG174 coax cables.

2.7 What Happened to the Inductance of the Transmission Line?

In this analysis of the instantaneous impedance and the characteristic impedance of a transmission line, there was no mention of any *inductance* associated with the transmission line. We introduced the concept of the capacitance per length of the transmission line and the speed of propagation of the signal on the transmission line.

Where does the inductance of the transmission line come in?

When the electrical properties of a physical transmission line are described in other textbooks, they are often introduced in terms of the lumped circuit (LC) elements. This is often referred to as the *n-section lumped circuit model*. An example is shown in **Figure 2.8**.

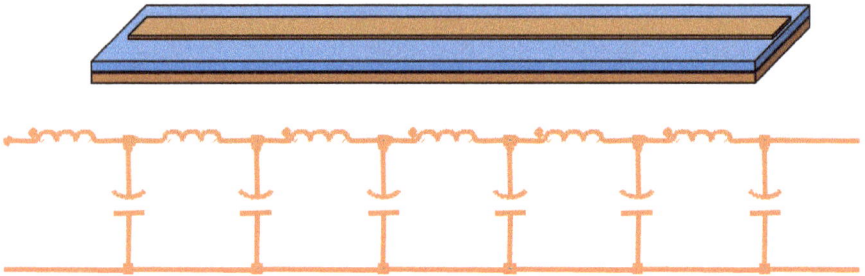

Figure 2.8 A common electrical model (bottom) for a physical transmission line (top) based on a lumped circuit LC ladder model.

The physical transmission line is approximated as a combination of n-capacitor, C, elements and n-inductor, L, circuit elements. This equivalent electrical circuit model is an approximation to the actual electrical behavior of a real physical transmission line. It can be a good approximation, in some situations, if used correctly. It is equally likely to be a poor approximation in other situations.

As is pointed out in Chapter 8, this is an approximation to a uniform transmission line, but is not a very good approximation *compared to* the ideal T-element model of a transmission line, which every SPICE simulator understands. The T-element circuit model is based on the physics model. This n-section lumped circuit model is introduced here only to clarify the role of the inductance associated with the transmission line.

In this model, there are three different inductance-related elements with the following parameters:

L_{total} = the total loop inductance between the signal and the return path

Len = the length of the transmission line

L_{Len} = the inductance per length in the transmission line

ΔLen = the small length of the transmission line to which the inductance elements in the model correspond

L = the small inductance of each element used in the equivalent circuit model

n = the number of sections into which the transmission line is broken

The connection between all of these is:

$$\text{Len} = \Delta\text{Len} \times n$$

$$L_{total} = L \times n \tag{2.8}$$

$$L_{Len} = \frac{L_{total}}{\text{Len}}$$

There are comparable sets of capacitance elements:

C_{total} = the total capacitance between the signal and the return path

Len = the length of the transmission line

C_{Len} = the capacitance per length in the transmission line

ΔLen = the small length of the transmission line for which the inductance elements in the model correspond

C = the small capacitance in each circuit element used in the equivalent circuit model

n = the number of sections into which the transmission line is broken

The connection between all of these is:

$$\text{Len} = \Delta\text{Len} \times n$$

$$C_{total} = C \times n \tag{2.9}$$

$$C_{Len} = \frac{C_{total}}{\text{Len}}$$

Using the equivalent lumped circuit model, we can derive a time delay of the transmission line, the speed of propagation, and a characteristic impedance. The process is to rewrite each circuit element with its connection between the voltage across it and the current through it:

$$\frac{\partial}{\partial x} V(x,t) = -L\frac{\partial}{\partial t}I(x,t)$$

and $$\tag{2.10}$$

$$\frac{\partial}{\partial x}I(x,t) = -C\frac{\partial}{\partial t}V(x,t)$$

These coupled, first-order differential equations are combined to create a linear second-order differential equation, the telegrapher's equation, for the voltage and current distributions on the transmission line:

$$\frac{\partial^2}{\partial t^2}V(x,t) = \frac{1}{LC}\frac{\partial^2}{\partial x^2}V(x,t)$$

and $$\tag{2.11}$$

$$\frac{\partial^2}{\partial t^2}I(x,t) = \frac{1}{LC}\frac{\partial^2}{\partial x^2}I(x,t)$$

The telegrapher's second-order, linear differential equation is solved with one solution being sine waves of current and voltage.

The mathematical details of deriving and solving these differential equations can be found in almost every other book on transmission lines and are therefore skipped here.

The impedance of the voltage and current waves is calculated as a characteristic impedance and the propagation delay through the circuit. It is the time delay (TD) of a sine wave entering the circuit until it comes out the other end.

The solution of the wave equation using sine waves results in:

$$Z_0 = \sqrt{\frac{L}{C}} = \sqrt{\frac{L_{total}}{C_{total}}} = \sqrt{\frac{L_{Len}}{C_{Len}}}$$

$$TD = \sqrt{L_{total}C_{total}} = n \times \sqrt{LC} = Len \times \sqrt{L_{Len}C_{Len}} \quad (2.12)$$

$$V = \frac{Len}{TD} = \frac{Len}{Len \times \sqrt{L_{Len}C_{Len}}} = \frac{1}{\sqrt{L_{Len}C_{Len}}}$$

These relationships are a direct result of solving the second-order linear differential equation.

It is from these methods that most engineers are taught the behavior of the characteristic impedance being the square root of L/C. But what does it mean?

While these equations result in the relationships between the characteristic impedance, time delay and speed of the signal, and the circuit elements of capacitance and inductance, it provides little physical insight into the behavior of signals on transmission lines.

What was presented in the previous section as a zen approach to analyzing the instantaneous impedance of a transmission line, with the distribution of small capacitor buckets, is NOT an equivalent electrical circuit model. It is a physics model.

We explicitly introduce the concept of the signal propagating at the speed of light in the material. We introduced the concept of small capacitor elements distributed some distance apart. Neither of these features are found in electrical circuit models.

From the physical model, we concluded the signal sees an instantaneous impedance that we used to define the characteristic impedance. We also calculated the time delay (TD) of the transmission line as:

$$TD = \frac{Len}{v} \tag{2.13}$$

Using these two different models, we derived the same terms but from very different approaches. The values derived from each approach are exactly the same. They just include different elements.

In the physics model, we explicitly introduced the concept of the finite propagation speed of the signal, v. In the circuit model, we derived the speed of the signal in terms of the L_{Len} and the C_{Len}.

Using this connection, we can convert the physics model into the electrical model:

$$Z_0 = \frac{1}{C_{Len} \times v} = \frac{1}{C_{Len} \times \frac{1}{\sqrt{L_{Len} C_{Len}}}} = \sqrt{\frac{L_{Len}}{C_{Len}}} \tag{2.14}$$

Where did the inductance go in the physics model? It was hidden in the assumption about the speed of the signal. We explicitly included the speed as a property of the interconnect.

The characteristic impedance of the transmission line is both the square root of the ratio of the inductance to the capacitance in the transmission line and it is the inverse of the capacitance per length times the speed of the signal. They are the same impedance written in slightly different forms.

While both approaches result in the same values for the important figures of merit of a transmission line, the physics approach offers some intuitive understanding of what environment the signal sees. The circuit model approach does not provide this same level of understanding, but has other applications, as is introduced in Chapter 8.

2.8   What is Special about 50 Ohms?

There is nothing fundamental or unique about 50 ohms. The only really fundamental impedance is the impedance of free space.

When a signal propagates in free space as light, it is a changing electric field and a changing magnetic field. It sees an impedance as it propagates.

The definition of impedance in circuits is based on

$$Z = \frac{V}{I} \tag{2.15}$$

This is really a circuit approximation of the fields. Voltage is an approximation for electric fields and current is an approximation for magnetic fields. The impedance a propagating electromagnetic field sees is related to the ratio of the electric to magnetic fields:

$$Z = \frac{E}{H} = \sqrt{\frac{\mu_0}{\varepsilon_0}} \approx \sqrt{\frac{1.257 \times 10^{-6}\ \mathrm{N \cdot A}}{8.854 \times 10^{-12}\ \mathrm{F \cdot m}}} = 377\ \Omega \tag{2.16}$$

We refer to this impedance as the characteristic *impedance of free space*. It is sometimes referred to as the characteristic impedance of the universe as it has fundamental significance in electrodynamics.

While it is important in electromagnetic applications, it has little importance when it comes to signals in the GHz range propagating on transmission lines built on circuit boards.

So why is 50 ohms so special and commonly used? The answer lies in the early use of coax cables in the communications and radar systems in the 1930s. This is described in Microwave Tubes by A.S. Gilmour. It is based on the geometry for lowest loss coax cables.

In the early days of radio, transmitters and receivers were not very efficient and every watt of power was precious. When transporting an RF signal from a TX to an antenna for broadcast, the goal was to use as low a loss cable as practical.

When a low-loss dielectric, such as polyethylene, with a Dk = 2.3, is used in a cable, the cable's losses are dominated by copper losses. The skin depth in copper is

$$\delta = \sqrt{\frac{\rho}{\pi f \mu_0}} \qquad (2.17)$$

where

δ = the skin depth in m

ρ = bulk resistivity of copper, 1.7×10^{-8} ohm-m

f = the frequency in Hz

μ_0 = the permeability of free space = $4\pi \times 10^{-7}$ H/m

The series resistance of the cable, above about 1 MHz, when limited by skin depth, is the series resistance of the center conductor and outer shield. A larger diameter inner conductor means a lower resistance cable. The series resistance of the center and outer conductor, assuming the current is skin depth limited, is:

$$R_{Len} = \frac{\rho}{\pi\delta}\left(\frac{1}{d} + \frac{1}{D}\right) = \left(\frac{1}{d} + \frac{1}{D}\right)\sqrt{\frac{\rho f \mu_0}{\pi}} \qquad (2.18)$$

where

R_{Len} = the resistance per length of the center conductor

d = the outer diameter of the center conductor

D = the diameter of the outer shield

ρ = the bulk resistivity of copper, 1.7×10^{-8} ohm-m

δ = the skin depth of copper

As the diameter of the center conductor increases, the resistance of the center conductor decreases, exactly as expected.

But the attenuation in a cable is about more than just the series resistance. The attenuation per length of a coax cable is:

$$\text{atten}\left[dB/len\right] = 4.34 \times \frac{R_{Len}}{Z_0} \qquad (2.19)$$

where:

atten[dB/len] = the attenuation of the signal in dB/length

R_{Len} = the resistance per length in ohms/length

Z_0 = the characteristic impedance of the cable.

The characteristic impedance of a coax cable is related to the dielectric material and the diameter of the inner conductor and inner dimension of the outer shield. For a coax cable, it is given by:

$$Z_0 = \frac{377\,\Omega}{2\pi\,\sqrt{Dk}}\ln\left(\frac{D}{d}\right)$$

(2.20)

where

Z_0 = the characteristic impedance of the cable

Dk is the dielectric constant of the material inside the cable

D = the inner diameter of the shield

d = the outer diameter of the center conductor

For the lowest loss cable, we want the smallest series resistance and the largest characteristic impedance. This means as large an outer diameter as practical. If we fix the outer diameter to be the largest practical size, we can ask, is there an optimal center conductor for lowest loss?

The larger the center conductor, the lower the series resistance and the lower the loss. But, a larger center diameter with a fixed outer diameter shield conductor means a lower characteristic impedance, which means a higher attenuation.

Using these relationships, we can fix the outer shield to a diameter of 1 cm, for example, and vary the inner diameter of the conductor and calculate the resulting loss and the characteristic impedance. By plotting the attenuation against the characteristic impedance, we can find the optimum impedance that results in the lowest attenuation for the special case of polyethylene dielectric.

Figure 2.9 is a plot of the attenuation and characteristic impedance. The impedance for lowest attenuation using polyethylene is about 50 ohms.

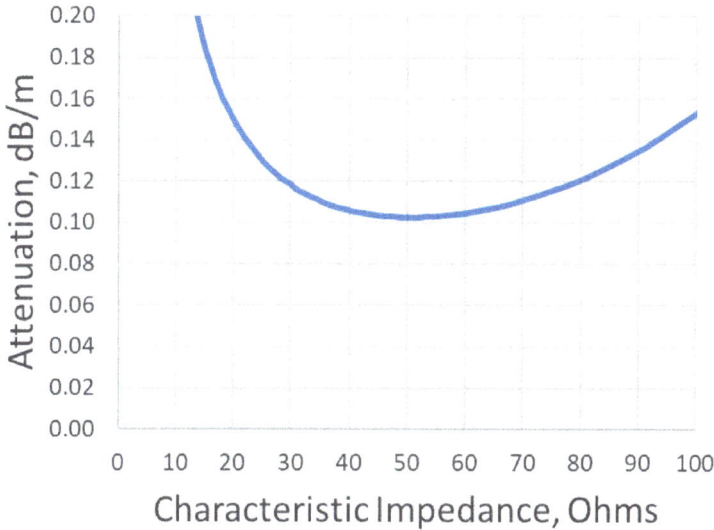

Figure 2.9 The attenuation verses the characteristic impedance of a coax cable with a fixed outer diameter with a polyethylene dielectric.

This is the origin of the use of 50 ohms as the characteristic impedance for cables.

Once the optimal characteristic impedance of a coax cable is fixed, the rest of the interconnect system, from the connectors, the receivers, the amplifiers, the drivers, and all the test equipment used to test all the components, should be the same 50 ohms impedance.

With the early industry moving toward 50-ohm components, it became more cost effective to engineer all systems to this 50 ohm value.

But 50 ohms is not necessarily the most cost-effective impedance environment for all electronic systems, especially wide bandwidth high speed digital systems fabricated not as coax cables in polyethylene, but as planar interconnections in FR4 type polymers.

Some electronic products use a different impedance environment. Depending on which performance parameter is most important, a higher or lower characteristic impedance for the system may have an advantage. For example:

- A *higher impedance* means a higher termination resistance and lower power dissipation

- A *lower impedance* generally means lower cross talk

- Narrower traces and *higher impedance* can result in higher interconnect density

- *Lower impedance* lines can be a better match to the typically lower impedance of through-hole-vias and results in less reflection noise

- Wider lines and slightly *lower impedance* can sometimes result in lower attenuation

But, if there is no strong compelling reason otherwise, a 50-ohm target impedance for single-ended traces and 100-ohm differential impedance is a reasonable starting place for high-speed digital systems.

2.9 The Paradox of Current Flow into a Transmission Line

The signal is the voltage difference between adjacent points on the signal and return conductors. As it propagates down a transmission line, there is also a current associated with it. It's the combination of this voltage and current that creates the instantaneous impedance the signal sees.

If the characteristic impedance of a transmission line is 50 ohms and a 1V signal is a launched into the transmission line, a current of 1V/50 ohms = 20 mA will be launched into the transmission line.

After the 20 mA current enters the signal conductor, when does the 20 mA current come out of the return path? This situation is illustrated in **Figure 2.10**.

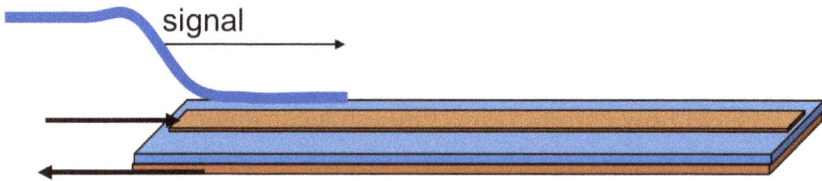

Figure 2.10 When the 1V signal is launched into the transmission line, and 20 mA is sent into the signal conductor when does the 20 mA come out the return conductor?

Keep in mind that there is an insulating dielectric between the signal and return conductor. Of course, free conduction charges cannot flow through the insulating dielectric; it's an insulating dielectric.

Based on the common principles taught in school about current flow, we would think that the 20 mA of current would enter the signal line and have to flow all the way down to the end of the transmission line before it could then connect to the return path and make its way back to the front of the transmission line and exit out of the return path.

But what if the far end of the transmission line is open? How does the current entering the transmission line at the front know that the far end is open and there is no connection to the return path? Would any return current come out the return conductor, ever?

The notions we learn early on of how current flows in a circuit are only suitable for what happens at DC. They do not guide our understanding in the high-speed, time domain world.

2.10 Displacement Current as a New Type of Current

To understand the behavior of currents in a transmission line, we have to introduce a new type of current, displacement current. This is the basis of how current flows through a capacitor.

One of the many contributions James Clerk Maxwell made to unifying the concepts of electromagnetics is realizing there are two types of currents. There is current composed of free conduction charges, which is what we usually think of when we think current, and there is a current that is due to a changing electric field, which he called *displacement current*.

This new current was needed to provide continuity to the current flow through a capacitor. Of course, at DC, there is no current flow through a capacitor, since there is an insulating dielectric between the two conductors that make up the capacitor.

The definition of the capacitance of a capacitor is that:

$$C = \frac{Q}{V} \quad \text{or} \quad Q = CV \qquad (2.21)$$

Capacitance is a measure of the efficiency for two conductors to store charge at the cost of voltage. A large capacitance means the conductor geometry is very efficient at storing a lot of charge for a small voltage.

To increase the voltage across a capacitor, we must add some more excess positive charge to the top conductor and more negative charge to the bottom conductor.

When negative charge is added to the bottom conductor, this is the same as pushing positive charge out of the bottom conductor.

When the voltage across a capacitor is increased, it looks from the outside like positive charges flow into the top conductor and positive charges flow out of the bottom conductor. This is illustrated in **Figure 2.11**.

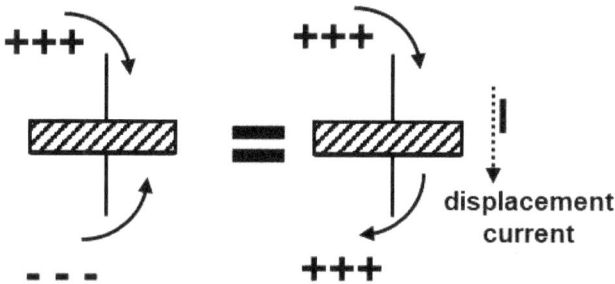

Figure 2.11 When the voltage across a capacitor is increased, positive charge goes into the top conductor and positive charge flows out of the bottom conductor. Displacement current keeps the current continuous in the circuit.

Do actual free charges move through the insulating dielectric between the top and bottom conductors? Of course not. There is an insulting dielectric between the conductors. But it sure looks that way.

To James Clerk Maxwell, when the voltage across a capacitor changed, it looked like current flowed into and out of the terminals of the capacitor. To maintain continuity of current, which was a cornerstone of his unified theory, he required a current to flow through the insulating dielectric.

In his vision, he saw all dielectric insulating materials composed of a sea of positive and negative charges, bound together by local electric forces. When an external electric field was applied, these charges were pulled slightly apart. He termed this behavior, *polarization*.

65

When the external electric field polarized the material, the charges were slightly displaced. This motion of the positive and negative changes when their separation changes is a current. Positive charges moving one way and negative charges the opposite is a net, positive current.

When the external electric field changes, the polarization changes and the movement of the bound charges, the displacement of the charges, is a current. When the electric field stops changing, the material is still polarized, but the bound charges are not moving.

He called this current from the changing polarization, *displacement* current.

It was easy to understand for a dielectric material which could be polarized. But how could this apply to a vacuum in which there is no material to polarize?

In his 1861 perspective of the universe, all space, even a vacuum, was filled with a medium referred to as the ether. It was the medium which permeated everywhere and through which light traveled. It was thought to be composed of small charged particles bound together, but polarizable, just like a dielectric material.

When an electric field is applied to this ether-dielectric, its positive and negative charges, still bound to each other, are polarized. When the electric field changes, displacement current flows through the ether as the bound charges' separation changes.

In Maxwell's view, a changing electric field was a displacement current just as real as a conduction current. Displacement current from changing electric fields affects magnetic fields just as much as conduction current. Displacement current provides the continuity of current with conduction current.

The displacement current density in a material is exactly equivalent to the changing electric field in the material:

$$J_d = \frac{\partial D}{\partial t} = \varepsilon_0 \frac{\partial E}{\partial t} + \frac{\partial P}{\partial t} \qquad (2.22)$$

where

J_d = the displacement current density

D = the electric displacement field in the material

ε_0 = the permeability of free space

E = the electric field

P = the polarization of the material

The first term is the displacement current from the changing polarization of free space, the ether in Maxwell's view. The second term is the displacement current from the changing polarization of the dielectric of the material, related to the dielectric constant of the material.

When you see the electric field lines that connect two conductors at different voltages, and the voltage between them increases, displacement current will literally flow along the electric field lines that will change. This is illustrated in **Figure 2.12**.

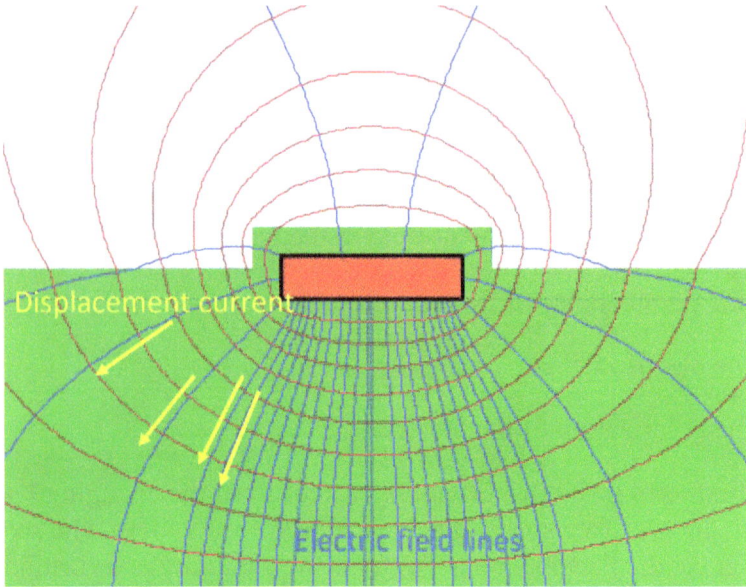

Figure 2.12 Displacement current flows along the electric field lines when the voltage between the conductors change.

The higher the electric field line density, the higher the displacement current density when the electric field changes. The faster the voltage changes, the larger the electric fields change with time and the larger the displacement current density.

In a capacitor, if 20 mA of current flows into the top conductor and 20 mA of current flows out the bottom conductor, 20 mA of displacement current will flow through the insulating dielectric.

As the applied voltage to a capacitor increases, the electric field inside the dielectric between the conductors increases and the displacement current flows.

This is precisely how the return current returns between the signal and the return conductor in a transmission line.

2.11 Return Current in a Transmission Line

When a voltage is launched in a transmission line, as the voltage at the beginning of the line increases, the electric field between the signal and return conductors increases and displacement current flows to connect the input current into the signal path to then instantly return out the return path. This current loop is shown in **Figure 2.13** at two instances in time.

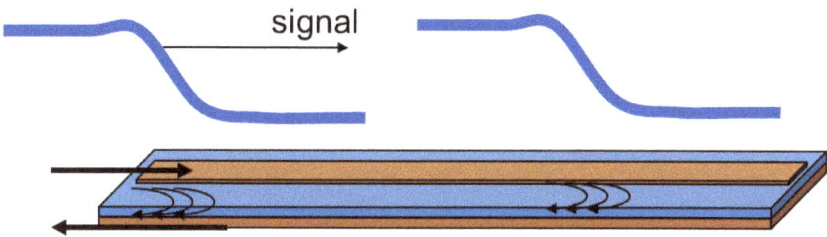

Figure 2.13 At any instant in time, displacement current flows between the signal and return path ONLY where the rise edge of the signal is.

As the wavefront moves down the transmission line, the only place that displacement current flows is where the changing voltage wavefront is located. As this voltage wavefront moves down the transmission line, a coincident displacement current wavefront also moves down the transmission line.

When 20 mA of current flows into the signal path, when does 20 mA of current flow out of the return conductor? Instantaneously. It flows between the signal conductor and the return conductor through the displacement current of the changing electric field that is created by the changing voltage.

This current wavefront has two directions associated with it. It has a direction of propagation, which is the same as the direction of propagation of the signal, and it has a direction of circulation.

When the signal is a positive edge, propagating from the left to the right, the current wavefront circulates in the clockwise direction, advancing as signal-return, signal-return, signal-return.

If the voltage were a negative-going voltage, also propagating from the left to the right, the direction of circulation of the displacement current would be opposite, from the return conductor to the signal conductor, circulating in the counterclockwise direction.

The observation that the current flow in a transmission line has two directions, and they are independent of each other, is a critically important principle. Understanding this behavior will resolve what may otherwise appear as paradoxes in transmission line behavior.

Watch this video and you can see the dynamic nature of the signal-return current wavefront propagating down a transmission line.

2.12 Where Does the Return Current Flow?

If the return conductor is a wide plane, like a ground plane, the actual path the return current takes will not spread out everywhere in the plane.

Above about 100 kHz, all the return current will flow directly underneath the signal current. The path the current takes is driven by the path that reduces the impedance the current loop sees.

At DC, the impedance the current sees is all about the DC resistance of the signal and return paths. The current will spread out in the return plane at DC finding the lowest resistance path.

But at any other frequency, the impedance the current sees is the series combination of the series resistance and the loop inductance of the signal-return path loop:

$$Z = R + j\omega L \qquad (2.23)$$

Above about 100 kHz, the inductive impedance is higher than the resistive impedance and the current will redistribute to take the path of lowest loop inductance.

The lowest loop inductance translates to:

- As far apart for currents in the same direction

- As close together for currents traveling in the opposite direction.

This is literally the same driving forces giving rise to currents flowing in the outer surface, or skin of the conductor, referred to as the *skin effect*.

It is difficult to calculate the current distribution for microstrip or other planar geometries. This requires a 2-D full-wave field solver. **Figure 2.14** shows an example of the current distribution in the same microstrip transmission line at three different frequencies. As the frequency increases, the current in the return path flows mostly under the signal line. It does not spread out.

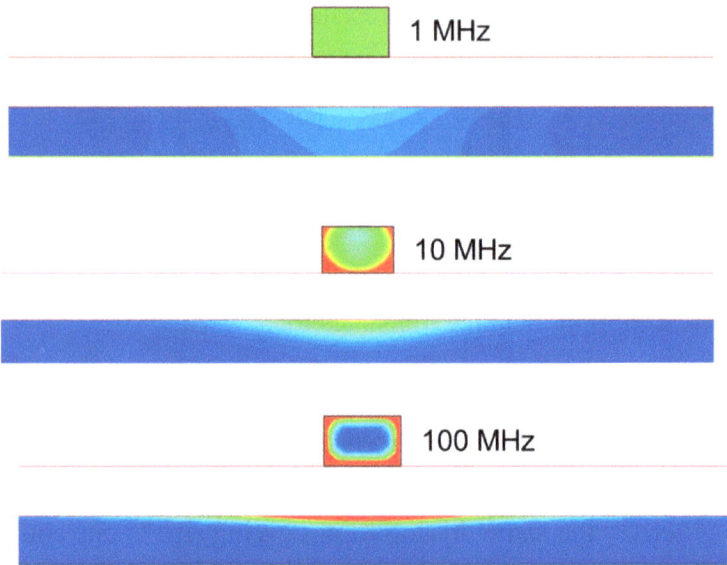

Figure 2.14 The calculated current density in a transmission line at three different frequencies. Dark blue means no current and red means very high current density.

This means that as the signal traces on a board route around the board, the return current distribution in the return plane will also route around and follow the signal path, staying directly underneath. This is illustrated in another simulation of the current distribution in a microstrip that makes a sharp bend in its route shown in **Figure 2.15**.

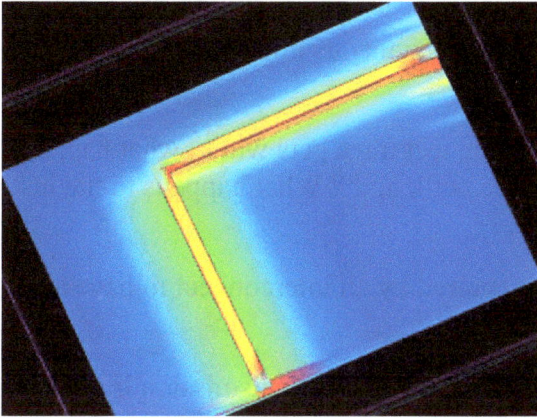

Figure 2.15 Current distribution in the return plane when a signal traces makes a sharp turn. Note the return current stays directly beneath the signal conductor.

2.13 Review Questions

1. Why is it a good idea to not call the return path ground?

2. What is generated between the signal and return conductors at the beginning of a transmission line when a signal is launched? What will happen?

3. Is the signal the moving electrons in the conductor or the electric field in the dielectric?

4. What is a good rule of thumb for the speed of a signal in an FR4 transmission line?

5. What is the difference between the voltage measured by an oscilloscope and the signal on the transmission line?

6. Using the wiring delay of a transmission line, what is the time delay of a 1-inch long interconnect in a large BGA package? A 3-inch-long interconnect on a small circuit board? A 30-inch-long interconnect in a backplane?

7. What does instantaneous impedance refer to? Why is this so important?

8. If an interconnect trace gets wider, what happens to the capacitance per length? What happens to the instantaneous impedance?

9. What is required by an interconnect to have a characteristic impedance?

10. In the physics model of a transmission line, where did the inductance of the interconnect go?

11. What is special about 50 ohms and why is it so common in high-speed digital applications?

12. When 20 mA of current flows into a signal conductor that is 12 inches long, when will the return current come out of the return conductor?

13. What is displacement current and where does it flow?

14. Where in the return path will the return current flow?

Chapter 3   Categorizing Transmission Lines

Every interconnect is a transmission line, with no exceptions. But not all transmission lines are the same. Using a few important criteria, we can distinguish five different families of transmission lines. Within each family many of the properties are similar.

The first step in understanding the properties of a transmission line is to identify the families to which it belongs.

3.1   Distinguishing Metrics

In principle, every interconnect is a transmission line and the principles of signal propagation and how the signal interacts with the interconnect are all identical.

In practice, the details in the specific behaviors of signals on interconnects depend on the specific features of the transmission line. There are five important distinguishing features of transmission lines that separate them into different families. The electrical properties of these interconnects depend on to which family or families the interconnect belongs.

Any transmission line can be classified as belonging to one of the two options for each family. To understand the properties of a transmission line, we need to first classify it according to which family it belongs.

There are 32 different combinations or categories to which a transmission line can be placed, based on the combinations of five categories, each with two options. Every transmission line can be described by one of the two options. The combination of all five defines the transmission line's classification. The five categories are:

1. Uniform or nonuniform

2. Single-ended or differential

3. Uncoupled or coupled

4. Lossless or lossy

5. Balanced or unbalanced

For example, a transmission line can be:

A uniform, single-ended, coupled, lossy, balanced transmission line. This classification defines its properties.

These properties are very different from a nonuniform, differential, uncoupled, lossless, unbalanced transmission line.

To understand an interconnect's behaviors, we must first determine its classification.

This book focuses specifically on lossless, balanced, single ended and differential, uncoupled and coupled, uniform and nonuniform transmission lines.

The major exception that is not covered in this book is lossy transmission lines.

3.2   Uniform or Nonuniform Transmission Lines

A uniform transmission line has the same cross section down its length. A nonuniform transmission line does not. The key property of a uniform transmission line is that the instantaneous impedance is constant down the length of the transmission line.

This means, as a signal propagates down the transmission line, it sees the same instantaneous impedance and propagates undistorted with no reflections. This means a uniform transmission line has one value of instantaneous impedance that characterizes it, which we call its *characteristic impedance*.

A nonuniform transmission line has a cross section between the signal and return path that varies down its length. This means the instantaneous impedance changes down the length of the transmission line. These changes will cause reflections and distort the signal as it propagates.

A nonuniform transmission line does not have one value of instantaneous impedance that characterizes it. It does not have a characteristic impedance.

We may approximate a nonuniform transmission line as a uniform transmission line, with the condition that the instantaneous impedance is constant to within some impedance range or percentage.

Or, we may divide a nonuniform transmission line into sections that are mostly uniform.

Figure 3.1 shows examples of uniform transmission lines on a circuit board with constant cross section and nonuniform transmission lines on a circuit board with a changing spacing between the signal and return path. With no return plane on the bottom layer, the spacing between the signal lines and whichever other line is the return varies down the length of the transmission line.

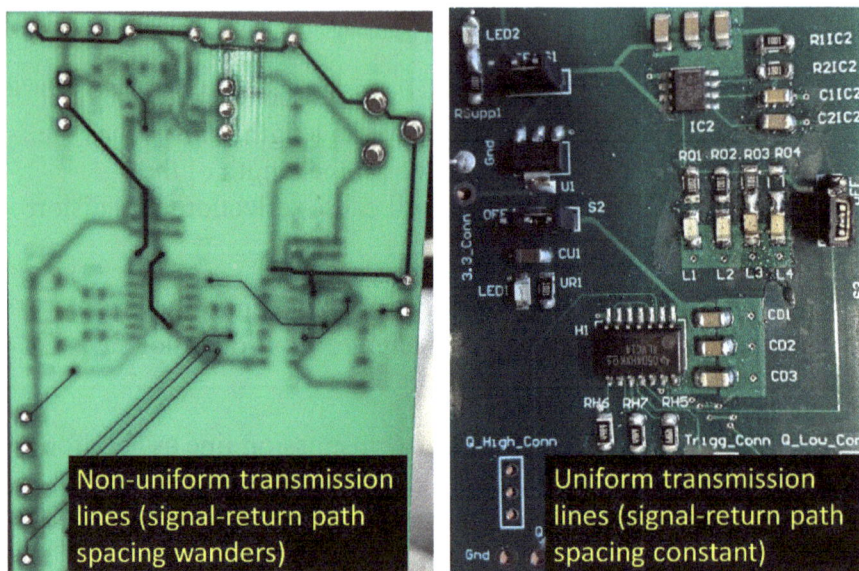

Figure 3.1 Example of nonuniform transmission lines on a circuit board, with no return plane on the bottom layer and uniform transmission lines on a circuit board with a continuous return plane on the bottom layer.

If you are designing a transmission line, engineering it as a uniform transmission line is the right goal. It will result in the best signal quality and least distortion of the signal.

3.3   Single-Ended or Differential

A *single-ended transmission line* has one signal path and a return path. There is only one type of signal that will propagate on this signal and return path, a single-ended signal. This is the voltage between the signal conductor and the return conductor.

A *differential pair transmission line* has two signal lines and a return path. There are two types of signals that can propagate on a differential pair: a differential signal, which is the difference

78

between the voltages on the two lines, or a common signal, which is the average voltage on the two lines, compared to the return path.

Two signal lines with their return paths can be considered as *two* single-ended transmission lines with some cross talk, or as *one* differential pair.

When we have two single-ended transmission lines and use them to transport a differential signal, we refer to the two transmission lines as a differential pair transmission line. Two single-ended transmission lines make up one differential pair.

These two descriptions of two transmission lines are perfectly equivalent. Both contain all the geometry and electrical information to translate from one description to the other without losing any information.

Given one description with its properties, we can convert between the two lines as two single-ended transmission lines with cross talk, or one differential pair. **Figure 3.2** shows an example of a single-ended microstrip transmission line and a differential pair microstrip.

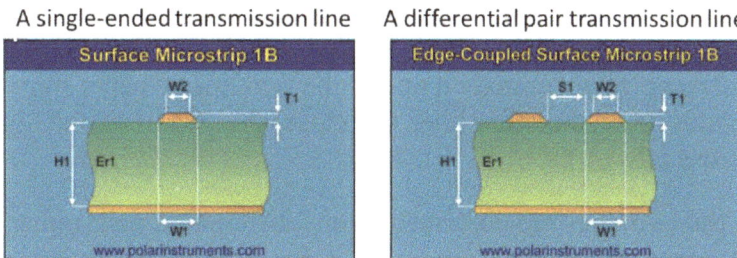

Figure 3.2 An example of a single-ended transmission line and a differential pair. These both happen to be a microstrip topology. Illustration courtesy of Polar Instruments.

A twisted pair cable such as a CAT5 cable, is composed of four pairs of two wires, tightly twisted together. Is each pair a single-ended or differential transmission line?

It can act as both. When one line is used explicitly as the return path, the twisted pair carries a single-ended signal and is a single-ended transmission line.

When the transmission line is used to carry a differential signal, it is a differential pair. But where is the return path for the p and n signal lines?

Suppose the twisted pair is sitting a few inches above a copper sheet. When the differential signal is sent over the p-line and the n-line, the return currents for the p-line and the n-line will spread out in the copper sheet and overlap. They will cancel out. The presence of the copper plane plays no role on the differential currents or the differential impedance.

This is not true for the common impedance of the differential pair. This impedance is very strongly dependent on the proximity of the return plane, which might be literally the floor.

3.4 Uncoupled or Coupled

When we have two single-ended transmission lines, we describe each one with a characteristic impedance and time delay. In addition, their proximity influences their *cross talk*. This is the amount of noise generated on one of the two lines when a signal propagates down the other line. The closer the two lines, the higher their cross talk.

When the two lines are described as one differential pair, rather than the cross talk between them, we refer to the degree of coupling. The closer the spacing, the higher the coupling.

The coupling between the two lines that make up a differential pair is one of the factors that influences their differential impedance. By adjusting other geometry terms, it is possible to make a differential pair with the same differential impedance either uncoupled or tightly coupled.

Should a differential pair be designed as tightly coupled or uncoupled?

When loss is not important in the performance of a differential pair, a tightly coupled differential pair is preferred as this will result in the highest interconnect density.

When loss is important, a design with the widest line width will have the lowest conductor loss. A wider line differential pair for a fixed target impedance can be achieved with an uncoupled differential pair.

When both conductor loss and interconnect density are important, the optimal coupling is generally somewhere between tightly coupled and uncoupled. This is often referred to as *loosely coupled*.

3.5   Lossless and Lossy Transmission Lines

Loss occurs when some of the signal, as it propagates down the transmission line, turns into heat from conductor loss and dielectric loss. As real, physical transmission lines, EVERY transmission line is lossy. There is always some conductor and dielectric loss in all real transmission lines.

In all real interconnects, the attenuation from the dielectric loss and conductor loss are frequency-dependent so that all transmission lines get more lossy with higher frequency.

The measurement of the amount of loss, and the analysis of the loss, and including it in a circuit simulation can be a challenging task. If the loss is low enough to not affect circuit performance, we can approximate the transmission line as lossless to make the analysis or interpretation of measurements easier.

One way of describing the loss in a transmission line is the transmission coefficient from one end to the other. This is usually measured in dB. A lossless transmission line has a transmission coefficient of 0 dB. All the signal gets through.

The more the loss, the smaller the transmission coefficient. When measured in dB, a more lossy transmission line has a larger, more negative dB transmission coefficient.

There is no such thing as a real lossless transmission line. Every real physical transmission line has some loss. But we might describe a transmission line as lossless if the transmission coefficient is larger than – 3 dB for frequencies below the bandwidth of interest.

Figure 3.3 shows some examples of the measured transmission coefficient of transmission lines with various dielectrics, conductors, and cross-section geometries.

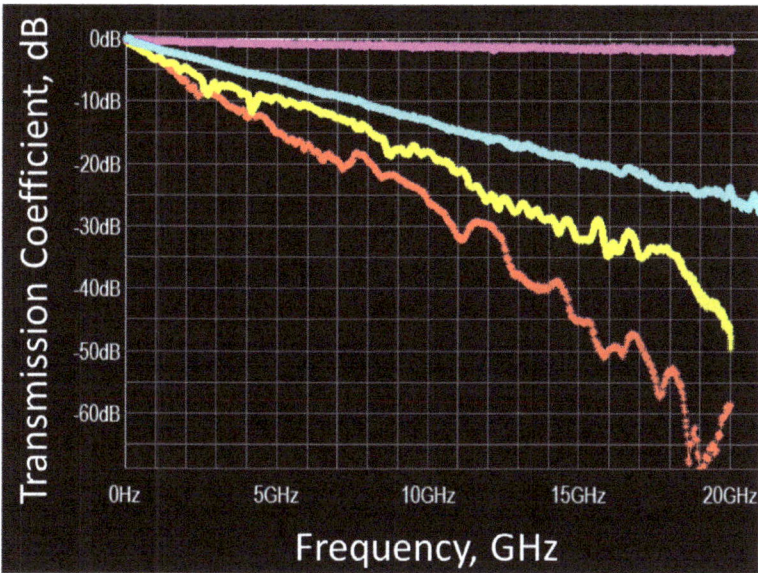

Figure 3.3 Examples of the measured transmission coefficient for various transmission line structures. Only the top measurement of a coax cable would be considered lossless up to 20 GHz.

The interconnect with the lowest loss in these examples is a coax cable. It is an indication of the future direction for interconnects, which need to be low loss at high bandwidth.

The problem lossy transmission lines create is not so much from the loss itself, it is from the frequency dependent of the loss. This is evident in the measured transmission coefficients shown in Figure 3.3.

Higher frequencies are attenuated more than low frequencies. If a high bandwidth signal with short rise time is sent through the lossy interconnect, the high frequencies are attenuated more than the low frequencies and the rise time of the signal will increase. The bandwidth of the signal decreases.

When the resulting rise time is comparable to the unit interval of the digital signal, the longer rise time will begin to affect the signal

83

quality of the collapse of the eye diagram. **Figure 3.4** is an example of the signal at the input of a lossy transmission line and the signal as it would appear coming out for two different data rates. In addition is the resulting eye diagram created by slicing the received data pattern into unit intervals and superimposing successive bit.

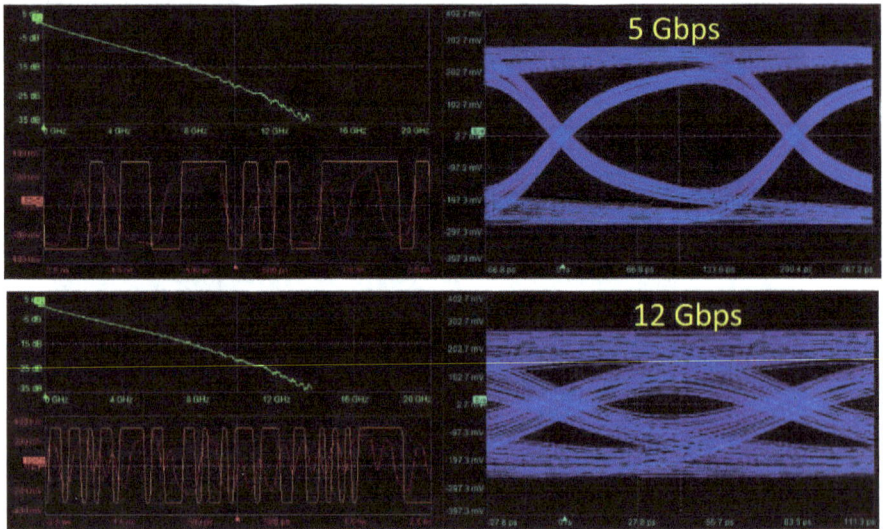

Figure 3.4 Top: A 5-Gbps data pattern entering a lossy transmission line and the resulting received signal and then turned into an eye diagram. Bottom: the same interconnect with a 12-Gbps data pattern. Note the collapse in the eye due to the frequency dependent loss.

In this example, it is the same interconnect, with a 5-Gbps and 12-Gbps PRBS. The higher data rate has a shorter unit interval. When the unit interval is shorter than the rise time, the data pattern influences the resulting signal at the receiver.

This distortion in the received signal is the most important impact on high-speed digital signals from losses in transmission lines used in backplane applications. While the goal is to engineer interconnects with less loss, invariably the features of wider conductor traces for lower copper loss or lower dissipation materials for lower dielectric loss all cost more. This is why the analysis of lossy transmission lines is so important. The cost-performance trade-off determines successful board designs that are affordable.

3.6   Balanced or Unbalanced

A single-ended transmission line is composed of two conductors, a signal and a return path. How similar the two conductors are is a measure of how balanced they are. When the conductors have an identical cross section, they are considered balanced. When they are different, they are unbalanced.

A microstrip or stripline transmission line is unbalanced. A twin rod or a twisted pair or ribbon cable is a balanced transmission line. These examples are shown in Figure 3.5.

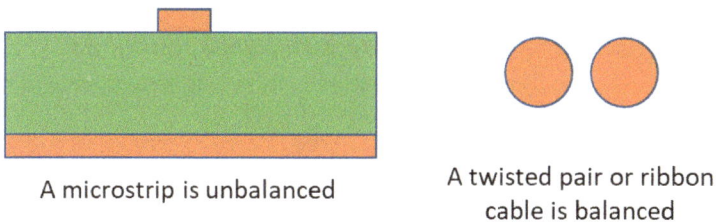

A microstrip is unbalanced

A twisted pair or ribbon cable is balanced

Figure 3.5 Examples of unbalanced and balanced transmission lines.

In a differential pair, the degree of balance refers to how similar the two signal lines, the n and the p-lines, are in the differential pair. A differential microstrip with identical signal conductors is balanced. A differential pair with the p-line on one layer and the n-line on another layer with a different line width and dielectric thickness is unbalanced.

When we include the floor, or external ground surface, as part of the system with a single-ended transmission line, we introduce a third conductor. This means we have three conductors just like in a differential. The concept of balance in a single-ended transmission line becomes exactly the same as in a differential pair.

A balanced and unbalanced differential transmission line can have just as good quality instantaneous impedance and show just as good signal propagation.

The importance of the balance in a differential pair transmission line arises when the degree of balance of the transmission line changes. This happens, for example, when a microstrip transmission line is connected to a ribbon cable that may be a balanced transmission line.

In this case, some of the differential signal will convert to common signal and this will contribute to radiated emissions.

While the concept of balance in a transmission line does not affect the signal quality, it can have a dramatic impact on failing a radiated emissions test.

While the instantaneous impedance may be matched, it is not the full story. When the balance changes, there will be a possibility of radiated emissions.

3.7 Review Questions

1. What are the five families of transmission line types?

2. What is the difference between a uniform and nonuniform transmission line?

3. Why doesn't a nonuniform transmission line have a characteristic impedance?

4.   What is the difference between a single-ended transmission line and a differential pair transmission line?

5.   What does the cross talk between two single-ended lines refer to?

6.   What is the advantage of a tightly coupled differential pair?

7.   Which type of differential pair can achieve a better differential impedance, a tightly coupled differential pair or a loosely coupled differential pair?

8.   What is a loosely coupled differential pair a good balance between?

9.   What are the two root causes of loss in lossy transmission lines?

10.   What is the impact on transmitted signals from lossy transmission lines?

11.   If lower dissipation factor materials result in less distortion of high-speed signals, why not always use the lowest dissipation factor laminate available?

12.   What is the difference between a balanced and unbalanced transmission line?

13.   What properties of a transmission line are affected by whether it is balanced or unbalanced?

14.   Why are uniform transmission lines preferred for carrying signals?

15.   What type of transmission lines does this book focus on?

16.   What is an example of a very important type of transmission line that is outside the scope of this book?

Chapter 4   Five Impedances of a Transmission Line

The basic, fundamental definition of the impedance of a two terminal device is that it is always, without exception,

$$Z = \frac{V}{I} \qquad (4.1)$$

where

Z = the impedance

V = the voltage across the device

I = the current through the device

A transmission line is not really a two-terminal device, but each end is. When we apply the principles of impedance, we will always be looking at the impedance between the signal and return path conductors, either at one or the other end, or somewhere in the middle.

The only metric we care about when looking between the beginning and end of only the signal conductor in a transmission line is the DC series resistance. In this book, we consider this resistance to be very small compared to the 50 ohms of the characteristic impedance of a typical transmission line. Generally, this is a good assumption. In the second book in this series, we consider the impact of the series resistance.

We will never look at the impedance from one end to the other of the signal conductor only in the transmission line. This is not an important metric. Instead, we will consider the dynamic nature of the signal between the signal and return conductors, propagating down the length of the transmission line.

Based on this definition of impedance, we will look at five different situations and see that while each is an impedance, the interpretation of the impedances, and what we might do with it, is very different. To minimize the confusion, it is important to get in the habit of adding the qualifier for which impedance we are referring.

The five different impedances are:

- The *instantaneous* impedance
- The *characteristic* impedance
- The *initial* or *surge* or *wave* impedance in the time domain
- The *input* impedance in the time domain
- The *input* impedance in the frequency domain

This observation that there are multiple types of impedance is a major source of confusion when describing transmission lines. One person may be thinking frequency domain input impedance, while another person may be hearing instantaneous impedance. These are two very different quantities. This ambiguity means confusion.

Get in the habit of adding the qualifier when describing the impedance. Determining which impedance is important in your application will also help you understand the principles.

4.1 The Instantaneous Impedance

The most important and fundamental impedance associated with a transmission line is the *instantaneous impedance* the signal sees each step along the way as it propagates down a transmission line.

This is described in Chapter 2. The instantaneous impedance the signal sees is inherently a time domain concept.

4.2   Characteristic Impedance

A transmission line with a uniform cross section has the same instantaneous impedance everywhere. The transmission line has one value of instantaneous impedance that characterizes it. This one value of instantaneous impedance that characterizes the transmission line is called the *characteristic impedance*.

By definition, only a uniform transmission line has a characteristic impedance. If it is not uniform, there is no one value of instantaneous impedance that characterizes it.

If you know the characteristic impedance of the line, you know what impedance a signal will see when it first launches into the line and what the signal will see at each step along the line.

4.3   The Surge or Wave Impedance

S*urge impedance* or *wave impedance* are older terms that relate to the input impedance of an infinitely long transmission line. It is exactly the same as the characteristic impedance or the instantaneous impedance.

This concept of surge impedance is similar to the input impedance a signal would see upon initially entering a transmission line. It adds no additional value to the concept of the instantaneous impedance.

The wave impedance is the input impedance a wave would see entering an infinitely long transmission line. It is an artificial concept as there is no such thing as an infinitely long transmission line. This would also apply to a transmission line terminated in an impedance equal to its characteristic impedance so there are no reflections from the end.

4.4 The Input Impedance in the Time Domain

When a signal is initially launched into a transmission line, it sees the instantaneous impedance of the transmission line at the beginning of the line. For a uniform transmission line, this is equal to the characteristic impedance of the line.

Once launched into the line, the signal will propagate down the line. While looking at the front of the line, the impedance, as due to the ratio of the voltage applied to the current going into the line, will be constant. The input impedance will appear to any driver to be constant and will look like a resistor.

The source would have no way of knowing about a termination at the end of the transmission line until a round-trip time has passed. If the length of the line is 1 meter in polyethylene, such as in a coax cable, with a speed of about 20 cm/nsec, the time delay of the line would be 5 nsec. It would take 10 nsec for the signal to enter the transmission line and reflect off the end for the source to know there is an end to the line.

If we only look for a time short compared to 10 nsec there is no measurement we could do to distinguish between seeing a uniform transmission line or a resistor with a resistance of the same value as the characteristic impedance of the transmission line.

This means that to any driver looking at the source impedance of the transmission line, it will look like and behave like a resistor for the first 10 nsec. After that, all bets are off, as there will be the complication of the reflections to deal with.

But if the end of the transmission line is open and we wait a day, the input impedance of any transmission line should appear as an open. After all, a transmission line open at the far end is just two conductors with no DC path between them. It should look like an open.

This means the input impedance of a transmission line is not constant over time. Initially and for a time shorter than the round-trip delay time, it will look like a resistor equal to the characteristic impedance. But eventually it will look like an open.

The way the input impedance gets from the characteristic impedance to an open is by multiple reflections, depending on the source impedance used to drive the current into the transmission line. This is illustrated in **Figure 4.1**.

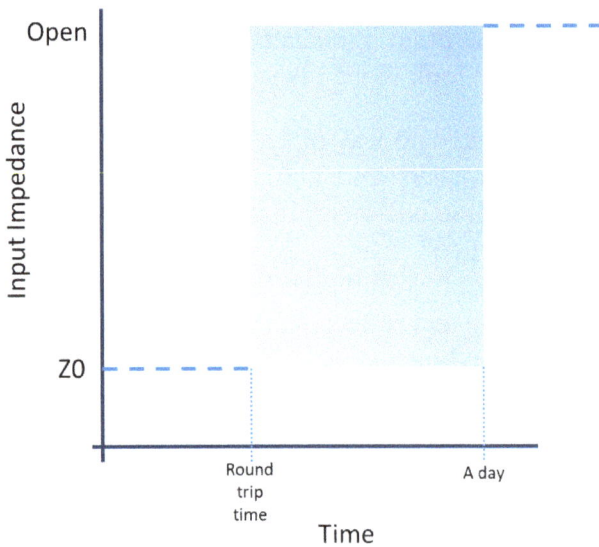

Figure 4.1 The input impedance of a transmission line changes over time.

During the round-trip time, the input impedance a driver sees looking into a uniform transmission line is the characteristic impedance of the line. The input impedance can be modeled as a resistor only during this period of time.

This circuit is shown in **Figure 4.2**.

For t < 2 x TD

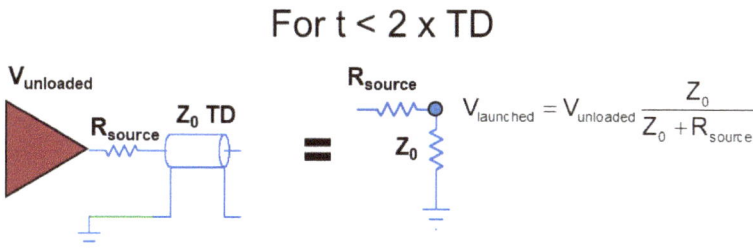

Figure 4.2 The input of a transmission line can be modeled as a resistor for a roundtrip time.

Initially and for a time equal to the round-trip time, a driver looking into a transmission line sees a resistor. This means the voltage launched into the transmission line is NOT the source voltage from the driver, but is a result of the voltage divider created by the Thevenin source resistance of the driver and the characteristic impedance of the line with the Thevenin voltage of the driver.

If the unloaded voltage of the driver, the Thevenin source voltage, is 1V and the Thevenin source resistance is 30 ohms, the voltage actually launched into the transmission line is the voltage divider of the 30-ohm source and 50 ohms of the line. This means the voltage appearing at the beginning of the line is:

$$V_{launched} = 1 \text{ V} \frac{50 \, \Omega}{30 \, \Omega + 50 \, \Omega} = 0.62 \text{ V} \qquad (4.2)$$

This is the voltage that appears between the signal and return conductors on the transmission line and propagates down the line.

If the Thevenin source impedance decreases, a larger fraction of the source voltage will appear across the front of the line and a larger signal will propagate down the line.

This is the signal that propagates down the transmission line and interacts with other impedances it may see, such as at the end of the line.

This principle of the voltage launched into a transmission line from a source is at the foundation of understanding transmission lines. The voltage launched depends as much on the Thevenin source resistance of the driver as it does on the characteristic impedance of the transmission line.

Watch this video and I will walk you through this very important principle.

4.5   Drawing Circuits with Resistors and Transmission Lines

Most engineers are familiar with circuit models with resistor elements. The voltages at any node can be calculated using simple network analysis or combinations of resistors in series and parallel.

But when a transmission line is part of the circuit, it is very confusing how the impedance of the line enters into the circuit. Do we consider the characteristic impedance of the transmission line as a resistor in series or in parallel with the rest of the circuit. Where does this impedance appear in the circuit? These two options for an equivalent circuit model of a transmission line as a resistor are shown in *Figure 4.3*.

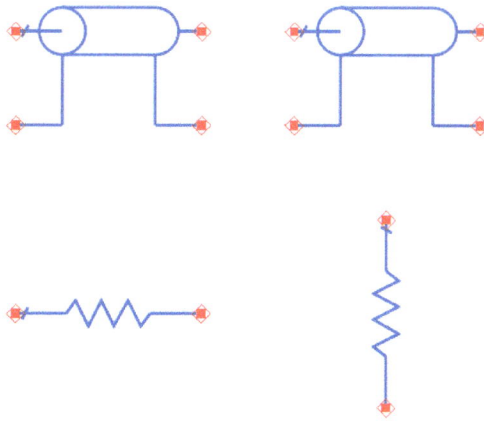

Figure 4.3 Which is the correct representation of a transmission line when viewed in a resistor circuit model element?

The answer is neither. A transmission line is a distinctly different circuit element than a resistor. It is a distributed circuit element, in that its properties are distributed down the length of the circuit element. This is in contrast to a resistor whose properties are all lumped in one point.

It is not correct to replace a real transmission line as a resistor element in a circuit, except under one very special condition. Generally, the characteristic impedance of a transmission line element will never appear as a circuit element in a circuit other than when part of a transmission line element model.

The one exception is during the initial response of the transmission line to a signal launched into it. At the initial instant a signal enters the transmission line, the signal will see the instantaneous impedance as a resistor element between the signal and return conductors.

There is no measurement that can be done that would distinguish the impedance between the signal and return conductors at the input of the transmission line from a resistor between the signal and return conductors with a resistance equal to the characteristic impedance of the transmission line.

However, the input to the transmission line will look like a resistor circuit element only for a finite, short period of time, shorter than the round-trip time delay of the transmission line. After this transient time, the transmission line will NOT look like a resistor circuit element.

This means we have to be very careful when thinking of a transmission line as a resistor. This analogy only works for a very short period of time. The rest of the time, the transmission line behaves like, well, a transmission line and you need to include the circuit element of an ideal transmission line to present the real physical transmission line.

4.6   Input Impedance in the Frequency Domain

The frequency domain is a very special domain. It is distinctly different from the time domain.

In the frequency domain, the only signals we can consider are sine waves. The projection of a sine wave in the time domain is of a specific waveform shape with three terms: a frequency, an amplitude, and a phase.

Described in the time domain, a sine wave is continuous in time: it has always been and will always be the same repetitive pattern. It is a steady-state waveform. In the frequency domain, there is no concept of time or transient behavior. There is only steady state. Each sine wave has always been and will always be.

When we consider the impedance of a transmission line in the frequency domain, it is always as the *input impedance* of the transmission line. This is the impedance the sine wave signal would see looking between the signal and return path of the transmission line at the beginning of the line. It is the steady-state behavior, including all the possible reflections that can happen, such as at the ends. Any transient as the wave is launched into the transmission line and undergoes multiple reflections has stabilized and we are seeing the steady state-behavior of the input impedance of the transmission line.

In the frequency domain, each sine wave frequency component that enters the front of the transmission line has always been and will always be. The voltage and current appearing at the front of the transmission lines includes the input wave and the voltages from all of the reflections at steady state.

This means that when we look at the front of the transmission line, each frequency includes the impact from the load at the far end of the transmission line.

The input impedance is defined as the ratio of the voltage to the current, but since both are sine waves, the resulting ratio of V/I is complex: it has a magnitude (the ratio between the voltage to the current amplitudes) and a phase (the phase difference between the voltage wave and the current wave.)

There is a separate impedance of the transmission line at each frequency. How the input impedance in the frequency domain varies with frequency depends on the characteristic impedance of the transmission line, its length, and its termination.

One of the first derivations in any graduate class on transmission lines is the input impedance of a lossless transmission line in the frequency domain. It is:

$$Z_{in} = Z_0 \frac{Z_L + jZ_0 \tan(\beta l)}{Z_0 + jZ_L \tan(\beta l)} \tag{4.3}$$

where

Z_{in} = the input impedance of the transmission line

Z_0 = the characteristic impedance of the transmission line

j = sqrt(-1)

Z_L = the impedance of the load at the far end

β = the propagation constant of the transmission line, also = $(2\pi f)/v$, where f is the frequency and v is the speed of a signal on the transmission line

l = the length of the transmission line

This is explicitly the input impedance of a transmission line in the frequency domain. The term inside the tan() function has a frequency as part of it:

$$\beta l = \frac{2\pi f}{v} l \qquad\qquad (4.4)$$

Since most graduate classes that introduce transmission lines are oriented toward RF applications, working in the frequency domain is natural. However, any intuition gained about the impedance of a transmission line in the frequency is difficult to directly apply to the time domain.

This impedance calculated in the frequency domain analytically is exactly the same value as is simulated with a circuit simulator such as SPICE. **Figure 4.4** shows the SPICE simulated impedance of a 12-inch-long transmission line with an FR4 laminate and the calculated input impedance using this analytical relationship. The results are exactly on top of each other.

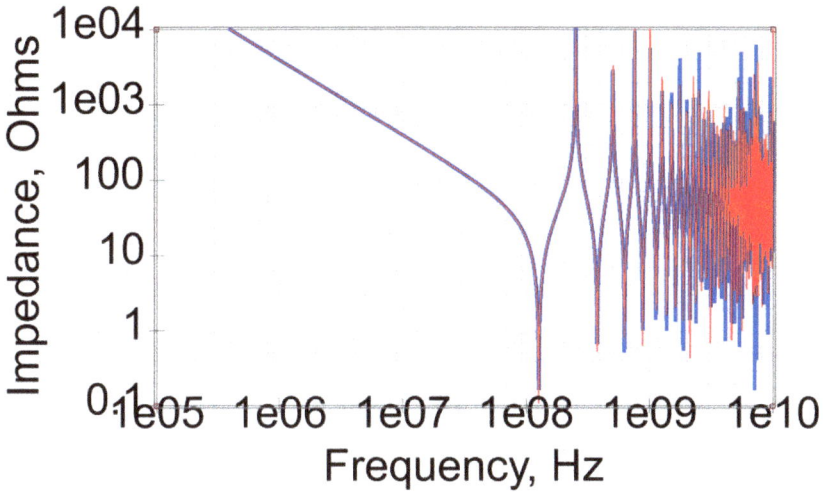

Figure 4.4 Simulated and analytically calculated input impedance of a transmission line, in the frequency domain. Simulated with QUCS, a SPICE-like simulator.

At low frequency, the input impedance of a transmission line open at the far end looks like the impedance of a capacitor. If the end of the transmission line were shorted, the input impedance of the shorted transmission line in the frequency domain looks like the impedance of an inductor.

This is the basis which we can approximate a transmission line as a capacitor or an inductor at low frequency. We refer to this similarity as the *lumped circuit approximation*. **Figure 4.5** shows this quality of the approximation.

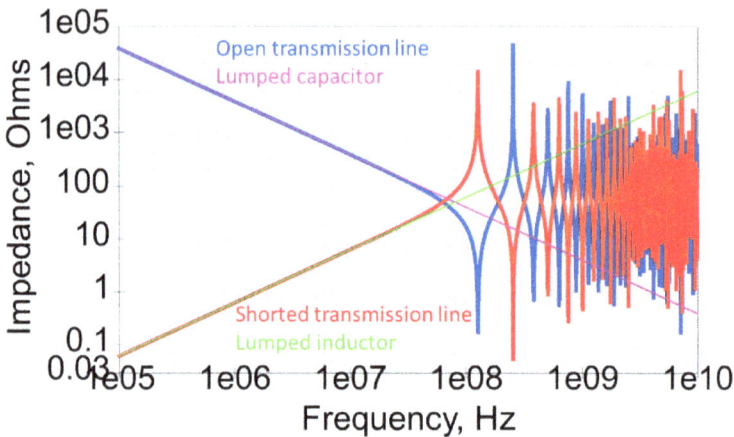

Figure 4.5 The simulated input impedance in the frequency domain of an open and shorted transmission line compared to a capacitor and inductor.

The highest frequency up to which a transmission line can be approximated by an ideal capacitor or inductor depends on the length of the transmission line. The longer the transmission line, the lower the frequency up to which the transmission line behaves like a lumped circuit element.

It is often difficult to translate this behavior at low frequency into the expected behavior in the time domain except for extreme conditions such as very long rise time signals and very short transmission lines.

Above this frequency, the input impedance of a transmission line in the frequency domain has a complicated behavior consisting of large fluctuations in input impedance, from very high impedance, like an open, to nearly zero impedance, like a short.

While this is the expected behavior of a transmission line in the frequency domain, it is difficult to translate this frequency domain behavior into the expected time domain behavior.

4.7 A Few Special Cases for the Input Impedance in the Frequency Domain

The input impedance of a transmission line varies with frequency with a very well-structured pattern. When the far end is open, the input impedance starts high, drops down, and then oscillates. When the far end is shorted, the input impedance starts low and increases with frequency until it begins to oscillate.

Using the relationship that describes the input impedance of a transmission line with frequency, we can explore a few special frequency values and the resulting impedance.

While this relationship is formidable looking as it involves complex numbers in the numerator and denominator, there are a few simple special cases worth considering. These are based on the phase term inside the tan() function.

This term is the phase angle of the wave in propagating down the length of the transmission line of length l.

The tan() function has a few special values:

- Case 1: $\tan(n \times \pi) = 0$

- Case 2: $\tan(\pi/2 + n\pi = 90 + n \times 180 \text{ deg}) = \text{infinite}$

- *Case 3: $\tan(45 \text{ deg} = \pi/8) = 1$*

Case 1: $\beta l = 0, \pi, 2\pi, 3\pi, ..., \tan(\beta l) = 0$

At the lowest frequency, and then as we increase frequency, there will be frequencies when the length of the line is an integral number of half wavelengths. Then:

$$Z_{in} = Z_0 \frac{Z_L + jZ_0 \tan(\beta l)}{Z_0 + jZ_L \tan(\beta l)} = Z_0 \frac{Z_L + 0}{Z_0 + 0} = Z_L \qquad (4.5)$$

This says, when the frequency is such that the phase length of the transmission line is really short, or an integral number of half wavelengths in other words, the round-trip delay is an integral number of whole wavelengths; the input impedance looking into a transmission line is the load impedance.

If the far end is open, the input impedance will look open. If the far end is a short, the input impedance will look like a short. If the far end is 50 ohm, the input impedance will be 50 Ohms. This is independent of the characteristic impedance of the transmission line.

Case 2: $\beta l = (0 + 1/2)\pi, (1 + 1/2)\pi, (2 + 1/2)\pi,..., \tan(\beta l) = $ infinite

As the frequency increases, there will reach a value where the phase length of the transmission line is exactly a phase of $\pi/2$, which is 90 degrees, or one quarter of a cycle. The round-trip delay is exactly half a cycle. This is the case when the length of the transmission line has an extra quarter cycle ($\pi/2$) delay in it.

In this case:

$$Z_{in} = Z_0 \frac{Z_L + jZ_0 \tan(\beta l)}{Z_0 + jZ_L \tan(\beta l)} = Z_0 \frac{Z_L + jZ_0 \infty}{Z_0 + jZ_L \infty} = Z_0 \frac{\dfrac{Z_L}{\infty} + jZ_0}{\dfrac{Z_0}{\infty} + jZ_L} = \frac{Z_0^2}{Z_L}$$

(4.6)

When the frequency corresponds to the length of the line being a quarter wavelength, the input impedance is the characteristic impedance squared over the far end load impedance.

When the far end is shorted, this quarter wavelength frequency corresponds to an infinite input impedance.

When the far end is open, this quarter-wavelength frequency corresponds to a shorted impedance.

This says we can easily find the frequency that corresponds to the length being a quarter wavelength by the first time the input impedance goes to either a short dip or an open peak.

Case 3: βl = π/8, tan(βl) = 1

As the frequency increases from zero, there will be a frequency where the length of the interconnect is exactly one eighth a wavelength, or half of π/2 phase. What an oddball phase to pay attention to. This frequency corresponds to exactly half the frequency of the transmission line length being one quarter of a wavelength.

At the frequency that is exactly half the quarter-wave frequency, where the dip or peak is located in the input impedance, for an open or short at the far end, the input impedance is:

$$|Z_{in}| = \left| Z_0 \frac{Z_L + jZ_0 \tan(\beta l)}{Z_0 + jZ_L \tan(\beta l)} \right| = \left| Z_0 \frac{Z_L + jZ_0}{Z_0 + jZ_L} \right| = Z_0 \sqrt{\frac{Z_L^2 + Z_0^2}{Z_0^2 + Z_L^2}} = Z_0$$

(4.7)

This is a very powerful observation. The phase angle of π/8 is 45 degrees. At the frequency where the transmission line length is half a quarter wavelength, the input impedance is exactly the characteristic impedance of the line. This is the case no matter what the load at the end of the line.

If you measure the input impedance of a transmission line with an open or a short, and you know the frequency where the length is one quarter a wavelength, take half this frequency. The impedance at this frequency is the characteristic impedance.

As an example, **Figure 4.6** shows the input impedance of three transmission lines. In each case, their far end is open. Their time delay is exactly 0.25 nsec. This means the frequency where this transmission line is one quarter a cycle is 1 GHz.

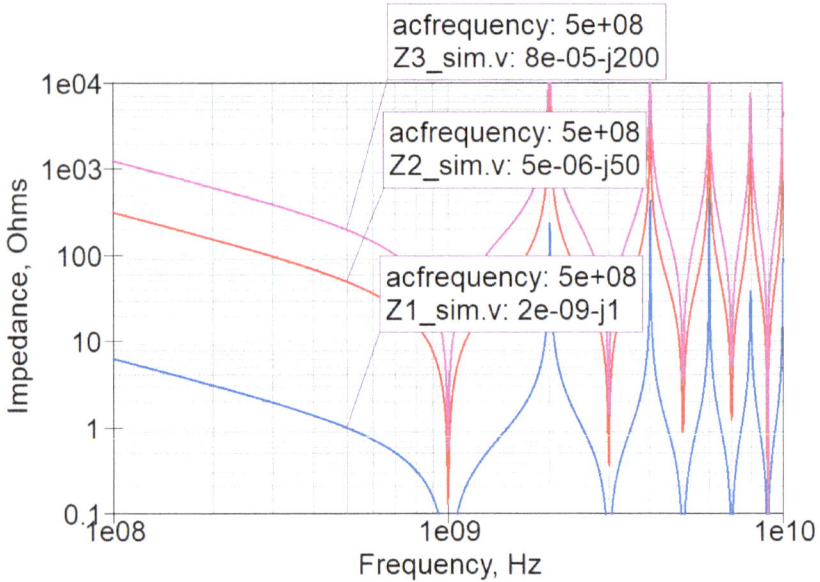

Figure 4.6 An example of the simulated input impedance of three transmission lines and their input impedance at a frequency where the line length is one eighth a wavelength.

Each transmission line was engineered with a different impedance: 1 ohm, 50 ohms and 200 ohms. Clearly their impedances at low frequency are very different. The lower impedance is for the transmission line that is 1 ohm and the highest impedance, the line that is 200 ohm.

The one quarter wave frequency is exactly 1 GHz in each case. The one eighth wave frequency is therefore 500 MHz. A marker was placed on the input impedance at 500 MHz for each line. We would expect that at this one eighth cycle frequency, the input impedance should be exactly the characteristic impedance of the line.

104

This is exactly what is found. The extracted magnitudes of the impedances are 1 ohm, 50 ohm and 200 ohms. The fact that the impedances are imaginary is not important. This relationship is about the absolute magnitude of the input impedance being the characteristic impedance.

This is a useful trick to use to quickly extract the characteristic impedance of any transmission line if the input impedance is measured in the frequency domain for the case of an open or shorted far end.

4.8 Which is Better, the Frequency or the Time Domain Impedance?

This input impedance of a transmission line in the frequency domain is a very different impedance than the input impedance of the same transmission line in the time domain.

They are both impedances of a transmission line but refer to very different situations and behaviors.

The frequency domain behavior with the wild impedance fluctuations is the expected impedance behavior of a real transmission line.

Not even experienced engineers can see the high-frequency input impedance of a transmission line in the frequency domain with these wild impedance fluctuations and use it to predict the time domain behavior of the input impedance of a transmission line or what happens to a signal as it propagates in the time domain.

For short rise time signals or high-frequency bandwidth signals, we really have to keep the impedance behaviors separate. While the information content is the same in both descriptions of the input impedance of a transmission line, their display and interpretations are very different.

This is why we need to learn to think in these two domains separately. It helps being schizophrenic, keeping these two world views separate. This is illustrated in **Figure 4.7**.

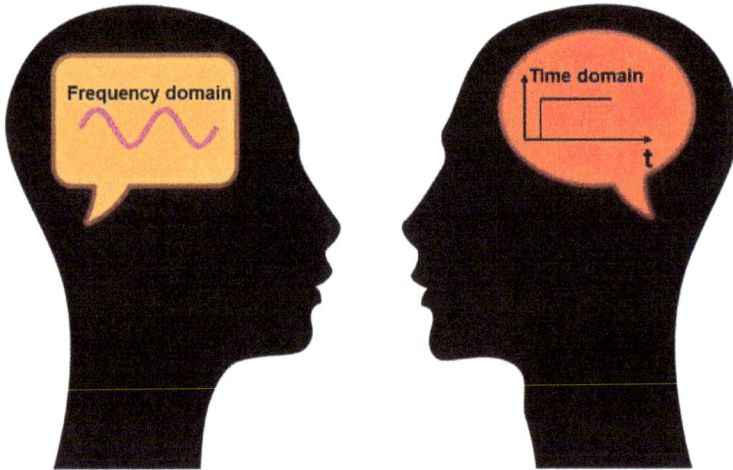

Figure 4.7 Being schizophrenic and thinking separately in the frequency domain and the time domain is important to understand transmission lines.

If you initially learned about transmission lines in the frequency domain, you have to add the separate behavior of the transmission line in the time domain to your understanding. You cannot apply your intuition about impedance in the frequency domain directly to the impedance in the time domain.

For most high-speed digital applications, understanding the impedance of a transmission line in the time domain is the most important perspective.

Depending on the question asked, we may get to the answer faster in the time domain or the frequency domain.

4.9   Review Questions

1. What is the fundamental, basic definition of impedance?

2. What are the five different impedances that can be associated with a transmission line?

3. What is the instantaneous impedance of a transmission line?

4. The characteristic impedance of a uniform transmission line is 60 ohms. What is the initial input impedance a signal will see entering the line? What is the instantaneous impedance the signal will see as it propagates down the line?

5. Why is it ambiguous when we just refer to "the impedance" of a transmission line?

6. When we do refer to "the impedance" of a transmission line, in the context of a digital signal propagating on the line, to which impedance are we probably referring?

7. What lumped circuit element does the front of a transmission line look like initially?

8. A 50 ohm transmission line is open at the far end. What impedance will a digital ohmmeter read if connected between the signal and return conductors? How does this relate to the 50 ohms of the transmission line?

9. Why is the source impedance of the driver important to know in order to calculate the voltage launched into a transmission line?

10. For how long a period of time can you think of a transmission line in a circuit as a resistor? And where would the resistor be placed?

11. A driver has a Thevenin source resistance of 10 ohms and a Thevenin source voltage of 3.3V. What is the voltage

initially launched into a 6- ohm transmission line? If its time delay is 1 nsec, for how long will the input of the transmission line look like a 60-ohm resistor?

12. In the frequency domain, what is the assumption about the sine wave signal entering the transmission line?

13. Why is there no reflection when a signal, traveling on a 75-ohm impedance transmission line encounters a 75-ohm resistor between the signal and return path?

14. At low frequency, what does the input impedance of a transmission line, open at the far end, look like? What if it were shorted at the far end? What will it look like electrically?

15. If the length of the transmission line, shorted at the far end, is exactly half a wavelength, what will the input impedance of a transmission line look like in the frequency domain?

16. If you have the measured impedance vs frequency of a transmission line open at the far end, you will see a minimum impedance at a frequency where the length of the transmission line is exactly a one quarter cycle. At what frequency would you look to read the characteristic impedance of the transmission line directly from the impedance curve?

17. What is different about the input impedance of a transmission line in the time domain than in the frequency domain?

Chapter 5   Why We Care About Impedance: Reflections

A signal propagating down a transmission line, seeing a constant instantaneous impedance, will propagate undistorted. It doesn't matter what the value of the instantaneous impedance is, whether 50 ohms, 20 ohms, or 90 ohms.

Distortions in the signal occur when the instantaneous impedance changes. It's a changing instantaneous impedance that causes reflections and multiple reflections can distort the signal at the receiver.

An example of the measured reflection noise at a receiver from a uniform transmission line with discontinuities on its ends is shown in **Figure 5.1**.

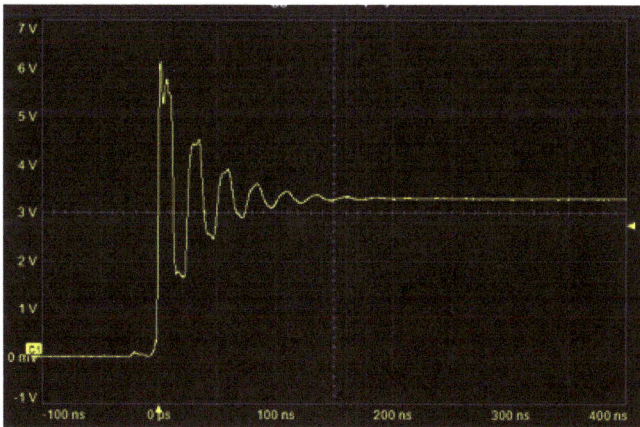

Figure 5.1 Measured voltage noise at a receiver from a driver with a clean step edge signal. The ringing is due to multiple reflections from discontinues in the transmission line circuit.

It is to prevent these sorts of distortions that we care about the instantaneous impedance a signal sees and want to engineer it to be as constant as practical. Then the goal becomes to reduce all the other impedance discontinuities in the circuit.

5.1 Reflections Keep the Universe from Blowing Up

If the signal sees the same instantaneous impedance each step along its path, it will propagate undistorted. Whatever rise time or edge shape goes into the transmission line will appear everywhere down the transmission line…unless the instantaneous impedance changes.

Whenever the instantaneous impedance changes, some of the signal will reflect and what continues will be distorted.

If there were no reflection, *the universe would blow up*. To prevent this from happening, a reflected signal is generated, and the transmitted signal is distorted in a special way.

The simplest circuit we consider initially is the transition between two uniform transmission lines with different characteristic impedances. The transmission line in region 1 has a characteristic impedance of $Z_1 = 50$ ohms and in region 2 it has a characteristic impedance of $Z_2 = 75$ ohms. This is illustrated in **Figure 5.2**.

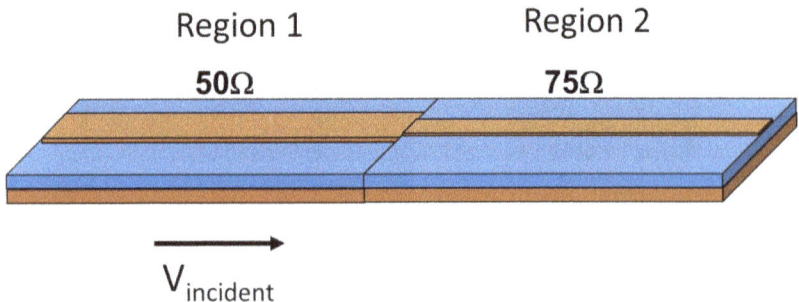

Figure 5.2 Two transmission lines connected at a boundary, with a signal incident from the 50-ohm side.

In region 1, a signal with a voltage of $V_{incident} = V_1$ is propagating toward the right. The current in the transmission line associated with the signal is also propagating from left to right and is circulating in the CW or positive direction. Its magnitude is I_1.

The same is defined in region 2.

In each region, by definition, the ratio of the voltage of the signal to the current magnitude is Z_1 and Z_2, with the definitions of:

$$\frac{V_1}{I_1} = Z_1 \quad \text{and} \quad \frac{V_2}{I_2} = Z_2 \tag{5.1}$$

These two statements have to be true, by definition.

We have engineered the two transmission line regions to have $Z_1 \neq Z_2$, so these ratios cannot be equal.

In addition to these two relationships, there are two other important relationships between the voltages and currents in both regions, called *boundary conditions*. These relate to properties of voltages on conductors and to currents flowing into and out of conductors.

The first boundary condition is that voltages between two adjacent points on a conductor must be continuous. In other words, the voltage between two adjacent points on a conductor cannot change in a sudden jump.

If there were a voltage difference between two adjacent points on a conductor, right next to each other, the electric field between these two points would be:

$$E = \frac{\Delta V}{x} \tag{5.2}$$

where

ΔV = the voltage difference between the two points

x = the separation between them.

This boundary condition means that if we make the distance between two points on a conductor smaller and smaller, the voltage difference between the two points must approach 0V. If it didn't, the electric field would get bigger and bigger and could get so large as to blow up the universe.

This boundary condition says that voltages across conductor boundaries must be continuous.

If we stand with one foot on line 1 and one foot on line 2 and bring our feet closer together so they are on adjacent points, the voltage on each side of the boundary must be the same. The condition is:

$$V_1 = V_2 \qquad (5.3)$$

The second boundary condition is that current that flows between the signal and return path into a node must equal the current that flows between the signal and return path out of the node. This is another way of saying charge is conserved.

This boundary condition says whatever charge flows in must flow out.

$$I_1 = I_2 \qquad (5.4)$$

If there were an imbalance in the currents flowing into and out of a junction, charge would build up. If we wait long enough, we could have an arbitrarily large amount of charge build up and the universe would blow up.

We've introduced four different relationships between the voltage and current on both sides of the boundary. They must all be true at the same time, yet if all we have is a signal entering the interface between the two regions and a signal leaving the interface, we run into an inconsistency that blows up the universe.

These relationships cannot all be true at the same time with just two signals. The impedances on either side are different, yet the voltages on either side of the interface must be the same and the currents going into and out of the boundary must be the same.

If all of these relationships were true, the universe would blow up.

To prevent it from blowing up, when a signal is incident to the interface between two impedances, a third signal is generated at the boundary, propagating in the opposite direction from the incident signal, reflected from the interface. This reflected signal is to keep the universe in balance.

Watch this video to see more details about the boundary conditions on the voltages and currents at the interface.

5.2   The Reflection and Transmission Coefficient

To keep track of the fact that we have two signals now in region 1, we will call the incident signal traveling from the left to right, $V_{incident}$, or V_i and the reflected signal, created at the boundary and traveling from the right to the left, $V_{reflected}$, or V_r. Likewise, we change the name of the voltage in region 2 to describe the fact that this is a transmitted signal. There is only one signal propagating in region 2, traveling from the interface to the right, V_2. For consistency, we can also call this signal the transmitted signal, labeled as $V_{transmitted}$, or V_t. This is illustrated in **Figure 5.3**.

113

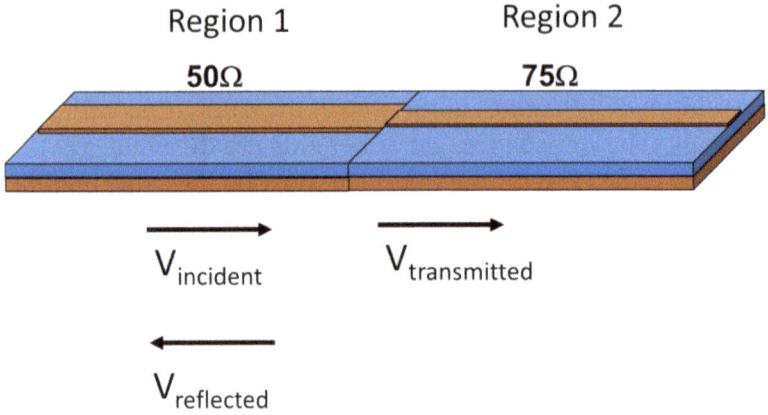

Figure 5.3 The incident and reflected signals in region 1 and the transmitted signal in region 2.

The reflected signal is a signal propagating in region 1. Regardless of the direction of propagation, all signals in region 1 will see an instantaneous impedance of Z_1.

This means that associated with the V_r signal is a current, I_r and in region 1, we must have:

$$\frac{V_r}{I_r} = Z_1 \tag{5.5}$$

What happens in region 1 at the boundary has changed. The voltage we measure between the signal and return path right at the boundary in region 1 is now the sum of two signals:

$$V_1 = V_i + V_r \tag{5.6}$$

The currents are a little more complicated.

Right at the boundary, the V_i signal is propagating from left to right and its current, I_i, is circulating in the CW, positive, direction.

Right at the boundary, the V_r signal is propagating from right to left and its current, I_r, is circulating in the signal-return direction, which is CCW, the negative direction. These current directions are illustrated in **Figure 5.4**.

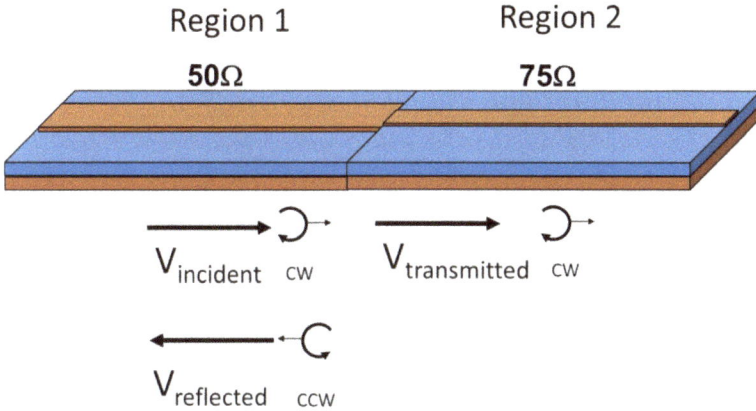

Figure 5.4 Direction of propagation of the signals and direction of circulation of the currents.

While the voltages at the boundary in region 1 add, the currents *at the boundary* in region 1 are in opposite directions and subtract. The net current flowing between the signal and return paths in region 1 *at the boundary* is

$$I_1 = I_i - I_r \qquad (5.7)$$

Now we apply our boundary conditions to the voltage across and current into the boundary:

$$I_1 = I_2$$
$$I_i - I_r = I_1 \qquad (5.8)$$

and

$$V_1 = V_2$$
$$V_i + V_r = V_1$$

(5.9)

Watch this video to see more details on the incident reflected and transmitted voltages and currents.

In addition, we have the three impedance relationships:

$$Z_1 = \frac{V_i}{I_i}$$

$$Z_1 = \frac{V_r}{I_r}$$

(5.10)

$$Z_2 = \frac{V_2}{I_2}$$

The starting place is the incident voltage and the two impedances, Z_1 and Z_2. These are the values we know. Using the boundary conditions, with a little algebra, we can derive the values of the reflected voltage, V_r, and the voltage that transmits into region 2, V_2, as:

$$V_r = V_i \frac{Z_2 - Z_1}{Z_2 + Z_1}$$

$$V_2 = V_i \frac{2 \times Z_2}{Z_2 + Z_1}$$

(5.11)

It has become traditional to define two new figures of merit that describe what happens at the boundary, the *reflection coefficient* and the *transmission coefficient*. They are a measure of the relative amount of the reflected signal and the transmitted signal compared to the incident signal.

We calculate the reflection coefficient as

$$R = \rho = rho = \text{reflection coefficient} = \frac{V_r}{V_i} = \frac{Z_2 - Z_1}{Z_2 + Z_1}$$

$$T = \text{transmission coefficient} = \frac{V_2}{V_i} = \frac{2 \times Z_2}{Z_2 + Z_1} \quad (5.12)$$

To see all the gory details on completing the algebra to derive these two coefficients, watch this video.

5.3 Using the Reflection Coefficient and the Transmission Coefficient

These two relationships are the basis of all behaviors of signals on transmission lines. In this first example, we looked at the case of one interface between two different uniform transmission lines.

This effect of a reflected signal generated, born propagating in the opposite direction of the incident signal, and the transmitted signal being distorted, describe even the most complicated interconnects. It's just sometimes hard to keep track of all the reflections and all the bounces the incident, reflected, and transmitted signals encounter.

Generally, keeping track of all the reflections can be complicated. In the next chapter we introduce the bounce diagram to keep track of all the reflections. Then we introduce simulation tools that can analyze even more complicated situations.

The simplest case we can analyze manually is when there is just one interface.

Example 1: Consider the boundary between two impedances, region 1 is 50 Ohms and the impedance of region 2 is 75 ohms, illustrated in **Figure 5.5**.

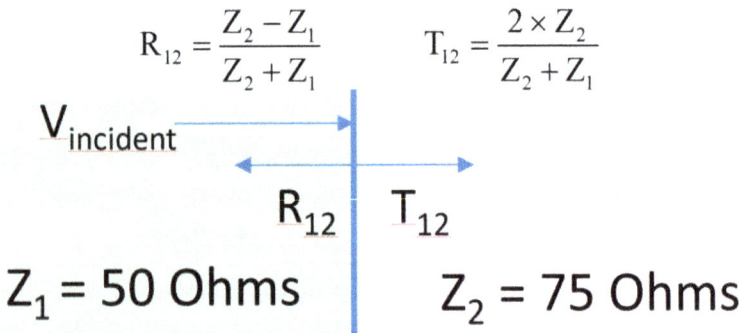

$$R_{12} = \frac{Z_2 - Z_1}{Z_2 + Z_1} \qquad T_{12} = \frac{2 \times Z_2}{Z_2 + Z_1}$$

$V_{incident}$

$R_{12} \quad T_{12}$

$Z_1 = 50 \text{ Ohms} \qquad Z_2 = 75 \text{ Ohms}$

Figure 5.5 The boundary between two impedances, 50 ohms and 75 ohms.

The reflection and transmission coefficients are:

$$R_{12} = r = \frac{Z_2 - Z_1}{Z_2 + Z_1} = \frac{75 - 50}{75 + 50} = \frac{25}{125} = 0.2$$

$$T_{12} = t = \frac{2 \times Z_2}{Z_2 + Z_1} = \frac{2 \times 75}{75 + 50} = \frac{150}{125} = 1.2$$

(5.13)

If the incident signal is 1V, then traveling in region 1, at the boundary, there are two waves. The incident wave is 1V and the reflected wave, traveling to the left at the boundary, is 1 V x 0.2 = 0.2 V. The net voltage we measure between the signal and the return path at the boundary is 1V + 0.2 V = 1.2 V.

On the right side of the boundary, the voltage transmitted that is propagating into region 2 is 1 V x 1.2 = 1.2 V.

On either side of the boundary is the same voltage, 1.2 V. This has to be the case. It is the boundary condition. Now we see how the reflected wave makes this happen.

The current in the incident signal, propagating from left to right, and circulating in the clockwise (CW) direction is:

$$I_i = \frac{V_i}{Z_1} = \frac{1\,V}{50\,\Omega} = 20\,mA \qquad (5.14)$$

After the incident signal hits the boundary and the reflected signal is created, the current in the reflected signal, propagating from the right to the left, is

$$I_r = \frac{V_r}{Z_1} = \frac{0.2\,V}{50\,\Omega} = 4\,mA \qquad (5.15)$$

This current is part of the reflected signal propagating from the right to the left. Since it is a positive signal, the current is circulating in the counterclockwise (CCW), negative, direction.

Right at the boundary, the net current is 20 mA in the CW, positive direction, and 4 mA in the CCW, negative direction. The net current right at the boundary on the left side is 20 mA – 4 mA = 16 mA.

In region 2, the transmitted signal is propagating from left to right. The current is circulating in the CW, positive direction and is:

$$I_2 = \frac{V_2}{Z_2} = \frac{1.2\,V}{75\,\Omega} = 16\,mA \qquad (5.16)$$

Incredibly, after all this algebra and calculations we find that the net current into the boundary from region 1 is exactly the same as the net current out of the boundary in region 2, exactly as it should be to keep the universe from blowing up.

Of course, if the incident voltage is 2 V and not 1 V, all the terms just scale.

The behavior of the incident, transmitted, and reflected signals is dynamic. It is not just about what happens at the interface. It is also about their propagation.

Watch this video to see the dynamic propagation of the signals in the transmission lines leading to and coming from the interface.

Example 2: Suppose the impedance in region 2 is 25 ohms and the incident voltage is 1 V. The same analysis applies, it's just with different values for the reflection coefficient and transmission coefficient. These values are:

$$r = \frac{Z_2 - Z_1}{Z_2 + Z_1} = \frac{25 - 50}{25 + 50} = \frac{-25}{75} = -0.33$$

$$t = \frac{2 \times Z_2}{Z_2 + Z_1} = \frac{2 \times 25}{25 + 50} = \frac{50}{75} = 0.67$$

(5.17)

The reflection coefficient is negative. This means that if a 1V signal travels from left to right into the boundary, a – 0.33 V signal will reflect, traveling from the right to the left. A negative voltage will reflect when the second impedance is less than the first impedance.

Right at the boundary, the voltage in region 1 is 1 V – 0.33 V = 0.67 V. This is exactly the same as the voltage traveling in region 2, the transmitted voltage, as it should be.

The current in the incident signal propagating from left to right, and circulating in the CW direction is:

$$I_i = \frac{V_i}{Z_1} = \frac{1\,V}{50\,\Omega} = 20\text{ mA} \qquad (5.18)$$

The reflected signal, of -0.33 V, traveling from right to left, has a current circulating in the return-signal direction, the CW, positive, direction. Its current is

$$I_r = \frac{V_r}{Z_1} = \frac{0.33\,V}{50\,\Omega} = 6.6\text{ mA} \qquad (5.19)$$

This is circulating in the same direction as the incident signal's current.

Right at the boundary, in region 1, the net current is 20 mA + 6.6 mA = 26.6 mA.

In region 2, the signal is propagating from left to right. Its value is 1 V × 0.67 = 0.67 V. The current is circulating in the CW direction. The current in the transmitted signals in region 2 is

$$I_2 = \frac{V_2}{Z_2} = \frac{0.67\,V}{25\,\Omega} = 26.8\text{ mA} \qquad (5.20)$$

This is exactly (to within the round-off error) the same current that is flowing between the signal and return path in region 1 of the boundary, as it should be.

Example 3: trivial case. Suppose the second region is also 50 ohms. Without doing any calculation, we know the answer. If the instantaneous impedance does not change, there will be no reflected signal and the transmitted signal should be the same as the incident signal.

The reflection and transmission coefficients in this case are:

$$r = \frac{Z_2 - Z_1}{Z_2 + Z_1} = \frac{50 - 50}{50 + 50} = 0$$

$$t = \frac{2 \times Z_2}{Z_2 + Z_1} = \frac{2 \times 50}{50 + 50} = \frac{100}{100} = 1$$

$$(5.21)$$

This is consistent with what we expect to see.

5.4 An Important Distinction Between the Signal and the Voltage

One of the most confusing aspects of the reflections of signals on transmission lines is the distinction between the signals that are propagating on the transmission line and the voltages that are measured at a specific node by a scope.

There is absolutely nothing we can do to prevent signals, once launched on a transmission line, from propagating. This is a fundamental property of all signals. Because of reflections, if we launch a signal into a transmission line, it may end up rattling around back and forth on the interconnect.

The problem is, we can't directly measure the propagation characteristics of signals on transmission lines. All we can measure with a scope, for example, is the net, total voltage between the signal and return path at some location. There is no direction information in the voltage the scope displays.

For example, **Figure 5.6** shows the measured voltage at the end of a transmission line. Is this voltage created by one, 6V peak-to-peak signal propagating into the scope, or two, 3V peak to peak signals, one propagating into the scope and one reflecting off the high impedance of the scope? You can't tell by looking at the measured voltage.

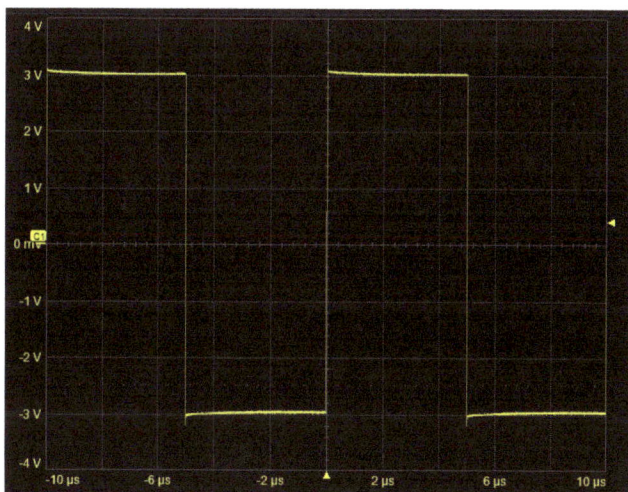

Figure 5.6 The measured voltage at the end of a transmission line. Is this one signal or two signals?

In fact, in this example, the input to the scope is 1 Meg, and the reflection coefficient for the signal coming from the source into the 1 Meg of the scope is nearly 1. This means at the scope input, there are really two signals traveling in opposite directions. The scope is measuring the sum of the incident + the reflected signals. But you can't tell this just by looking at the screen.

All we can measure with a scope is the net, total voltage, not the direction of propagation of the signals.

We have to use our engineer's eye to reverse engineer that the voltage signal we see on the scope is really composed of the wave traveling in one direction and another wave traveling in the opposite direction.

Be aware of these two separate terms, the *voltage* measured at some location, which has no direction information and is always the total, net voltage between the signal and the return path, and the *signals* that are propagating in specific directions.

5.5   Important Termination Special Case: A 50-Ohm Resistive Load

A termination is the resistance at the end of the transmission line. It terminates or ends the transmission line. If is sometimes assumed to be 50 ohms but can be any impedance value.

There are three important special case terminations to understand: when the termination is matched to the characteristic impedance of the transmission line, when it is an open, when it is a short.

The behavior of a signal when transitioning between a 50-ohm transmission line and a 50-ohm resistor is almost trivial but has a few subtleties. An example of this transition is illustrated in **Figure 5.7**.

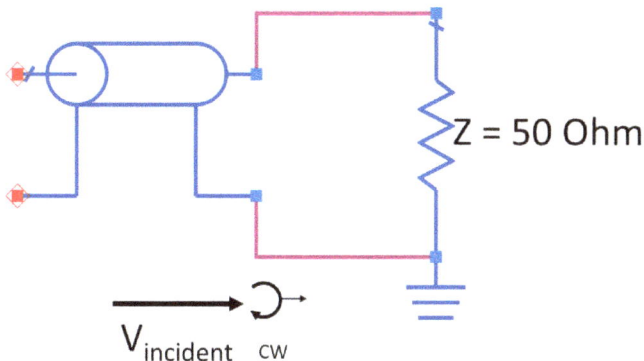

Figure 5.7 Special case 1 is a signal propagating on a uniform 50-ohm transmission line and encounters a 50-ohm resistor between the signal and return path.

The signal does not care what the nature of the instantaneous impedance is. All it cares about is the value of the instantaneous impedance.

As long as the instantaneous impedance between the signal and return path is constant as the signal propagates along its path, it will continue undistorted. If the instantaneous impedance changes for whatever reason, there will be a reflection.

When the signal encounters a resistor between the signal and return conductors, the signal will see an instantaneous impedance equal to the resistance of the resistor.

If the signal sees a 50-ohm instantaneous impedance on a transmission line and then encounters a resistor also of 50 ohms, connected between the signal and return paths, it will see no change in the instantaneous impedance.

There is no reflection at the 50-ohm resistor, so the signal keeps going. As soon as it hits the 50-ohm resistor, the voltage in the signal will appear across the resistor, driving the same current through the resistor as is propagating down the transmission line.

The energy being transported by the signal down the transmission line gets turned into heat in the resistor. Whatever rise time signal enters the transmission line, will appear as the voltage across the 50-ohm resistor.

Using the relationships above, we can calculate the reflection and transmission coefficients when the signal hits the boundary between the 50-ohm transmission line and the 50-ohm resistor.

The reflection and transmission coefficient from a 50-ohm load are:

$$r = \frac{Z_2 - Z_1}{Z_2 + Z_1} = \frac{50 - 50}{50 + 50} = 0$$

$$t = \frac{2 \times Z_2}{Z_2 + Z_1} = \frac{2 \times 50}{50 + 50} = 1$$

(5.22)

This says in algebra what we already expect to see:

- The reflection coefficient is 0: nothing reflects
- The transmission coefficient is 1: everything transmits

5.6 Important Termination Special Case: An Open

Suppose we have a simple 50-ohm transmission line with an open at the far end. This is illustrated in Figure 5.8 using a resistor to symbolize the infinite impedance at the far end.

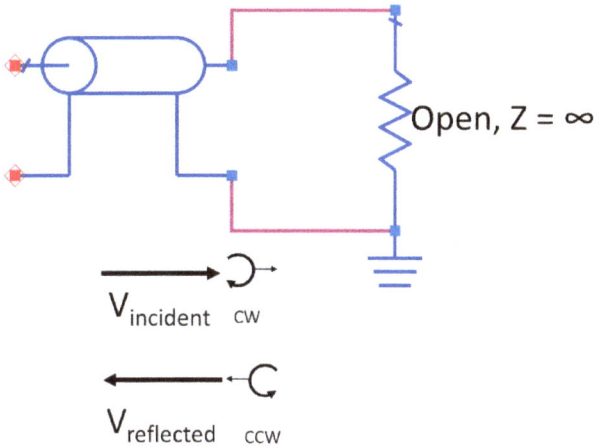

Figure 5.8 Reflections from an open.

From the signal's perspective, it doesn't care what the nature of the impedance is that it encounters. All it cares about is the instantaneous impedance. When it steps on the resistance at the far end, what reflects is only about the change in instantaneous impedance. The instantaneous impedance of a resistor is just its resistance.

The reflection and transmission coefficient from the open are:

$$r = \frac{Z_2 - Z_1}{Z_2 + Z_1} = \frac{\infty - 50}{\infty + 50} = 1$$

$$t = \frac{2 \times Z_2}{Z_2 + Z_1} = \frac{2 \times \infty}{\infty + 50} = 2$$

(5.23)

If the incident signal, propagating toward the right, is 1 V, the reflected signal from the open is also 1 V, but traveling toward the left. The transmitted signal that appears across the open impedance is 2 V. Of course, it has nowhere to go, and since the impedance is open, there is no power dissipation.

If we were to measure the voltage across the open, we would find either the transmitted signal, 2 V or the sum of the incident and reflected, 2 V. They both give the same result, of course.

While it is common to describe what happens at the open as "the signal doubles," thinking this way confuses the signal that propagates with the voltage that is measured and hides the real mechanism.

Don't think "the voltage doubled" at the open.
Think "the signal reflected with a reflection
coefficient of 1 and the voltage measured is the
sum of the incident and reflected signals." This
keeps straight the distinction between propagating
signals and net voltages.

The current into the open is more subtle. In region 1, the current at the boundary is the sum of the currents. The incident current is from the 1V signal in the 50-ohm impedance of the transmission line in region 1. This current is

$$I_1 = \frac{V_1}{Z_0} = \frac{1\,V}{50\,\Omega} = 20\,mA \qquad (5.24)$$

As the incident signal propagates, it is a 1V signal and a 20-mA current propagating from left to right circulating in the CW direction. As the incident signal propagates, it brings with it a 20-mA current flowing in the transmission line. Until it hits the open, it has no idea there is an open at the end of the line.

When the incident signal hits the open and reflects, the reflected 1V signal propagates to the left, but its current is circulating in the CCW direction. The net current flowing between the signal and return path at the boundary is the sum of these two currents.

128

Since the signals have the same voltage, and they see the same impedance, the magnitudes of the currents are the same. Since they are circulating in the opposite directions, they completely cancel out and there is no net current in region 1 at the interface.

Of course, this has to be the case. There is an open on the other side of the boundary, and you cannot have current through an open.

As the reflected signal propagates back to the source, each step along its way, it leaves in its wake a CCW circulating current. This will cancel with the CW circulating current from the incident wave that is also on the transmission line. As the reflected wave propagates to the left, where it has passed, there is no net current in the transmission line.

5.7 Important Termination Case: A Short

Next, consider the case of the far end having a short. The reflection and transmission coefficients are:

$$
\begin{aligned}
r &= \frac{Z_2 - Z_1}{Z_2 + Z_1} = \frac{0 - 50}{0 + 50} = -1 \\
t &= \frac{2 \times Z_2}{Z_2 + Z_1} = \frac{2 \times 0}{0 + 50} = 0
\end{aligned}
\tag{5.25}
$$

When the 1V signal hits the short, the reflected signal is -1 V. Right at the interface, the net voltage is the sum of the two signals, which is 0 V. They cancel out. And, of course, the voltage across a short has to be 0 V, it's a short.

Before the incident signal has reached the short, it is propagating along the 50-ohm transmission line exactly the same as the case of the open at the far end. The signal has no idea what the end termination is until it encounters it.

After the reflected -1V signal is created and propagates to the left, it leaves in its wake a -1V voltage. The net voltage we would measure with a scope probe is 1 V + -1 V = 0 V. As the reflected wave propagates to the left on the transmission line, it leaves in its wake a net 0 V.

The currents associated with the signals are analyzed the same way as with the open. It's just that the negative-reflected signal propagating to the left has a current circulating in the CW direction. It is circulating from the return to the signal as it propagates to the left. This is illustrated in **Figure 5.9**.

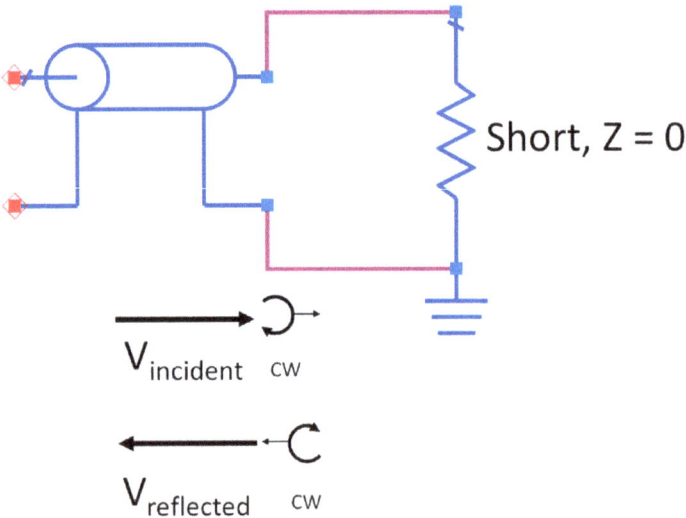

Figure 5.9 The negative reflected signal has a current circulation from the return to the signal path in the CW direction.

The incident current, propagating to the right, is 20 mA, circulating in the CW direction. After it reflects, the reflected current is also 20 mA, propagating to the left, circulating in the CW direction.

Right at the boundary, the current flowing between the signal and return paths is the sum of these currents. There is 20 mA in the

CW direction and 20 mA in the CW direction. This means the current is 40 mA in the CW direction.

As the reflected signal propagates toward the left, it leaves in its wake a net current in the transmission line, flowing down the signal and back up the return path, of 40 mA.

Watch this video and I will walk you through the reflections from an open and short with the resulting dynamic voltages on the transmission line.

In the case of the open at the far end, we saw that the voltage appearing across the open is 2 V and the current into it is 0 mA. In the case of the short at the far end, we saw that the voltage was 0 V and the current was 40 mA.

We expected the current through the open to be 0 mA. After all, it's an open. We expected the voltage across the short to be 0 V. After all, it's a short.

But we sent only 1 V into the transmission line toward the open, and we measured 2 V across the open. Where did the extra 1 V come from? We only started with 1 V.

We sent only 20 mA of current down the transmission line. We got 40 mA through the short. Where did the extra 20 mA come from?

Understanding the root cause of these paradoxical questions is at the heart of understanding transmission line circuits.

5.8   Resolving the Paradox: Where Did 2V and 40 mA Come From?

In the previous section, we ended up with two paradoxes. We send 1V down the transmission line and we see 2V across the open. We send 20 mA of current down the transmission line and we end up with 40 mA through the short. Where did the extra voltage and current come from in each case?

The answer lies in looking at the complete circuit. This includes the *source voltage* and the *source resistance*, in addition to the transmission line and its load.

Signals are dynamic. They are constantly propagating. When we looked at the special case of reflections from the open or short at the end, we looked at a snapshot in time, shortly before and shortly after the reflection occurred. But there is also what happens over a longer period of time and to analyze this, we have to know the important features of the source: its *voltage* and *resistance*. The complete circuit is shown in Figure 5.10.

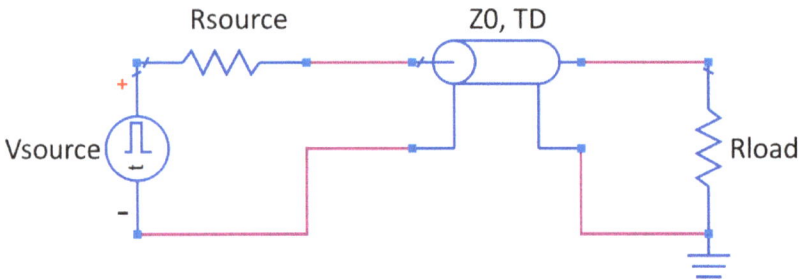

Figure 5.10 A simple transmission line circuit with a source and a resistive load at the end.

In this example with a 1V signal launched into a transmission line, let's engineer this with a 50-ohm source resistance. In order to launch a 1V signal into the transmission line, the source voltage has to be 2V. The circuit, for the case of the open or short at the far end, is show in **Figure 5.11**.

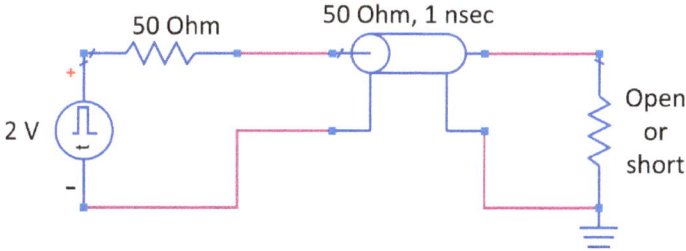

Figure 5.11 Complete circuit to explore the reflections from an open or a short.

Consider first an open at the far end. When the 2V source voltage turns on, it sees the voltage divider at the connection between the source resistance and the input resistance of the transmission line. At this voltage divider, half of the source voltage is launched into the transmission line. The 1V signal that is launched travels down the transmission line and reflects from the open at the far end.

A 1V reflected signal propagates back toward the source, leaving a net voltage of 2V in its wake. A current of 20 mA circulating in the CCW direction is propagating back to the source, and wherever they overlap, cancels out with the forward propagating 20 mA current circulating in the CW direction.

The impedance of an ideal voltage source is 0. You can convince yourself of this by considering the definition of impedance and the properties of an ideal voltage source. The voltage at its output is constant, independent of any current from it. If the current changes, the voltage change is 0V and,

$$Z = \frac{\Delta V}{\Delta I} = \frac{0}{\text{anything}} = 0 \tag{5.26}$$

The resistance of an ideal voltage source is 0 ohms. The current draw could be anything, but the output voltage would stay the same. It behaves like a 0-ohms impedance.

When the reflected signal reaches the source, it will see a resistance from the end of the transmission line to the return path through the 50-ohm resistor and the 0-ohm impedance of the ideal voltage source. This is still 50 ohms, which is the same impedance the signal sees on the transmission line. There is no change in impedance and no reflection. The result is the superposition of the 1V incident and 1V reflected wave at the entrance to the transmission line, which is 2V between the signal and the return.

The 20-mA current of the reflected signal circulating in the CCW direction will combine with the incident current circulating in the CW direction. The net current goes right into the 50-ohm resistance and the voltage source will be 0 mA.

After the reflected signal reaches the source resistor, there is no reflection and there are no further changes. The reflected signal is effectively terminated at the source.

In this steady-state situation, there are two waves constantly traveling down the transmission line. One is circulating 20 mA of CW current in the line, and the other is circulating 20 mA of CCW current in the transmission line. Everywhere they overlap, the currents cancel out and there is no net current.

Everywhere these two signals overlap there is a net 2V between the signal and return paths on the transmission line. You can't tell this is created by two waves traveling in the opposite direction.

This is the configuration we would expect to see based on our transmission line analysis.

At steady state, the 2V source is connected to an open at the far end through this transmission line. There should be no current in the transmission line, and since it is connected to a 2V source, we should see 2V between the signal and return path everywhere.

Where did the 2V come from that we see at the end of the transmission line, when we sent in just 1V? It came from the 2V of the source voltage.

In the case of the short, the same situation is at the launch. The reflected signal is a -1V signal propagating to the left. Everywhere it overlaps the incident signal, it leaves the net voltage, which is 0V in its wake.

The current in the reflected wave is 20 mA, circulating in the CW direction, propagating to the left, back to the source. Everywhere in its wake, it leaves a net current of 40 mA circulating in the CW direction.

When the -1V signal reaches the source resistor, it also sees no change in impedance and there is no reflection. Instead, the -1V signal subtracts from the 1V signal already at the entrance to the transmission line, leaving 0V.

Across the 50-ohm resistor at the source is a 2V drop. The current into this transmission line, flowing down the signal path of the transmission line and back through the return, is

$$I = \frac{2 \text{ V}}{50 \text{ }\Omega} = 40 \text{ mA} \qquad (5.27)$$

This is exactly what we get using the transmission line analysis. Where did the 40-mA come from? If came from the source. When the source is shorted through its 50-ohm output resistance, it will supply 40 mA into the 50-ohm load.

Watch this video and I will walk you through the transmission line analysis and the steady-state simple source and load analysis to show they are the same.

5.9 Another Paradox: The Signal Launched into the Transmission Line

When describing the initial voltage launched into a transmission line, we used the model of the voltage divider between the Thevenin source resistance and the input impedance of the transmission line.

Why don't we think of the same interface in terms of reflections? The signal comes from one impedance environment and encounters the impedance environment of the transmission line. Why don't we use the reflection and transmission analysis to calculate the voltage launched into the transmission line?

Consider the special case of a Thevenin source impedance of 10 ohms driving a 50-ohm transmission line. This is illustrated in **Figure 5.12**.

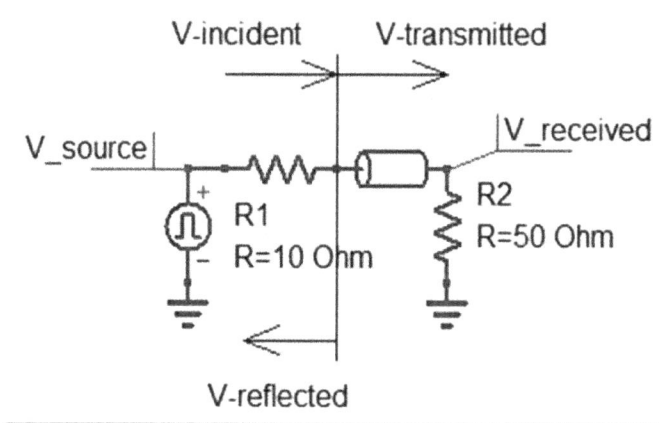

Figure 5.12 Initial set up of the driver launching a signal into a 50 Ohm transmission line.

In the first approach, we consider the voltage divider circuit translating the input impedance of the transmission line as a resistor. This will be the case for a time shorter than the round-trip time of the transmission line. The equivalent circuit of this system is shown in **Figure 5.13**.

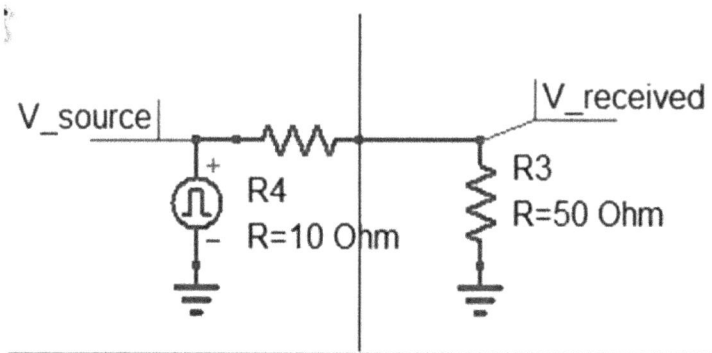

Figure 5.13 The equivalent circuit of the Thevenin source resistance driving the input resistance of the transmission line.

Using the voltage divider approach, we expect the voltage launched into the transmission line to be

$$V_{received} = V_{source} \frac{50}{10+50} = V_{source} \frac{5}{6} \qquad (5.28)$$

But, if we were to naively apply the transmission coefficient analysis, we would have estimated the transmission coefficient of the signal going from the 10-ohm resistor into the 50-ohm transmission line as

$$T = \frac{2 \times Z_2}{Z_1 + Z_2} = \frac{2 \times 50}{10+50} = \frac{100}{60} = 1.67 \qquad (5.29)$$

This analysis would have resulted in an estimate of the received signal of

$$V_{received} = V_{source} \times t = V_{source} \times 1.67 \qquad (5.30)$$

This result is very different than what we estimated based on the voltage divider approach. What gives? Which is the correct approach? Why don't they give the same results?

This is the seeming paradox.

When we apply the transmission line analysis to derive the reflection and transmission coefficient, we are using the model of signals propagating on a transmission line, seeing a change in the instantaneous impedance.

The assumption in the derivation of the reflection and transmission coefficient was that the signal starts in a transmission line environment and sees a change in the instantaneous impedance. But this is not the environment in which we applied the transmission coefficient calculation in this problem.

To use the transmission coefficient analysis and analyze the signal launched into the transmission line in terms of reflection and transmission coefficients, we have to redraw the circuit model slightly to illustrate how to think of this problem as a dynamic signal propagating down a transmission line.

We can think about the source impedance being a 10-ohm source resistor and a short length of 10-ohm transmission line. Its length is irrelevant. It could be a micron; it could be 10 inches. There is no difference in the result. **Figure 5.14** shows what the equivalent circuit really is.

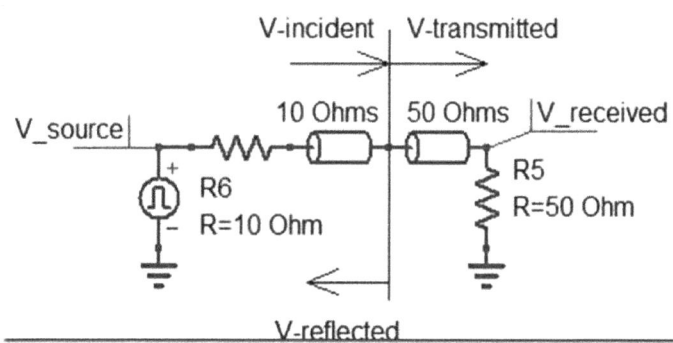

Figure 5.14 The equivalent circuit described in terms of transmission lines at the interface.

Adding the 10-ohm transmission line is perfectly equivalent to the original schematic, and explicitly shows the path for the signal propagating from the source to the 50-ohm transmission line.

In this circuit, we can analyze the voltage launched from the 10-ohm resistor into the 10-ohm transmission line, or when there is a signal in the 10-ohm transmission line, traveling to the left, it sees a terminated line with no reflection.

In this new circuit, it's clear that the incident signal to the interface between the two transmission lines that the signal sees is not going to be the source voltage. It is going to be half $\times$ V_{source}. The signal has to get through the voltage divider between the 10-ohm source impedance and the 10-ohm transmission line.

Based on this correct view of the problem, the transmission coefficient is still going to be the same, 10/6, but the incident voltage is ½ V_{source}, so we predict the received signal will be

$$V_{received} = V_{incident} \times T = \frac{1}{2}V_{source} \times \frac{10}{6} = V_{source} \times \frac{5}{6} \quad (5.31)$$

This is exactly what we calculated using the voltage divider method.

Using the correct perspective — thinking that signals dynamically propagating on transmission line — we see the transmission line view or the pure resistor voltage divider view gives exactly the same answer.

5.10 Review Questions

1. What is the problem that results when the instantaneous impedance a signal sees changes?

2. What is the first goal in interconnect design to reduce reflection noise? What is the second goal?

3. What is the consequence if there were no reflections generated at an interference between two different characteristic impedance transmission lines?

4. What are the two boundary conditions at any interface that keep the universe from blowing up?

5. When a signal is propagating from the left to right direction, incident on an interface between two transmission lines, what are the two signals that exist in the left side of an interface? What are the signals that exist on the right side of the interface?

6. What are the two boundary conditions that voltage and current must meet?

7. What is the difference between the signal propagating in a transmission line and the voltage measured by a scope?

8. A 2V signal is incident from 50 ohm to 75 ohms. What voltage reflects? What is the voltage a scope would measure at the interface, right before and right after the reflection is created?

9. A 2V signal is incident to the interface from the other side, from the 75-ohm region propagating to the 50-ohm impedance region. What is the reflection coefficient? What is the voltage on either side of the interface just after the reflection?

10. When a 2V signal travels from 50 ohm to 75 ohms, the voltage at the interface is different than if the signal travels from 75 ohm to 50 ohms. Why?

11. A 1V signal travels from a 100-ohm impedance to an open. What is the final voltage at the open and the current through the open?

12. A 1V signal travels from a 100-ohm impedance to a short. What is the final voltage at the short and the current through the short?

13. When a 1V signal encounters an open, what is the reflection coefficient? What is the transmission coefficient? What is the voltage transmitted into the open?

14. What is the initial input impedance a source will see looking into a 50-ohm transmission line? A 25-ohm transmission line?

15. When a 2V source with a 50-ohm source resistance drives a 50-ohm transmission line, what is the initial current launched

into the transmission line? What is the current into the transmission line after all the reflections have died out, if the far end termination is open, shorted, or 50 Ohms?

16. When a 2V source with a 10-ohm source resistance drives a 50-ohm transmission line, what is the initial current launched into the transmission line? What is the current into the transmission line after all the reflections have died out, if the far end termination is open, shorted, or 50 ohms?

17. When we send a 1V signal into a transmission line open at the far end and measure a 2V signal at the open, where did the extra 1V come from?

18. What's wrong with analyzing a resistor driving a transmission line using the concept of a reflection coefficient?

Chapter 6   Analyzing Reflections with the Bounce Diagram

At every interface a signal encounters, some of it will reflect and some of it will transmit. These two new signals may interact sometime later with other interfaces.

At each interface, the behavior of the signals is described by the reflection coefficient and transmission coefficient introduced in the last chapter. However, where there is more than one interface, keeping track of the multiple reflections and transmitted signals can get very complicated very quickly.

When the rise time of the incident signal is not very short compared with the time delay of the transmission lines, the interactions of the reflections with the rising edge of the signal can add another level of complication.

We can take advantage of an important analysis method to keep track of and visualize these multiple reflections.

6.1   A Typical TX-RX Circuit

The simplest circuit with two interfaces is that of a driver, driving a transmission line circuit with an open at the far end. It consists of:

- A driver, described by a source voltage, a source resistance, often referred to as the Thevenin voltage and Thevenin resistance of the source, and a rise time

- A uniform transmission line with a characteristic impedance and time delay

- A receiver with an input impedance

143

This equivalent circuit is shown in **Figure 6.1**.

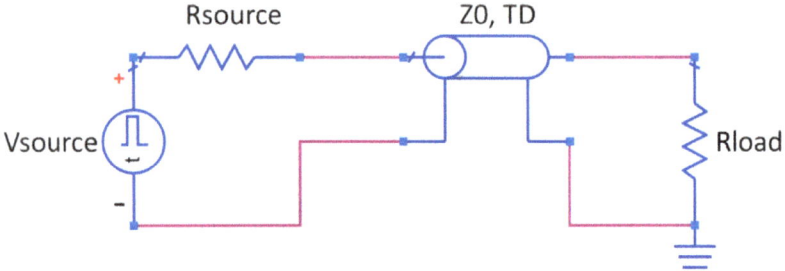

Figure 6.1 Very common transmission line circuit with a driver, transmission line, and receiver.

The input impedance of many receivers, the load resistance, is very high, often in the megohm range. It is effectively open. The output impedance of many transmitters, the source resistance, is very low, often in the range of 5-10 ohms.

In this environment, the behavior of the signal at the receiver is very distinctive.

Due to the low output impedance of the source, most of the source voltage is launched into the transmission line. This signal travels down to the receiver, where it sees an open and a reflection coefficient of 1.

The receiver sees a voltage equal to twice the launched voltage: one amount from the incident signal and one amount from the reflected signal.

The reflected signal reaches the low impedance of the source, sees a lower impedance and reflects. The reflection coefficient when going from a higher impedance to a lower impedance is always negative. A negative voltage reflects from the source, and heads back to the receiver.

When this negative signal hits the receiver, it sees the high impedance and reflects again, heading back to the source.

At the receiver, we had a large positive voltage before, then the negative reflected voltage pulls us down.

When the negative reflected signal hits the source again, it reflects and changes sign, comes back to the receiver as a positive signal, and increases the voltage we see there.

This process repeats with smaller and smaller amplitude reflected signals, alternating between higher or lower additions.

The net result is seeing oscillations at the receiver which we sometimes call ringing. This behavior is due to the multiple reflections bouncing back and forth between the large impedance change from the transmission line and the high impedance of the receiver and the low impedance of the source. An example of the measured voltage at the receiver with this ringing is shown in **Figure 6.2**.

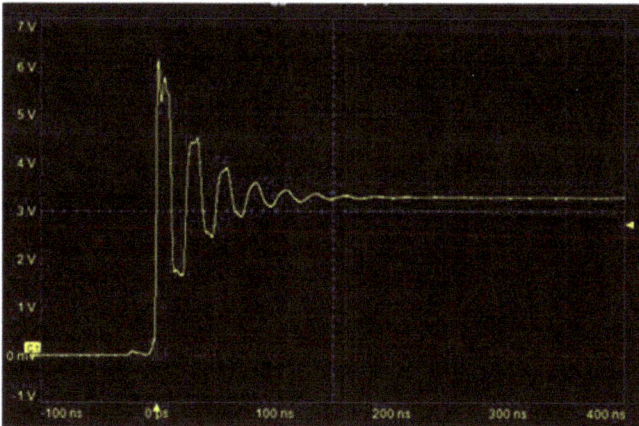

Figure 6.2 Typical voltage measured at the receiver with a low impedance source.

Watch this video to see the dynamic nature of the signals reflecting in this example.

6.2 The Bounce Diagram: An Example

Keeping track of all the reflections in this circuit example can quickly get very complicated. There are two interfaces with reflections at each one and a modified signal that transmits through each interface each time the signal reaches the interface.

To keep track of the dynamic nature of this signal propagating, we use a *ladder* or *bounce* diagram. The bounce diagram is a shorthand way of following the signal as it propagates and interacts at each interface. It is tedious but straightforward.

As an example, we take the circuit above and turn it into a bounce diagram. We start by defining the boundary conditions of the circuit and the reflection and transmission coefficients generated by each interface.

In this circuit, the source impedance is 10 ohms, the transmission line is 50 ohms, with a time delay of 1 nsec and the termination is 1 megohm. The starting place is shown in **Figure 6.3**.

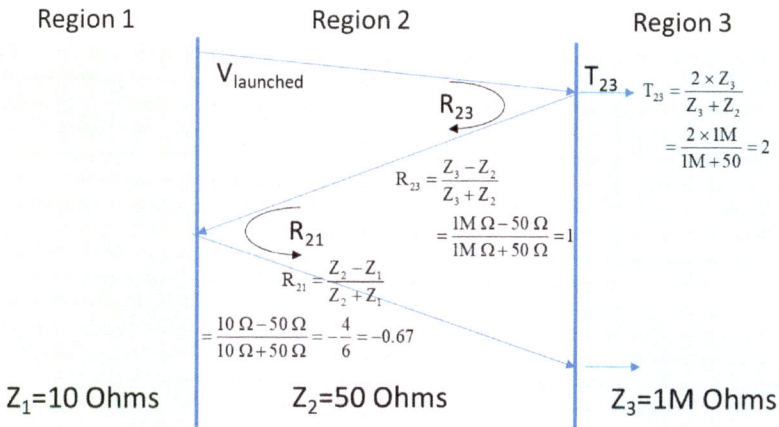

Figure 6.3 Setting up the boundary conditions for the bounce diagram based on calculating the reflection and transmission coefficients.

We use the bounce diagram to keep track of all the reflections at all the interfaces, and their time delays. Once the conditions are set up, we calculate the voltages that will appear at the output, where we would measure the actual voltage. This is where the distinction between the voltage that is measured and the signal that propagates is so important.

Using these values of reflection and transmission coefficient, we follow an incident voltage of 1V as it propagates through the transmission line and reflects from each end. This analysis is shown in **Figure 6.4**.

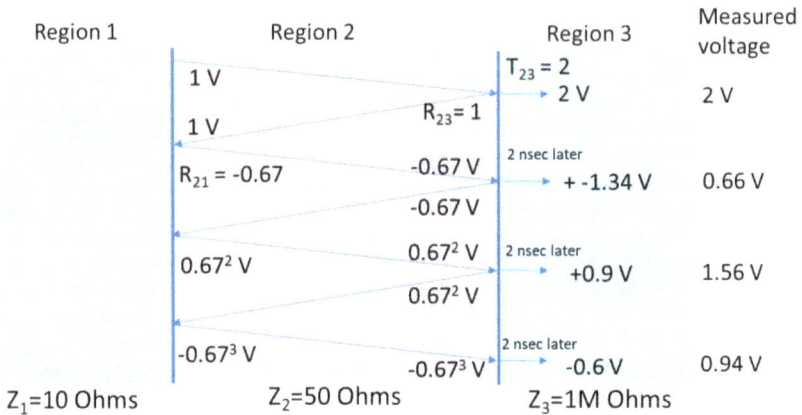

Figure 6.4 The bounce diagram for this simple case.

This example of a low impedance source, a 50-ohm line and an open at the far end is the simplest possible transmission line circuit.

It predicts the voltage appearing at the far end, as starting at 2V, then after 2 nsec, dropping to 0.66V, then 2 nsec later, rising to 1.56V, then 2 nsec later, dropping to 0.94V. This is the ringing-like behavior we see in the measured voltage at the receiver.

Setting up and using the bounce diagram for even this simple case is very tedious. Every engineer should go through this process at least once in their career. It helps visualize the reflections the signals undergo and illustrates the dynamic nature of the propagating signals and how they interact at each interface.

In any general transmission line circuit, the bounce diagram will help map out the reflected and transmitted signals at every interface. But other than in the simplest of circuits, the insight returned is rarely worth the effort required to map each reflection, especially when there are multiple transmission line segments with different time delays.

148

After you have calculated a bounce diagram once, to help understand the dynamic mature of signals, we generally solve all real-world problems using a transient simulator such as SPICE. These simulators will keep track of all the reflections and transmissions and include the impact of a finite rise time with a finite time delay of the transmission line.

We show examples of such simulations using QUCS, HyperLynx, and ADS in Chapter 8.

6.3   Simulating the Dynamic Nature of Reflections

It is sometimes difficult to visualize the propagation of signals on transmission lines. How the signal behaves at the boundary is very dynamic. The instantaneous voltage along the transmission line changes as the incident signal makes contact and the reflected signal is generated and the transmitted signal continues on.

It is very hard to visualize this dynamic nature of reflections, yet it is the dynamic nature that is so important.

To aid in seeing the dynamic nature of reflections, we will use a simple, freely available simulation tool you can download from here.

This tool simulates the dynamic nature of the signals as it encounters various impedance changes and displays the voltages moving on the transmissions line.

Watch an introduction to using this tool.

Figure 6.5 shows the setup for this tool when it is first opened. In this default configuration, we can:

- Change the source rise time, voltage level, and impedance

- Change the receiver impedance

- Change the Z0 and TD of the transmission line

- Observe the signal enter the transmission line, propagate to the end, and reflect from the impedance changes for the two ends

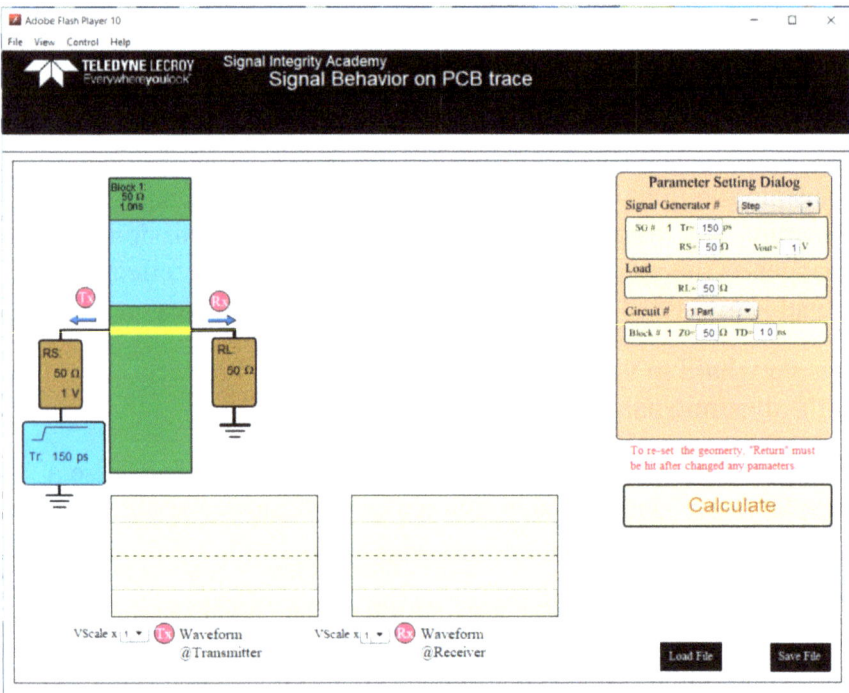

Figure 6.5 The default condition for simulating reflections using this TDR tool.

The default condition is set up to simulate a 150 psec source rise time, a 50-ohm source impedance, a 5-ohm transmission line 1 nsec long and a far end termination of 50 ohms.

As a starting place, we don't have to make any changes. We just push the enter button. This sets up the geometry, if there had been

150

any changes, and then we press the Calculate button. This action sets up the simulation and creates a play button. Press play and we see the signal propagating on the voltage scale in real time.

The yellow line is a top view of the transmission line, as though it were a microstrip. The green is the dielectric over the assumed solid return plane.

The blue space is where we plot the instantaneous voltage between the signal and return path directly beneath it. In the animation, we see this signal propagating. **Figure 6.6** is a snapshot of this signal midway through its travels.

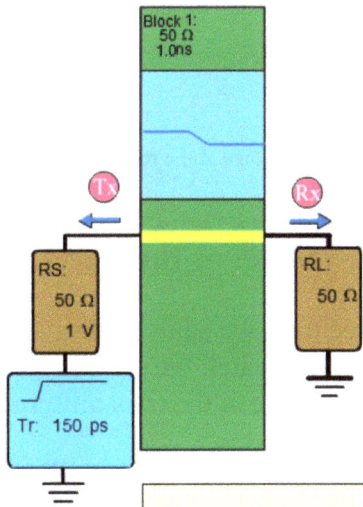

Figure 6.6 Snapshot of the signal as a voltage transition propagating down the transmission line, stopped in midflight.

The default condition illustrates the case of a signal propagating through an environment with no reflections. While this illustrates the dynamic nature of the signal, it is boring.

We'll make two small modifications to turn this problem into the simple case of a low impedance source and a high impedance receiver.

In the right-hand panel, change the RS to 10 ohms and the RL to 1,000. This is the highest resistance that can be used. This is now identical to the simple problem we analyzed in the previous section.

We know what to expect. When we look at the receiver, we should see ringing.

In addition to watching the signal propagate, we can also display the resulting voltage that would be measured at the receiver. This simulated voltage is shown in **Figure 6.7**.

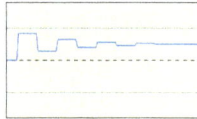

Figure 6.7 The simulated voltage at the receiver after multiple reflections.

6.4   Special Case: Short at the Far End

Using this simple, free tool, we can simulate the behavior of many transmission line systems, watch the signals propagate, and see the time evolution of the voltage at the transmitter and the receiver.

For example, using the default conditions, we just change the source resistance to 10 ohms and the RL to 0 ohms. This is the case of a short at the far end.

Before pushing the play button, it is always important to think about what to expect. This is Rule #9.

In the case of a short at the far end, the reflection coefficient will be -1. The voltage across the short should always be 0V, because it is a short.

But the voltage at the transmitter will be the initial voltage launched into the line from the 1-ohm source impedance into the

50-ohm transmission line and decreasing each time the reflected signal reaches it. **Figure 6.8** shows the evolution of the voltage on the transmitter and the receiver.

Figure 6.8 An example of the simulated voltage at the transmitter and receiver when the RS = 10-ohms, and the RL = 0 ohms.

6.5 Circuits with an Interface between Two Transmission Lines

With this tool, we can build transmission line circuits with up to three elements, each with its own Z0 and TD. As an example, we built a 2-element transmission line with 50 ohms connected to 75 ohms. **Figure 6.9** captures the voltage distribution on the two transmission lines right after the reflection occurred.

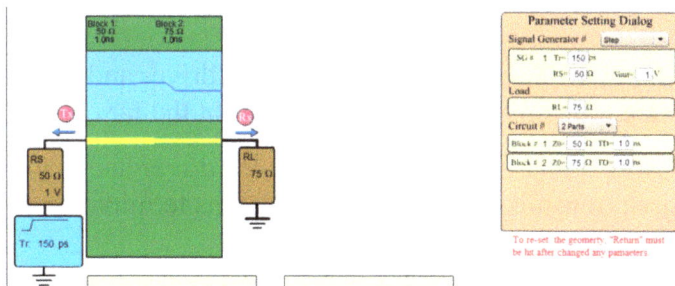

Figure 6.9 Two transmission lines showing the voltage distribution right after the reflection occurred.

In this example, the source impedance was matched to the transmission line it feeds and the load resistance was matched to the transmission line to which it is connected. This prevented any further reflections. This way we can just analyze the reflections at the single interface.

153

6.6 Try These Examples of Transmission Line Circuits

This free tool is a great way of exercising your understanding and building engineering intuition about how signals interact with transmission lines.

Before you try any of the following simulation examples, apply Rule #9 and anticipate what you expect to see before you simulate.

- Using a 10-ohm source impedance driver and a 50-ohm, 1 nsec transmission line open at the far end.

- Using a 100-ohm source impedance driver and a 10-ohm, 1 nsec transmission line open at the far end.

- Using a 50-ohm source impedance driver and a 50-ohm, 1 nsec transmission line open at the far end.

- Using a 50-ohm source impedance driver and a 50-ohm, 1 nsec transmission line with a 50-ohms termination at the far end.

- Using a 500-ohm source impedance driver and a 50-ohm, 1 nsec transmission line with an open at the far end.

- Using a 500-ohm source impedance driver and a 50 Ohm, 1 nsec transmission line with a 50-ohms termination at the far end.

6.7 Review Questions

1. What fundamentally causes ringing when viewing the output from a fast driver by a scope?

2. What does a bounce diagram describe?

3. What are two common features of a real transmission line system a bounce diagram does not do a good job of calculating?

4. If the source resistance is lower than the transmission line impedance, is the voltage launched into the transmission line higher or lower than half the source voltage?

5. When simulating the reflections from an interface between two transmission lines, why did we terminate each end in its characteristic impedance?

6. When the source impedance that drives a transmission line is much higher than the impedance of the line, and the receiver is a high impedance, what is the shape of the signal at the receiver?

7. When the source impedance that drives a transmission line is much lower than the impedance of the line, and the receiver is a high impedance, what is the shape of the signal at the receiver?

8. When there are three transmission line segments, each with different impedances, why is it nearly impossible to solve using a bounce diagram, but trivial to set up in a TDR simulator?

9. What are two examples of SPICE compatible simulation tools that can be used to simulate a transmission line circuit?

Chapter 7   Practical Applications of Transmission Line Properties: What Every Scope User Needs to Know about Transmission Lines

Above signal bandwidths of 20 MHz or rise times shorter than about 20 nsec, the properties of transmission lines influence virtually every measurement with an *oscilloscope*, often referred to as a *scope*. To illustrate how important transmission lines are to understanding scope measurements, we start with a commonly misinterpreted measurement.

7.1   A Commonly Misinterpreted Effect

On the front panel of every scope is a connection with clip-on terminals used to *compensate* 10× probes. The signal at these terminals is typically a 1-kHz square wave.

Since the connectors on the front of the scope are usually terminal posts, we can't just connect a coax cable directly to this signal source. We have to use a mini-grabber adaptor from the terminals to the BNC of the coax cable. An example of these terminals on the front of a Teledyne LeCroy HDO 8108 scope is shown in **Figure 7.1**.

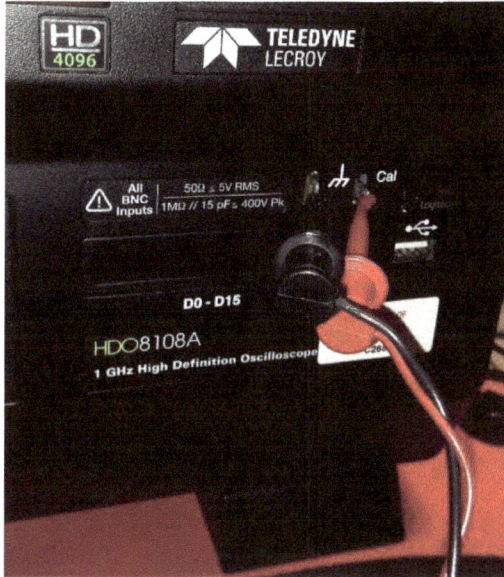

Figure 7.1 The compensation terminals in the front a scope with a mini grabber connected to a coax cable.

In this experiment, I connected a 3-foot-long 50 ohm RG174 coax cable from the mini-grabbers to the scope channel input. The signal we see with the scope is a 1-kHz square wave with about a 1V peak-to-peak value. This is almost universal in every scope. An example of this measured signal is shown in **Figure 7.2**.

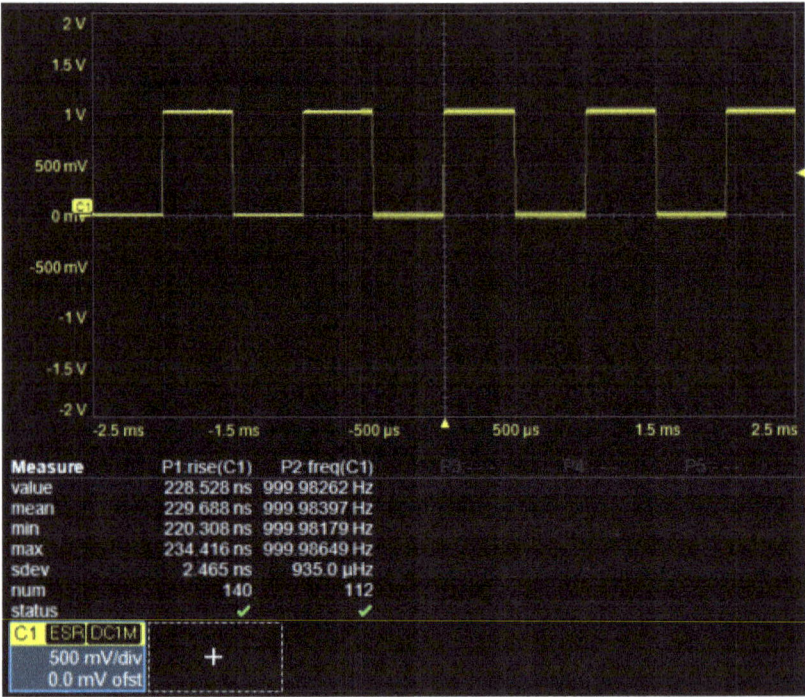

Figure 7.2 The measured voltage from the compensation source in a Teledyne LeCroy HDO 8108 scope. Note the measured values for the rise time and the frequency.

In this measurement, I used the internal scope parameter calculators to measure and calculate the 10%-90% rise time of each edge and the frequency of each cycle. These are displayed in the measurement boxes below the trace window.

It is also important to note that the scope bandwidth was 1 GHz and the sample rate was 10 GS/sec, with 12-bit vertical resolution. This means we can easily resolve rise times as short as 0.3 nsec.

The measured rise time of the compensation signal is 230 nsec, with a 3-foot length of coax cable.

This bandwidth is way less than the 1-GHz bandwidth of the scope, and with 10 GS/sec, there are more than 2,000 measurements along the rising edge, so this measurement is well within the range for this scope. This is always an important analysis to perform when interpreting any measurement.

What do you expect to happen to the rise time if we increase the length of the coax cable?

It is the expectation of most engineers, based on experience trying this experiment, that the rise time will increase.

Sure enough, as I spliced in 3-foot lengths of RG174 coax cable in the path from the compensation source to channel 1 of the scope, I measured a longer rise time. **Figure 7.3** shows these scope traces with 3-foot, 6-foot, 9-foot, and 12-foot lengths of RG174 coax cable. In fact, the rise time has increased with each increase in cable length.

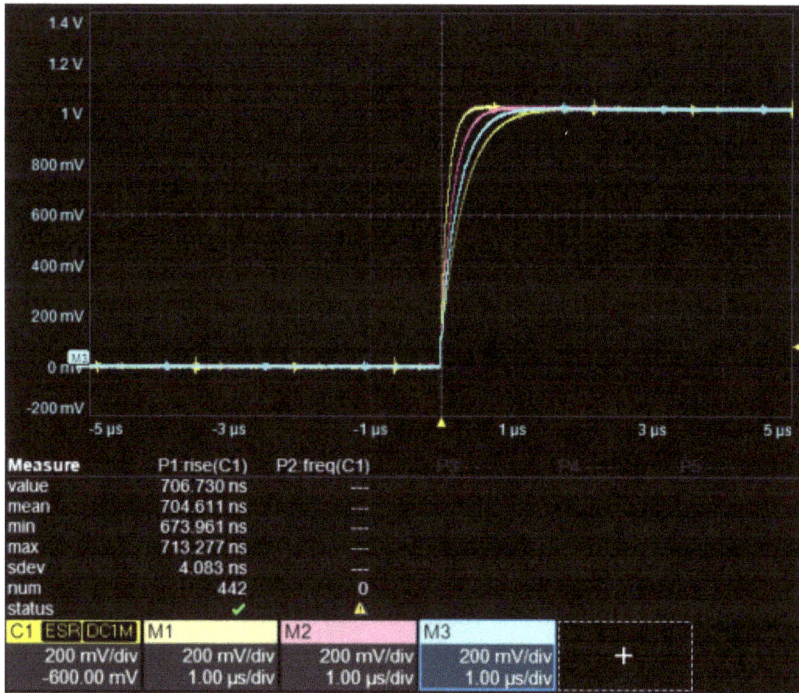

Figure 7.3 The measured rise time of the compensation signal with increasing lengths of coax cable. The rise time increases with each cable length increase.

This experiment shows that the rise time of the received signal went from 230 nsec with a 3-foot length to 704 nsec with a 12-foot length. This measurement certainly seems to support the observation that a longer cable results in a longer rise time.

Watch this video and I will walk you through this simple experiment.

Why is this? Why does the rise time increase with cable length?

7.2 The Wrong Root Causes of Rise Time Increase with Cable Length

Many engineers think the reason the rise time increases is related to one of three effects:

1. *Attenuation in the cable.* The longer the length, the more the attenuation, which cuts out high frequency more than low frequency.

2. *Charging of the cable.* The longer the cable, the more capacitance in the cable and the longer the RC charging time.

3. *The load on the source.* The longer the cable, the larger the capacitive load on the source, and the longer the rise time from the source.

None of these explanations really pin down the precise root cause to the extent we can use the information to fix the problem.

If we want to fix a problem, it is critical to have the correct root cause.

For example, if the root cause is attenuation in the cable, the way to reduce the rise time is to use a cable with lower loss, which usually means a more expensive cable.

If the root cause is capacitance in the cable and an RC charging, then the way to get a shorter rise time is to use a lower capacitance cable. We can achieve this with a shorter cable and a cable with a higher characteristic impedance, which means lower capacitance per length.

If the root cause is the impact from the driver on the cable's capacitive load, then we want to use a low capacitance cable; that is, a high characteristic impedance cable, and a really low output impedance driver source.

These root causes would lead us to use as a high a characteristic impedance cable as we can find or a high-performance, low-loss VNA metrology grade cable, either of which would be very expensive and have no impact on the rise time of the signal and actually make the signal quality much worse.

If we want to understand the origin of the longer rise time and engineer a technique to measure short rise times, we need to have the correct root cause.

This simple experiment raises a number of fundamental questions:

- Why does the rise time increase with the cable length?
- What is the cable doing to make the rise time longer?
- What is the intrinsic rise time of the source? Is it really changing with cable length?
- If we wanted to measure a rise time shorter than 230 nsec, how would we do it?

The answers to all of these questions are all rooted in the behavior of transmission lines and how signals interact with transmission lines, sources, and terminations.

7.3 Modeling Any Transient Source with Three Figures of Merit

The measurements we made are just as much about the device under test (DUT) we are measuring as they are about the scope, cables, and its setup.

In the last chapter we saw how important it is to know what is going on in the transmitter to interpret the measurements at the receiver and the entire circuit.

Any DUT acting as a transient voltage source has three important figures of merit that describe it:

- Its Thevenin source voltage
- Its Thevenin source resistance
- Its 10%-90% rise time

As a first-order approximation, every voltage source can be modeled as a Thevenin equivalent circuit, as shown in **Figure 7.4**.

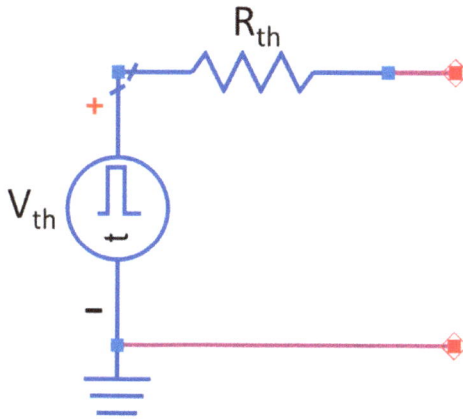

Figure 7.4 The first order model of any DUT voltage source.

In this model, the *Thevenin source voltage* is an *ideal voltage source*. This means it creates a constant voltage and has a 0-ohm output impedance. No matter what current is drawn, its voltage is always absolutely constant. It never changes.

This is why the impedance of an ideal voltage source is always 0 ohms:

$$Z = \frac{\Delta V}{\Delta I} = \frac{0}{\text{anything}} = 0\,\Omega \qquad (7.1)$$

Every source has a *Thevenin source resistance*. This is modeled as a series resistor. As current flows through the resistor, connected to some external load, the voltage at the output terminals, which we measure, will decrease due to the voltage drop across the Thevenin resistor.

We measure the voltage of the Thevenin source by measuring the voltage at the terminals of the DUT with no load, like with a 1-meg input impedance to a scope. This measured voltage is the Thevenin source voltage.

To measure the Thevenin source resistance, we apply a known load and measure the resulting voltage, like using the 50-ohm input impedance of the scope.

For example, using the compensation reference source of the scope as the DUT, I measured the open circuit voltage with 1-meg input to the scope and the voltage from the DUT with a 50-ohm load.

These measurements are shown in **Figure 7.5**.

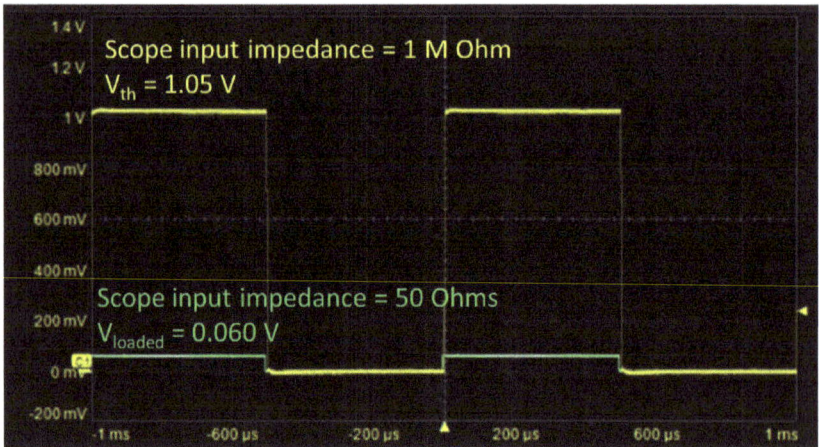

Figure 7.5 The measured voltage from the DUT with a 1-meg ohm input impedance and 50-ohm input impedance to the scope.

The equivalent circuit for the case of a load attached to the DUT is shown in **Figure 7.6**.

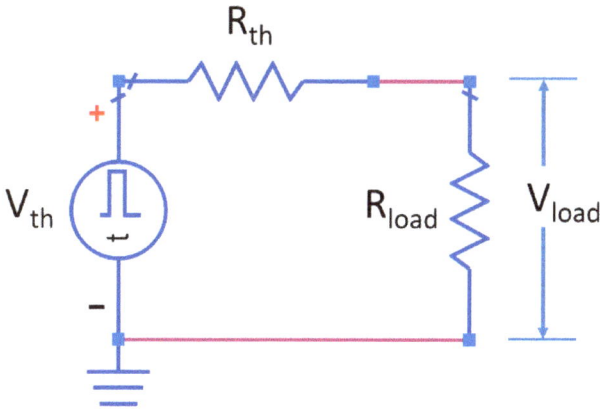

Figure 7.6 Equivalent circuit with a load to the DUT.

In the case of the load attached to the DUT, what the scope measures, the V_{load} voltage, is the voltage divider of the Thevenin voltage by the load resistor and the Thevenin resistance.

Using this circuit, the voltage divider is:

$$V_{load} = V_{th} \frac{R_{load}}{R_{load} + R_{th}} \qquad (7.2)$$

With a little algebra, the Thevenin resistance is:

$$R_{th} = R_{load} \left(\frac{V_{th} - V_{load}}{V_{load}} \right) \qquad (7.3)$$

In this measurement of the compensation source in my HDO8108 scope, I calculate a Thevenin resistance for the DUT as

$$R_{th} = R_{load} \left(\frac{V_{th} - V_{load}}{V_{load}} \right) = 50\,\Omega \left(\frac{1.052\ \text{V} - 0.060\ \text{V}}{0.060\ \text{V}} \right) = 825\,\Omega \qquad (7.4)$$

In this particular scope, the source resistance of the compensation circuit is 825 ohms. This is a relatively high resistance. Part of the reason why a resistance other than 50 ohms is used in this circuit is that the actual output resistance is not important for the application of adjusting the compensation for a 10× probe.

The use of a high resistance is a safety feature. If the user were to accidently connect the ground lead of the 10x probe to the signal source, shorting the output to ground, a high resistance keeps the current flowing in the return path of the probe very small, just a few mA. As experienced as I am with scopes, I have made this mistake more than once.

Now that we know the Thevenin source resistance of the compensation source, we can analyze the behavior of this circuit in terms of transmission lines.

7.4 Analyzing the Long Rise Time in Terms of Transmission Lines

The equivalent circuit, when the scope input is 1 meg, is shown in **Figure 7.7**.

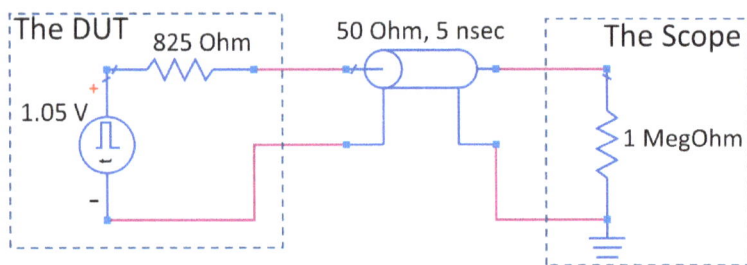

Figure 7.7 The equivalent circuit of the compensation source as DUT and scope as load.

When the source turns on, due to its high output impedance, it launches a small fraction of its internal Thevenin source voltage into the 50-ohm transmission line attached. The voltage launched into the transmission line is

$$V_{launched} = V_{th} \frac{Z_0}{R_{th} + Z_0} = 1.05 \, V \frac{50 \, \Omega}{825 \, \Omega + 50 \, \Omega} = 0.060 \, V \qquad (7.5)$$

The DUT should launch 60 mV into the transmission line. This signal will propagate down the transmission line until it encounters the 1-meg input.

The reflection coefficient is

$$r = \frac{Z_2 - Z_1}{Z_2 + Z_1} = \frac{1 \, M\Omega - 50 \, \Omega}{1 \, M\Omega + 50 \, \Omega} = 99.99\% \qquad (7.6)$$

This means that at the scope input, the first signal we should see is the sum of the incident 60 mV, and the reflected 60 mV, or a total of 120 mV.

Of course, this reflected 60-mV signal will make its way back to the source. When it hits the source, it will see the 825-ohm impedance of the Thevenin source resistance and reflect. The reflection coefficient from the 50-ohm cable to the 82-ohm source resistance is:

$$r = \frac{Z_2 - Z_1}{Z_2 + Z_1} = \frac{825 \, \Omega - 50 \, \Omega}{825 \, \Omega + 50 \, \Omega} = 89\% \qquad (7.7)$$

The reflection of the 60-mV signal from the source results in 89% of it reflecting back to the scope. This is a 53-mV signal heading back to the scope.

It will take the reflected signal one-time delay, or about 5 nsec to make its way down to the source, and another time delay, another 5 nsec, to make its way from the source back to the scope. This is a total of 10 nsec after we see the incident signal to see the additional 53-mV signal hit the scope and reflect.

The total signal we should see at the scope is 60 mV × 2 from the incident signal and 53 mV x 2 from the second reflection, or 226 mV. Each time the signal reflects from the 1 meg of the scope and heads back to the source, a smaller voltage will reflect from the source.

This behavior of a high impedance source with a 50-ohm transmission line open at the far end was one of the circuits analyzed in the last chapter. The simulation resulted in a ramp up of the voltage at the receiver, exactly as we anticipate. This simulated voltage is shown in **Figure 7.8**.

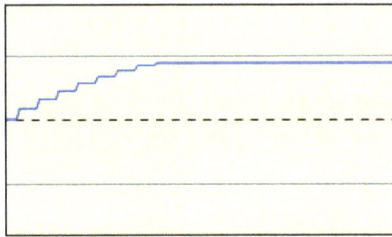

Figure 7.8 Simulated voltage at the receiver with a 850 Ohm source resistance, 50-ohm transmission line, and high impedance at the receiver.

To see this behavior in the scope on the 1-meg input, I had to change the scales. While measuring the same compensation signal from the scope DUT with just a 3-foot length of RG174 cable, using a 1-meg input resistance to the scope, I zoomed in on the time base to see the 10-nsec response and zoomed out on the voltage scale to see the 125 mV of initial signal. This result is shown in **Figure 7.9**.

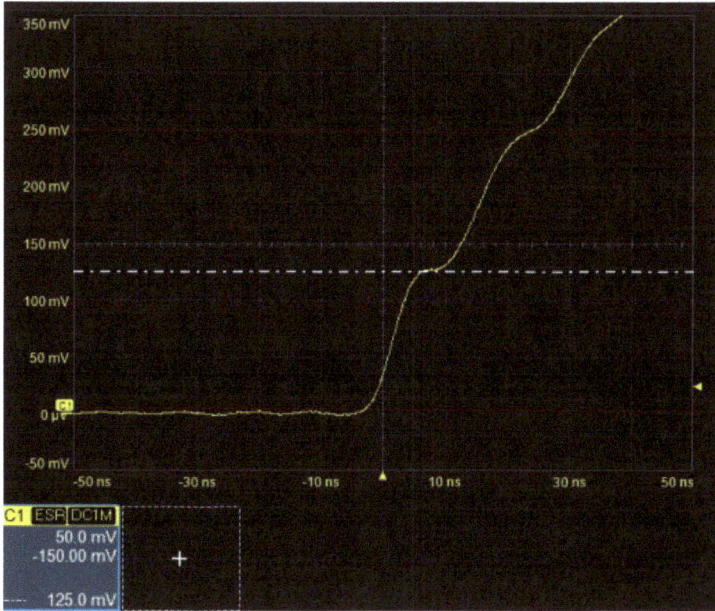

Figure 7.9 The measured compensation signal zoomed in on the voltage and time scales.

We measure precisely what we expect to see. In fact, the initial signal into the scope is only 125 mV, which increases on each reflection, with a time of about 10 nsec between reflections.

The long rise time we see with the 1-meg input is not due to attenuation in the cable. It is not due to an RC charging and it is not due to the cable loading the source and changing the source.

> *The long rise time is really the superposition of individual reflection steps between the high impedance of the scope input and the high impedance of the DUT.*

When the cable is longer, the voltage of each step is exactly the same, since the impedance discontinuities and reflection coefficients at the ends have not changed.

What has changed is the time between the reflections because the time delay has increased. **Figure 7.10** shows the measured input signal when the time base is on 100 nsec per division, for the case of the 12-foot-long cable. Each reflection takes about 40 nsec. The time delay of the cable is about 12 inches × 1.5 nsec/ft = 18 nsec and the round-trip time is 36 nsec, close to the 40 nsec we measure.

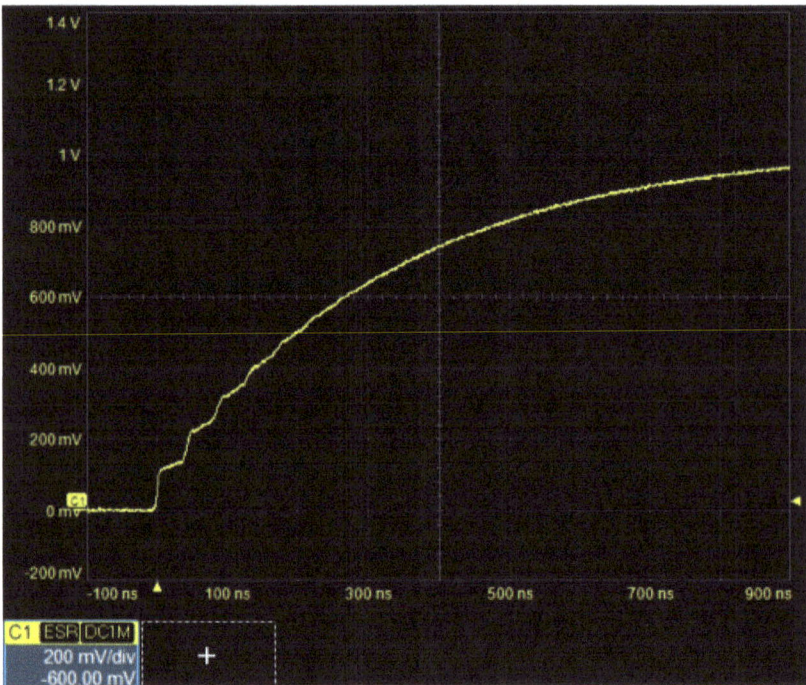

Figure 7.10 The measured signal from the DUT with a 12-foot-long cable and 1-meg input to the scope.

This signal received at the scope with a 1-meg input resistance and a higher than 50-ohm source impedance looks suspiciously like an RC charging time constant. There is a reason for this.

We saw in Chapter 4, and will explore in more detail in the next chapter, a transmission line can be **approximated** at low frequency by a simple capacitor. The value of the capacitance is

$$C = \frac{TD}{Z_0} \tag{7.8}$$

Another way of looking at the driver-transmission line-scope-system is as a resistor charging the capacitance of the transmission line. We know the real mechanism of the ramp-up is reflections back and forth between the high impedance at the ends, but we can approximate the transmission line as a capacitor and the system as an RC charging model.

When the output impedance of the driver is very high, and there are many reflections to reach the final value, this is a pretty good approximation. The RC model predicts a similar rise time as the multiple reflections.

But when the output impedance of the source is 50 ohms, there are no reflections and the RC rise time is unrelated to the rise time of the signal at the scope. These examples are illustrated in **Figure 7.11**.

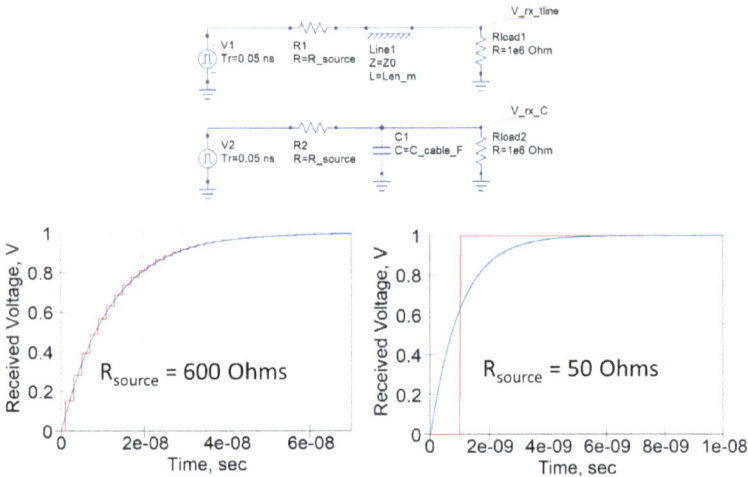

Figure 7.11 Top: Two circuits to simulate the received signal at the end of an ideal transmission line and a lumped capacitor. Bottom: The simulated received signal of each model with a high output impedance and 50-ohm output impedance.

The transmission line model for the cable interconnect is a good model across all time frames and output impedances of the source. Thinking of the cable as a capacitor is only a useful model when the source impedance is high and we look for a time that includes many reflections.

When we use 50 ohms in the scope, there are no reflections when the signal passes from the 50-ohm cable to the scope. The signal turns into heat and this initial voltage launched into the cable is measured by the internal scope amplifier.

The voltage we measure across the 50-ohm resistor is the voltage initially launched into the transmission line *without the artifacts of reflections*.

The signal we measure when the scope uses a 50-ohm input impedance is the intrinsic rise time of the signal from the source. **Figure 7.12** shows this initial measurement. Once this signal reaches the resistor, there are no more reflections, it is turned into

172

heat and the signal will stay constant until the voltage out of the source changes.

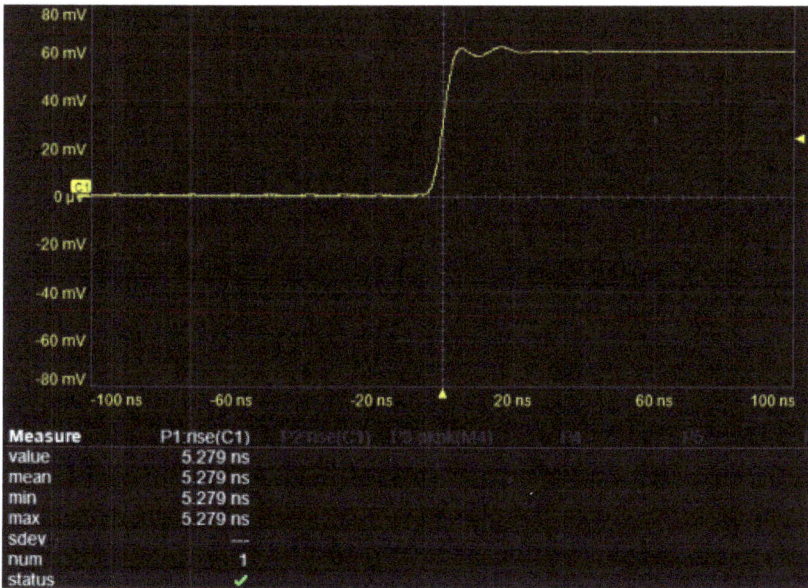

Figure 7.12 The measured signal from the source, with 50-ohm input to the scope. The cable length is 3 feet.

The 10-90 rise time of the signal from the compensation source is about 5.3 nsec. This is a far shorter rise time than the 700 nsec we measured with the 12-foot cable, or even the 230- nsec rise time we measured with the 3-foot cable using a 1-meg input to the scope.

This is the general method we should always use to measure the intrinsic rise time of the source: use a 50-ohm termination in the scope that will prevent any reflections from the 50-ohm cable and measure the intrinsic signal launched into the cable by the source.

Using this measurement setup we can evaluate the impact from the attenuation of the cable. These cables are RG174. This is a thin, flexible, 50-ohm cable. It is designed for flexibility, not for low loss. It is not considered a high-performance or high-speed cable.

Using this best measurement practice, when we use a 3-foot length of RG174 cable, we measure a rise time of 5.3 nsec, as shown in **Figure 7.12**. When we use a 12-foot length of RG174 cable, we measure a rise time of 6.9 nsec, as shown in **Figure 7.13**.

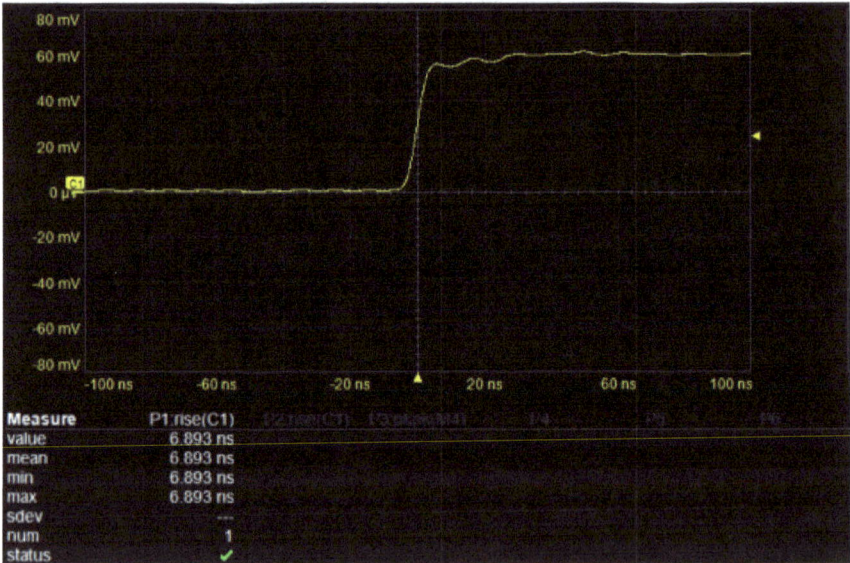

Figure 7.13 The measured signal from the source, with 50-ohm input to the scope. The RG174 cable length is 12-feet.

In fact, the cable *DOES* attenuate the high frequency components of the signal and cause the rise time to increase. But the impact is small. The increase in rise time from a 3-foot to a 12-foot cable is only 1.6 nsec. And this is in the cheapest and lowest bandwidth cable available.

In this particular scope, the rise time of the compensation signal was about 5.3 nsec. Generally, it will vary from scope to scope. Higher bandwidth scopes generally have shorter rise time compensation signals.

7.5   The Third Figure of Merit for all Sources: Intrinsic Rise Time

The way to measure the intrinsic rise time of any source is to connect it to a 50-ohm cable and use 50 Ohms in the scope. This way, there are no reflections from the cable and the measured signal by the scope is the actual signal launched by the source into the cable.

Other than the small impact from the attenuation in the cable, the rise time is independent of the length of the cable and is an intrinsic property of the source.

The rise time is the third important figure of merit of any source. It should always be measured using 50-ohm input impedance of the scope. This prevents the artifact of reflections.

The reason every scope has a 50-ohm input impedance option is based on the assumption that the user will be using a 50-ohm characteristic impedance cable to connect to the DUT. The 50-ohm input termination on scope will prevent any reflections from the signals traveling in the 50-ohm cable connecting the DUT to the scope and eliminate this source of measurement artifact.

Depending on the source impedance of the DUT, the apparent rise time could be increased to as much as 300 nsec or more due to reflections. If the typical signal we measure has a rise time longer than this, it doesn't matter what the scope termination is. Reflections will happen, but they will be smeared out during the rise time of the DUT.

There is one small potential danger when measuring a signal with a 50-ohm input impedance to a scope. Inside the scope is literally a precision, laser-trimmed, 50-ohm resistor soldered to a board.

It can dissipate only a 0.5 watt of power. If the resistor consumes more than 0.5 watt, the resistor will heat up, potentially being either thermally damaged or getting so hot it will melt off the board.

In either case, the entire scope has to be sent back to the factory to replace the input stage.

To prevent this problem, every user must be careful not to apply too much voltage to the 50-ohm resistor. The power consumption in a resistor is

$$P = \frac{V_{rms}^{2}}{R} \tag{7.9}$$

For the special case of $P < 0.5$ watt, the rms voltage must be less than:

$$V_{rms} < \sqrt{PR} = \sqrt{0.5 \times 50} = \sqrt{25} = 5 \text{ V} \tag{7.10}$$

As an important safety tip, do not apply more than a 5V rms signal to the front of any oscilloscope to prevent damage.

Watch this video to see the confirmation tests that it is reflections causing the longer rise time and how to measure the intrinsic rise time of the source. In this WPHD 804, high-performance scope, the rise time is even shorter than in the HDO8104 scope in the previous examples.

7.6   When the Source Impedance is 50 Ohms

In the previous example, we looked at the case when the source impedance was much higher than 50 ohms. The response measured by the scope with a 1-meg input resistance was an increasing rising edge, approximating an RC charging.

This was due to the multiple reflections between the high impedance of the source and the high impedance of the scope termination.

When the source impedance is exactly equal to 50 ohms, the behavior is very different.

Many function generators and the auxiliary output signals from many scopes all have a source impedance of 50 ohms. This is easy to measure by comparing the output voltage from the source with a high impedance load, like a 1-megohm scope input and a 50-ohm load.

The measured voltage with 50-ohm scope input termination will be exactly half the measured voltage with the scope set for 1 Mohm, which measures the unloaded source voltage. When the Thevenin resistance is 50 ohms, the 50-ohm cable makes a 2:1 voltage divider. Half of the Thevenin voltage from the source is launched into the transmission line.

With 1 meg in the scope, the one-half voltage reflects at the scope. The scope measures the incident one-half voltage and the reflected one-half voltage, which equals the full source voltage.

When the reflection from the scope's 1 meg hits the source, the signal sees the 50 ohms of the cable and the 50 ohms of the source. There is no impedance change and no reflection.

All the scope sees for as long as we measure is the full Thevenin voltage value.

When the scope is set for 50-ohm termination, the one-half voltage that is launched into the cable from the source reaches the 50-ohm of the scope, sees no reflection, and this one-half voltage is measured by the scope. Since there are no reflections, this is the constant, steady voltage that is measured by the scope.

This behavior of showing the full voltage with 1 Mohm input and one half the Thevenin voltage with 50-ohm scope input impedance is shown in **Figure 7.14**.

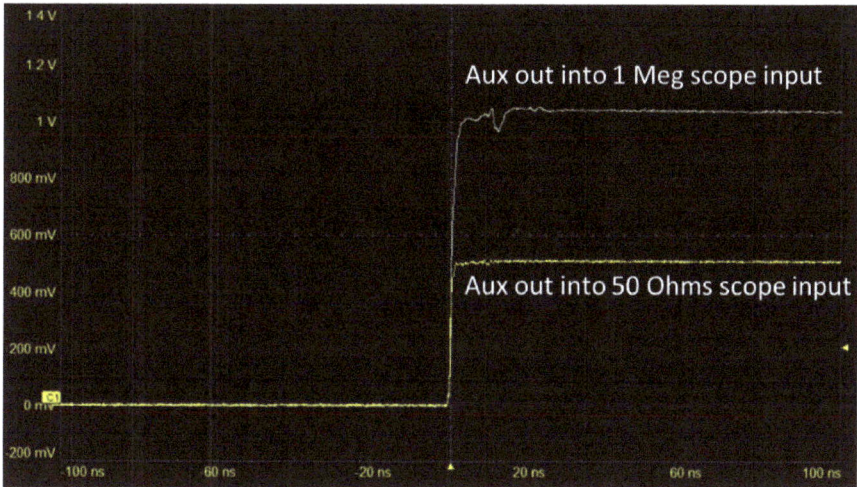

Figure 7.14 The measured voltage from the aux output of the scope on 1 meg and 50-ohm input.

Many function generators have a very confusing scale setting. When a 1 V peak -to-peak voltage is selected, the actual output voltage is often measured as 2 V. What gives?

Many function generators assume the user will connect the function generator to a 50-ohm cable. The voltage the user sets in the function generator is the voltage that would be launched into the 50-ohm cable.

What you do with that voltage signal after it enters the 50-ohm cable is up to the user.

If a 1 V peak to peak signal is selected, a 1V peak-to-peak signal will be launched into the 50-ohm cable. If the receiver at the far end of the cable is high impedance, like the scope set on 1 Mohm input, the 1V peak-to-peak signal will reflect from the 1 Mohm input. The scope will measure the 1V incident and the 1V reflected signal, seeing a 2 V signal.

179

Even though the user selected a 1 V signal, the scope and any high impedance load would see a 2 V signal. This is due to the reflections from the 50-ohm cable and the high impedance lead.

If the load were 50 ohms, there would be no reflection and the load would see the 1 V peak-to-peak signal selected on the function generator.

Now we see why.

7.7 When the Thevenin Source Resistance is Much Lower than 50 Ohms

Many high-speed digital drivers have a low Thevenin resistance, well below 50 ohms. This is the case with all power supplies, which can have an output resistance as low as 0.1 ohm or less.

The principles underlying what the scope will measure when connected to a low impedance source that suddenly turns on or off is the same as we have already considered when the Thevenin source resistance is low. It's just the details that are different.

The signal launched into the 50 ohm transmission line from a low impedance source will be very close to the Thevenin voltage. This is because the voltage divider has most of the voltage drop across the 50 ohm of the input of the coax cable.

This voltage travels to the end of the transmission line and hits the input to the scope. When the scope is set for 50 ohm, the signal is terminated, there are no reflections, and the scope measures the intrinsic rise time of the source.

When the scope is set for 1 Mohm input, the incident signal will reflect with a reflection coefficient of nearly 1. Due to the two waves, the incident and the reflected, the voltage at the scope will be the nearly twice the Thevenin open-circuit voltage.

The reflected voltage will then travel back to the source. But when it reaches the source, the reflection coefficient from the 50-ohm coax cable to the low impedance of the Thevenin output resistance will be negative. The incident voltage will reflect back to the scope with the opposite sign.

This negative voltage will head back to the scope, reflect from the 1-megohm resistance, and head back to the source again. It will hit the source, change sign, and head back to the scope. Each time the signal reflects from the low impedance of the source, it will change sign. This means the signal at the scope will alternate between positive and negative values. This will appear as ringing.

Using the free simulator, this situation can be set up and analyzed. The resulting signal at the receiver is shown in **Figure 7.15**. The behavior of the ringing is exactly as expected.

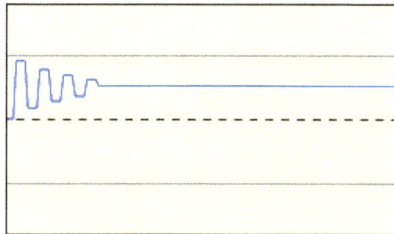

Figure 7.15 The first few reflection cycles of the signal at eh receiver with a low the impedance source and a high impedance at the load.

The time for a signal to travel from the scope back to the source is just the time delay (TD) of the cable. The time between hitting the scope, traveling back to the source, reflecting, and then traveling back to the scope to be detected is two time delays.

The time between peaks in the ringing is the time between the first positive signal, then a round trip to appear as a negative voltage, then a round trip to appear as another positive signal, or 4 × TD.

If the rise time of the source is larger than 4 × TD, the ringing will happen, but it will be smeared out. For a 3-foot-long cable, the time delay is about 5 nsec, and the peak-to-peak delay is

$$\text{Time}_{pk-pk} = 4 \times 5 \,\text{n} \sec = 20 \,\text{n} \sec \qquad (7.11)$$

This means the effective ringing frequency will be 1/20 nsec = 50 MHz.

This is the origin of the rough rule of thumb that if the rise time is greater than 4 × the TD, the ringing will be mostly smeared out and may not be important. In the above example with a 3-foot cable, if the rise time of the signal is > 20 nsec, or the bandwidth of the signal is < 50 MHz, the ringing will mostly be smeared out.

Two sources are selected to illustrate this behavior. One source is a hex inverter set to trigger at 1 kHz. This output Thevenin resistance is about 10 ohms. **Figure 7.16** shows the ringing noise at the scope input set for 1 Mohm.

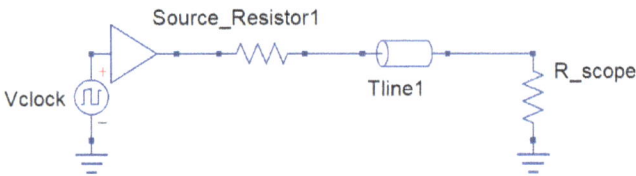

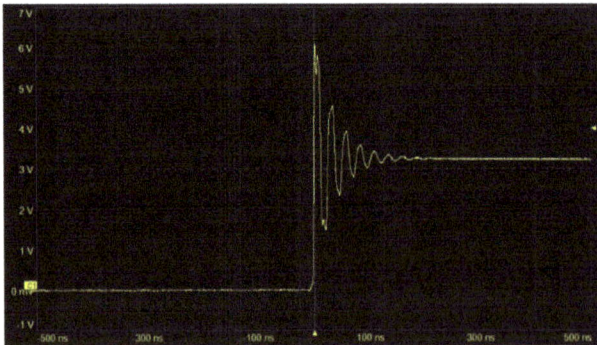

Figure 7.16 Measured voltage at scope with 1 Mohm input from a 10-ohm output driver switching on.

182

The second example is a relay that connects a 5 V power rail into the coax cable. The Thevenin source resistance is much lower. This means the ringing will last for a longer time. The circuit and measured signal at the scope is shown in **Figure 7.17**.

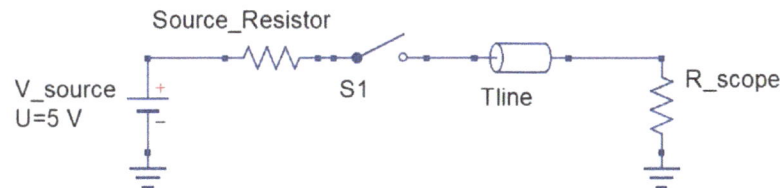

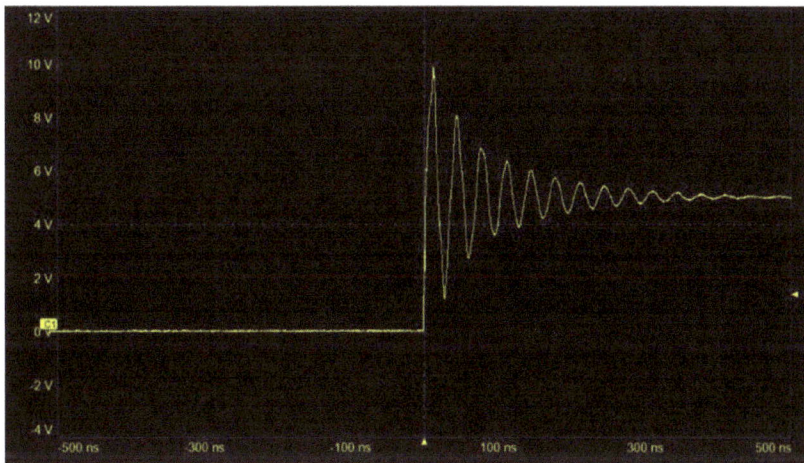

Figure 7.17 Using a relay switching a 5V power rail into the 50-ohm cable and 1 Mohm of the scope. The ringing lasts for a longer time.

7.8   When to Use 50 Ohms or 1-Mohm Input to Scope

It should be obvious now that the reason we use 50 ohms in the input to the scope is because the scope designers are assuming we are going to use 50-ohm impedance cables to hook up signals to the scope. We will want to terminate any signals at the scope propagating from the source into the 50-ohm cable. This prevents reflections at the scope and assures the measured voltage is the voltage launched into the 50-ohm coax cable by the source.

It's up to us to interpret what happens to the voltage after it is launched into the 50-ohm coax by the source.

If the impedance of the source is anything other than 50-ohms, using a 1-megohm scope impedance will result in reflections, which is a type of artifact. This is a feature of how we perform the measurements, not about the source signal.

If the rise time of the source is longer than about 20 nsec, the reflections between the ends of the cable will still happen, but they will be smeared out during the rising or falling edge of the signal and may not be noticeable.

One way around this problem of reflections in the cable adding an artifact is to use a 10 × passive probe. In the very best case, when using a coaxial connection to the DUT, the bandwidth of the 10 × probe can be as high as 500 MHz. The special features of the cable inhibit reflections by damping them out.

An example of a 10 × probe along with its equivalent circuit model is shown in **Figure 7.18**. The series L and C that is part of the probe contributes to a two-pole low-pass filter with a pole frequency of about 100 MHz.

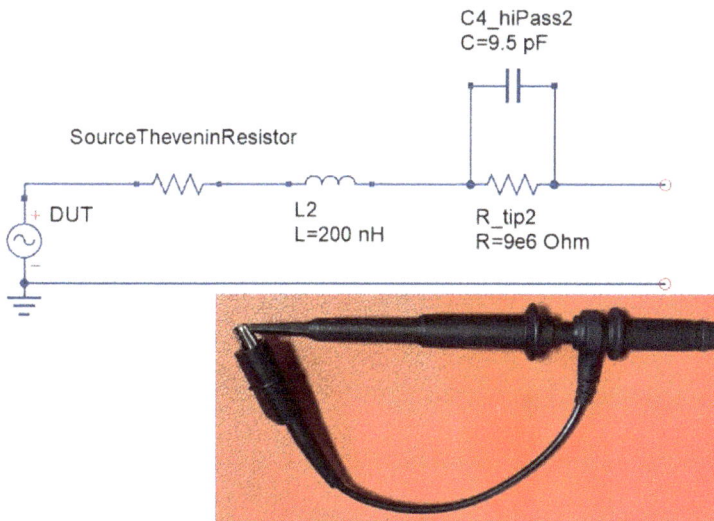

Figure 7.18 A typical 10 × probe with a long return path lead and its equivalent circuit model.

If the source impedance of the DUT is very low, there may be a large peak in the transfer function at 100 MHz. This will contribute to a ringing artifact if there are signal frequency components at 100 MHz. The transfer function of this simple circuit model with a low impedance source resistance is shown in **Figure 7.19**.

185

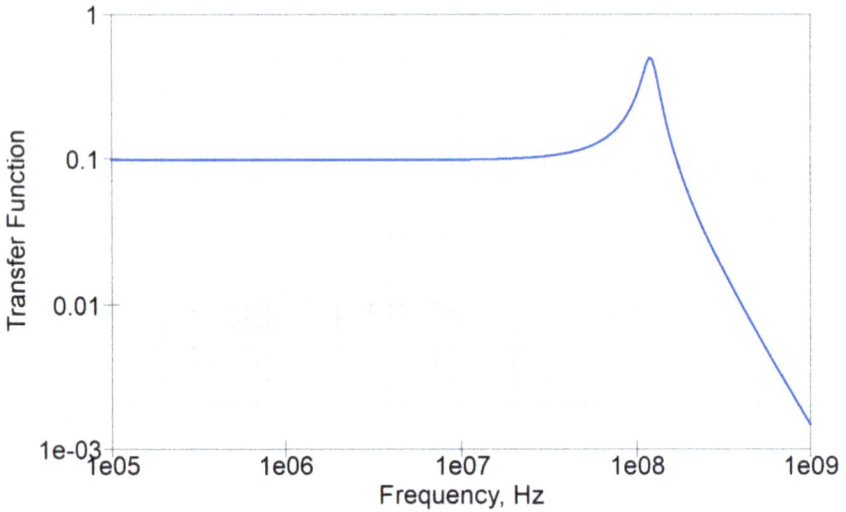

Figure 7.19 The transfer function of a 10 × probe connected to a low impedance source. Note the peak at about 100 MHz. This results in a ringing artifact.

An example of the measured noise picked up in a 10 × probe with the tip and ground shorted together, showing a ringing frequency of 100 MHz, is shown in **Figure 7.20**.

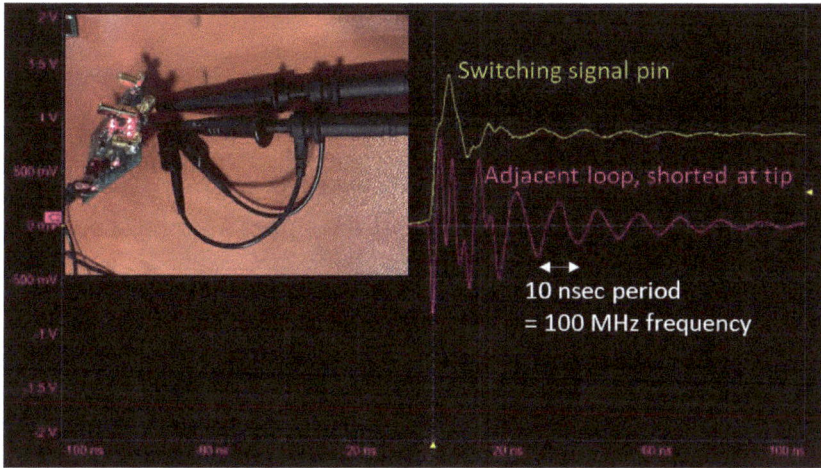

Figure 7.20 One probe measures a fast edge with high-frequency components. The second probe picks up the transient RF noise and passes the 100-MHz frequency components to the scope.

For frequency components of the signal below 100 MHz, the 10 × probe is a very good solution for probing signals.

This suggests a general strategy for probing signals using passive probes or cables based on not loading the source or introducing ringing or reflection artifacts:

- If you are trying to measure voltage levels less than 50 mV, do not use an attenuating probe of any sort, or the signal-to-noise ratio will be too low.

- Below 100-MHz signal bandwidth, use a 10 × probe with minimal tip loop inductance and 1 Mohm at the scope.

- Between 100-MHz and 500-MHz signal bandwidth, it is possible to use a 10 × probe, but with a coaxial connection to the DUT and 1 Mohm at the scope.

- Above 100 MHz, it is preferred to use a coax cable with 50 ohms at the scope.

187

7.9 Review Questions

1. What are the typical features of the signal on the compensation port of a scope?

2. Does the measured rise time of the signal into the scope increase with cable length?

3. If you want to efficiently fix a problem, what do you need to know about the problem?

4. If the root cause of the longer rise time from the compensation signal is capacitance in the cable, what two features would we want to change about the cable to reduce the rise time?

5. What does DUT stand for?

6. What are three figures of merit for every transient voltage source? If we know these three terms, we know the important features of the source.

7. What is the real root cause of the long rise time from the compensation signal?

8. What are three possible tests we can do to verify this explanation?

9. If we want to measure the intrinsic rise time coming from a source, how do we do that?

10. In a 12-foot-long, low-cost coax cable, what is the increase in rise time we might expect to see? How does this compare to the rise time initially measured in the scope with a 1-meg input impedance?

11. Why is 50 ohms one of the options as the input impedance of a scope?

12. If the maximum power dissipation the 50 ohm resistor inside a scope can handle is 0.5 watt, what is the max RMS voltage that should ever be applied to a scope set for 50-ohms input?

13. When the output impedance of the source is 50 ohms, what is the voltage measured by the scope set on 50-ohms input? On 1-meg input?

14. When the output impedance of the source is very low, what is the signature of the signal measured by a scope with its input on 50 ohms? 1 meg?

15. Why do some function generators output a 2V peak-to-peak signal when you set them for a 1V peak-to-peak signal?

16. When the source impedance is very low, the scope sees a voltage pattern that looks like ringing. What is the time interval between adjacent peaks related to? For a 3-foot coax cable, what is this period and what is this ringing frequency?

17. What is an advantage of a 10 × probe?

18. What are two disadvantages of a 10 × probe?

19. When should you always be using a direct coax cable connection between the DUT and the scope?

Chapter 8 Electrical Models of Transmission Lines

Real transmission lines are physical structures. All real, physical interconnects are transmission lines and every transmission line has a signal and a return conductor. They are defined by geometrical features such as line widths, trace thicknesses, lengths, dielectric thicknesses, and some material properties, such as dielectric constant, Dk, and dissipation factor, Df.

But electrically, we are limited in the variety of ideal equivalent electrical circuit models we can use to describe them.

8.1 Measured Electrical Behaviors of Transmission Lines

The electrical behavior of a real, physical transmission line structure can be measured, for example, with an impedance analyzer, a time domain reflectometer (TDR), or even a vector network analyzer (VNA). **Figure 8.1** is an example of the setup to perform a measurement on a real physical transmission line structure on a test board from CCN Labs using a Teledyne LeCroy WavePulser 40iX Network Analyzer.

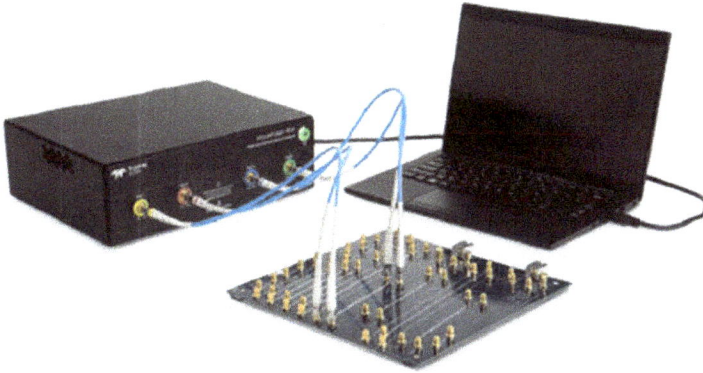

Figure 8.1 Example of the measurement setup for a transmission line on a circuit board with a Teledyne LeCroy WavePulser 40iX Network Analyzer.

We can also build a physical circuit with an active driver producing a real, time-dependent voltage waveform and measure the voltage at some location on the transmission line with a fast oscilloscope. **Figure 8.2** is an example of the circuit board with a driver chip, driving a transmission line in the form of a coax cable, measured with a Teledyne LeCroy WavePro HD high-bandwidth oscilloscope.

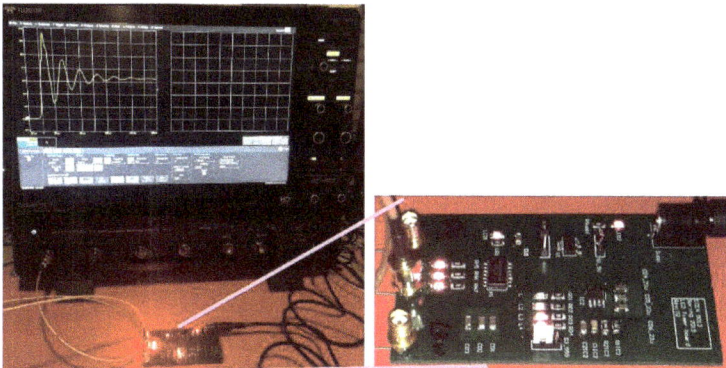

Figure 8.2 Example of a fast rise time driver on a circuit board measured with a fast scope. Inset is a closeup of the board with the driver IC as the small rectangle near the LEDs.

In these cases, the electrical properties of the transmission line structures influence the features of the voltages measured.

While we can use some measurements, such as S-parameters, as a behavioral or empirical black box model of the transmission line without any context of what goes on inside the black box, the most useful description of the electrical behavior of a transmission line is in terms of how it compares to an ideal equivalent circuit model.

8.2 Equivalent Electric Circuit Models of Transmission Lines

An equivalent electrical model of a transmission line approximates a real transmission line in terms of a collection of ideal circuit elements. The elements of a transmission line model from which we can select are defined in terms of the simulation engine in which the model will be used in.

The most common circuit simulator is SPICE (Simulation Program with Integrated Circuit Emphasis). Various forms of SPICE, both as open-source and commercial versions, have been around since 1972.

All versions of SPICE understand, among other ideal elements, these passive ideal components from which we can build transmission line models:

R: resistors

L: inductors

C: capacitors

M: mutual inductors

T: ideal lossless transmission lines

The R, L, C, and M elements are often referred to as *lumped* circuit elements, since all of their electrical properties can be assumed to be *lumped together* at one node in a circuit. They behave as point-like elements with no spatial extent.

The T-element is a *distributed* element in that its electrical properties are *distributed* over the length of the transmission line. Its spatial extent is an important feature of its electrical behavior.

All versions of SPICE understand active sources such as DC voltages, square wave or step voltage sources, and sine wave voltage sources, among others.

If you can draw the circuit connections or circuit topology of these ideal circuit elements and their parameter values, a SPICE simulation engine can calculate the voltages and currents at any circuit node in the time domain or the frequency domain.

> *The goal in constructing an ideal circuit model of a real transmission line is to construct the model so that any simulation of a measurement setup would accurately match the actual measured voltages or currents in the time or frequency domains.*

The outcome we want to achieve is to find the right combination of circuit elements in the right order (the circuit topology), with the right values (parameters), so that the simulated electrical behavior matches the measured electrical behavior to the level of accuracy required.

Generally, the higher the accuracy required, the more complex the model will have to be. However, it is remarkable how well simulations of simple, ideal models match the measured behavior of real physical transmission line structures.

As an example, **Figure 8.3** shows the measured voltage at the scope from a simple buffer circuit driving a coax cable and measured with a scope with a high input impedance. Using Keysight's Advanced Design System (ADS), a simple model was created using ideal circuit elements that simulated a waveform at the receiver very similar to what was measured.

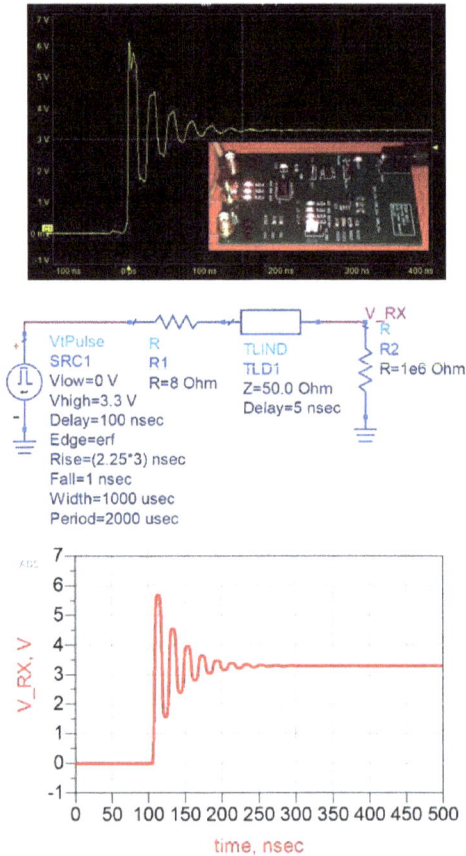

Figure 8.3 Top: Measured voltage at the scope from a buffer circuit on the circuit board in the inset. Middle: Ideal circuit model of this system using Keysight's ADS. Bottom: Simulated voltage of the ideal circuit model matching closely the measured voltage behavior.

194

In this example, the agreement between the simulated voltage signature and what was actually measured is very good. This is partly because this is such a simple circuit and partly because the ideal transmission line model used is such a good approximation of the behavior of a real transmission line.

8.3   Models and the Real World

It is important to keep these two worlds separate: the *physical world* of real transmission lines with real measured properties and the *ideal world* of ideal electrical circuit elements that can be used in a simulator.

There are really three different worlds in which we solve problems:

The mathematical world. This is where anything is possible as long as it follows the very precise rules of mathematics.

For example, the shape of a sine wave is precisely defined. The derivative or integral of a function is precisely defined. It is not a requirement that these mathematical structures should play any role in the real world. It is just remarkable that they do.

The ideal model world. This is the world in which we perform calculations and simulations. The elements from which we build our models are ideal elements with very well-defined properties and parameters. These properties and parameters are specific *figures of merit* of the ideal components.

For example, an ideal capacitor has one figure of merit, its capacitance. This term is perfectly constant with frequency. The impedance of an ideal capacitor will drop off with frequency and keep dropping off at ever-higher frequency.

Every model we use to perform a calculation or in a simulation is an *ideal* model. They just have different levels of complexity.

Every model element is ideal, in the sense that it behaves according to very specific rules that the simulator understands. By combining ideal circuit elements, very complex behaviors can be simulated, but still using only combinations of ideal circuit elements.

The real world. This is the world in which we live and in which we perform measurements and observe behavior. This is the world for which we would like to be able to predict performance and control behaviors. We can only do measurements on real structures.

These three different worlds are illustrated in **Figure 8.4**.

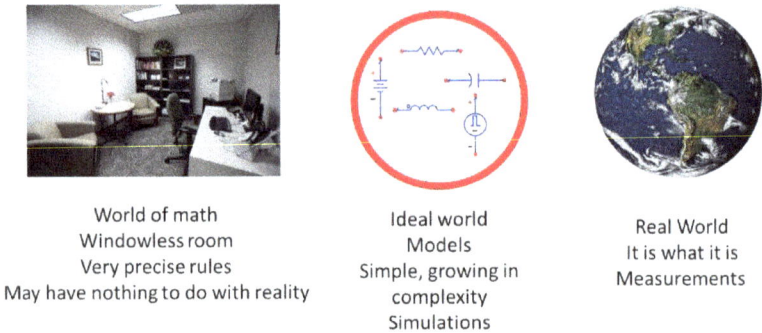

World of math	Ideal world	Real World
Windowless room	Models	It is what it is
Very precise rules	Simple, growing in	Measurements
May have nothing to do with reality	complexity	
	Simulations	

Figure 8.4 The three world views in which we will address problems.

It is important to keep straight the difference between the ideal circuit elements that have very well-defined mathematical descriptions and the real, physical elements that we measure.

Unfortunately, sometimes these two very different objects have the same name.

We call a *real physical capacitor*, a *capacitor*. But this is not the same as the *ideal circuit element* also called a *capacitor*. For example, the measured impedance of a *real capacitor* matches the simulated impedance of an *ideal capacitor* only at low frequency, then they diverge. This is shown in **Figure 8.5**.

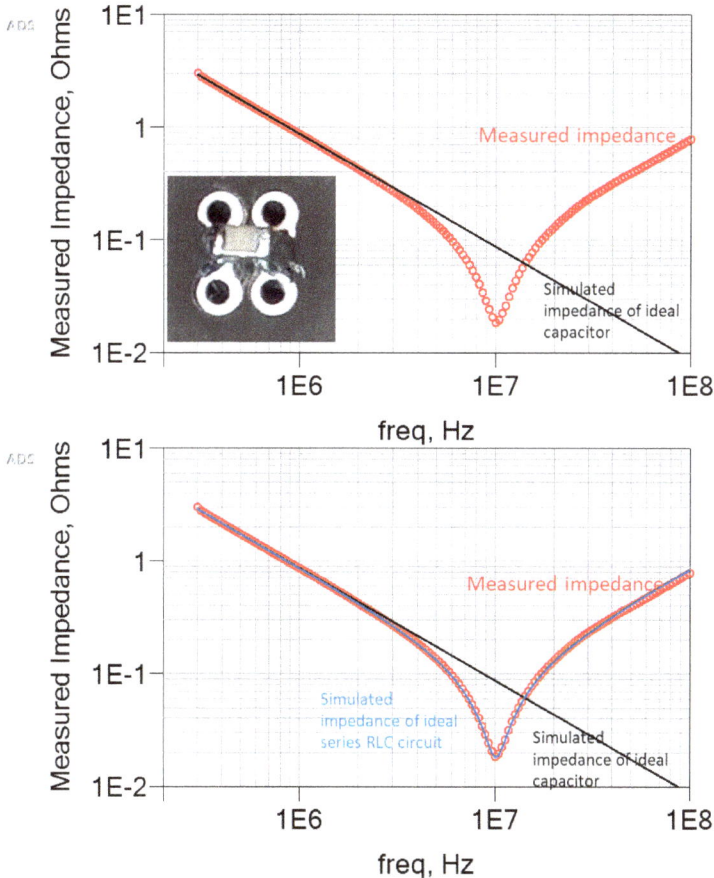

Figure 8.5 Top: Measured impedance of a real 0603 MLCC ceramic capacitor, shown in the inset, as the red circles. Black line is the simulated impedance of an ideal capacitor. Bottom: The same measured data with the simulated impedance of an ideal series circuit of ideal R, L, and C elements. The match is so good it is hard to see the simulated line.

If we increase the complexity of the ideal circuit model to include an ideal R and L element in series with the ideal C element, the resulting simulated impedance profile matches the measured impedance incredibly well to the full bandwidth of the measurement.

197

We need to be careful when referring to a "capacitor" to identify which type we mean. To avoid confusion, it is a good habit to add the preface, "real" or "ideal" when referring to any electrical elements.

8.4   Real and Ideal Transmission Lines

This distinction between real and ideal elements applies to transmission lines as well. A *real* transmission line is a trace on a board with its return path.

An *ideal* lossless transmission line is a *mathematical* model of a structure that has a uniform instantaneous impedance and a time delay. This ideal circuit element is understood by all SPICE-compatible simulators. **Figure 8.6** shows examples of these two very different transmission lines.

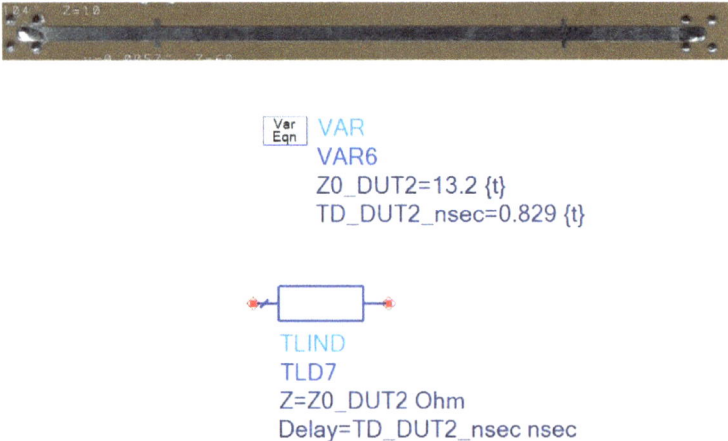

Figure 8.6 Top: A photo of a real transmission line on a circuit board. Bottom: An equivalent circuit model of an ideal transmission line in Keysight's ADS simulator.

When we describe the characteristic impedance of a real transmission line as 13.2 ohms, for example, we are really saying, "the measured impedance profile of this real transmission line matches the simulated impedance profile of an ideal transmission line with a characteristic impedance of 13.2 ohms."

The simulated reflected signal from an *ideal* transmission line element and the measured reflected signal from a *real* transmission line element can be remarkably close, if the circuit topology and parameter values of the ideal model are chosen correctly.

For example, **Figure 8.7** shows the equivalent circuit model of a TDR source and three ideal transmission line circuit elements in series. Each have a Z0 and TD parameter that define them. Their simulated impedance profile from a reflected step edge is compared with the actual measured impedance profile from a reflected step edge of a short transmission line on a PCB. The far end of the real transmission line and the ideal circuit are open.

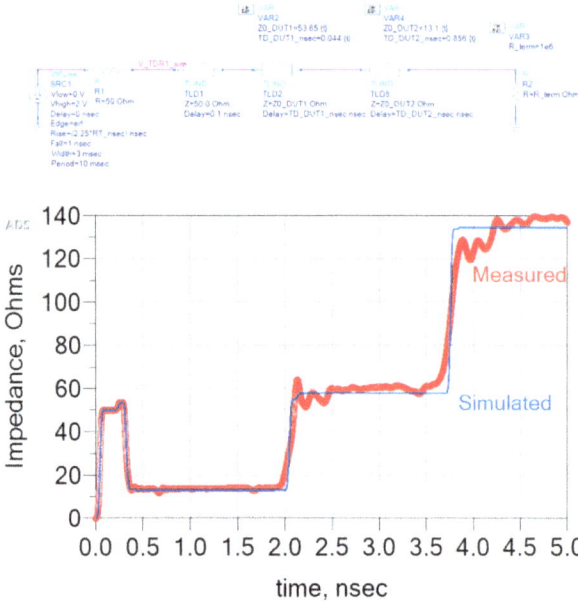

Figure 8.7 Top: The equivalent circuit diagram of a simple ideal transmission line circuit. Bottom: The measured (circles) and simulated (line) impedance profile showing excellent agreement.

In this physical system, the TDR was connected to the SMA on the circuit board to the relatively wide and low impedance trace on the board.

To model this, the first ideal transmission line represents the cable connection from the TDR to the SMA.

The second ideal transmission line, labeled as DUT1, represents the SMA connector attached to the circuit board. Its parameter values of Z0 = 53.65 ohms and TD = 0.044 nsec were chosen specifically to give the best match between the simulated impedance to the measured impedance profiles.

The third ideal transmission line element represents the trace on the board. Its parameter values, which gave the best match between simulated and measured performance, were Z0 = 13.1 ohms and TD = 0.856 nsec.

The match between the measured and simulated impedance profile is excellent. It is not perfect, though.

This suggests it could be improved, for example, by adding the short length of a 50-ohm SMA connector at the far end of the transmission line before the open.

These models are ideal lossless transmission line models. If the losses were included in a more advanced ideal circuit element, the agreement would be even better.

An ideal lossy transmission is still an ideal model. It is just a little more complex to include either the conductor loss or the dielectric loss, or both. Its behavior can more closely match the behavior of a real transmission line.

When referring to an ideal transmission line model, it's important to qualify what type of ideal transmission line it is. There are different levels of complexity, all of which can be ideal.

We describe the figures of merit of the real, physical transmission line in terms of the equivalent circuit model and its parameters that match the measured behavior.

We characterize a real transmission line using a TDR measurement by interpreting the measurement in terms of what ideal circuit element, with what parameters, would match the measured behavior of this real transmission line. We could then replace the real structure with the ideal structure and evaluate how it would behave in any circuit or situation.

8.5 Introducing a Simple, Open-Source SPICE-Like Simulator: QUCS

The examples in the last section illustrated how important it is to be able to simulate transmission line circuits. This is a process every engineer should be comfortable doing. A simple-to-use simulator to analyze the electrical properties of transmission line circuits in the time or frequency domain is an essential tool that should be in every engineer's toolbox.

The simplest to use, open-source (free) SPICE-compatible simulator with an excellent user interface is Quite Universal Circuit Simulator (QUCS). It runs on any OS and can be downloaded here.

Later in this chapter are directions on how to get started using QUCS.

It has an ideal lossless T element that is defined in a slightly unusual way. It is defined in terms of the characteristic impedance and a physical length, assuming air is the dielectric between the signal and return.

This means the length is

$$\text{Len}[m] = \text{TD}[\text{n sec}] \times 0.3[m/\text{n sec}] \qquad (8.1)$$

To define a transmission line 1 nsec long, a length of 0.3 m with air as the dielectric would be used as the parameter.

Every simulation in QUCS has at least four parts:

1. A simulation engine that defines the conditions of the simulation

2. A circuit with ideal circuit elements and some nodes labeled for which the simulated voltage will be displayed

3. Some parameters defined by equations or variables

4. Plots of voltage or currents in the time or frequency domain

An example of a QUCS circuit with a simple transmission line circuit is shown in **Figure 8.8**.

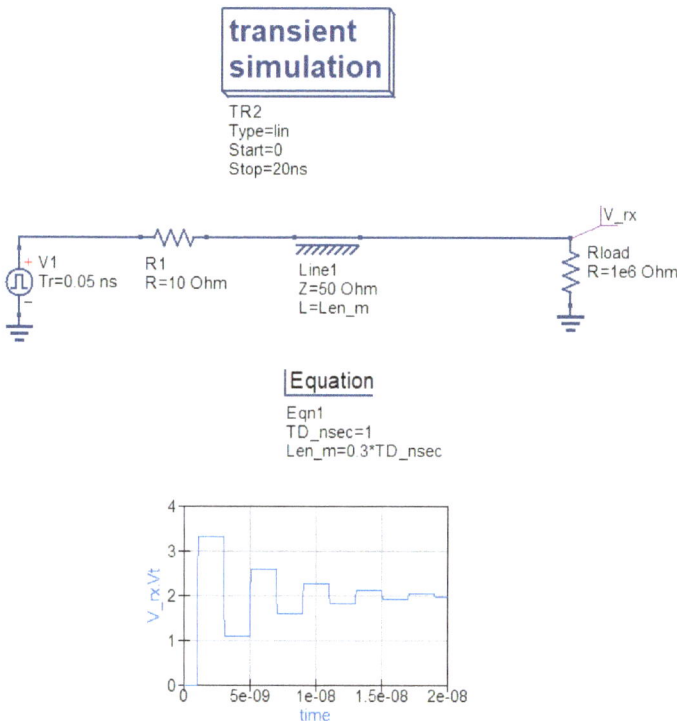

Figure 8.8 A simple QUCS simulation page with the simulation of a signal at the end of the transmission line circuit showing ringing.

QUCS can simulate the voltages or currents at any node in a circuit in the time domain or the frequency domain. A time domain simulation is a *transient* simulation. A frequency domain simulation is an *AC* simulation.

Click here to download the circuit files for this chapter.

Watch this video and I will show you how to get started with QUCS and do this very simulation.

8.6   The Tline Element and Real Transmission Lines

The best ideal circuit element model for a single-ended, lossless transmission line, is the ideal T-element. This has two parameters:

- Z0 = the characteristic impedance

- TD = the time delay.

The T-element is a fundamentally new and different circuit element than any ideal R, L, or C element. It is NOT a combination of L and C elements.

In the early days of introducing transmission line elements into SPICE, the early 1990s, the W-element was introduced by Metasoft in their version of SPICE, called HSPICE. This was a shorthand way of creating an n-section LC model. The W-element defined how many LC sections and the value of the total L and total C. The software would automatically generate the n-segments of LC elements. Of course, the more elements used, the longer the computation time.

The W-element was not really a new circuit element. It was a compact way of generating and editing an n-section LC circuit.

The T-element is a completely different way of describing a transmission line that has nothing to do with LC circuit elements. This model is not equivalent to combinations of LC elements. It is not a shorthand representation of an n-section LC model.

The T-element model defines an ideal transmission line in terms of an instantaneous impedance at the input to the element and a propagation delay from the signal entering the T-element to exiting the T-element. All simulators that use a T-element also understand the reflections that occur at every interface when the instantaneous impedance changes.

All the features of a voltage launched into the transmission line, reflections from the front, reflections from the back end, and reflections between multiple T-elements and time delays are fully taken into account by this ideal circuit element and simulators that use it.

This ideal, lossless transmission line model is a very good approximation to a real transmission line that has low loss.

Another way of comparing the electrical performance of a real transmission line and the simulated performance of the ideal T-element is with the input impedance in the frequency domain.

In the special case of the far end of the transmission line left open, the input impedance in the frequency domain will start high and drop down with frequency. At some point it will head back up and oscillate. This is the behavior of a real transmission line and an ideal T-element.

Figure 8.9 shows an example of a real transmission line built on a polytetrofluroethylene (PTFE, or Teflon) low-loss substrate. Its input impedance from one end was measured with a network analyzer, with its far end open. The measured impedance is compared with the simulated impedance of an ideal, lossless T-element transmission line model using Z0 = 50 ohms and TD = 0.41 nsec.

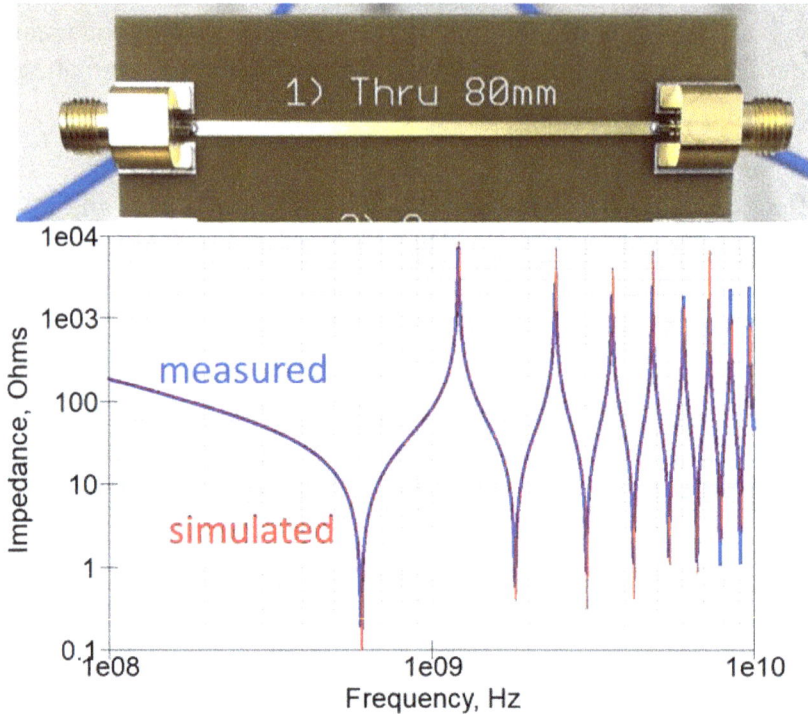

Figure 8.9 Example of a real transmission line built on a low-loss substrate, and the measured input impedance with a network analyzer compared with the simulated impedance of an ideal T-element.

The agreement between the measured, real transmission line and the simulated, ideal lossless T-element transmission line is remarkably similar, up to the 10-GHz bandwidth of the measurement. This ideal T-element is an excellent approximation to a real low-loss transmission line. As simple as this model is, it is a high bandwidth model. This is why it is such a valuable model.

The T-element model is the preferred model for all lossless transmission line circuits. It is a good approximation from DC to bandwidths above 10 GHz, providing the real transmission line is low loss.

Watch this video and I will show you how you can display the measured impedance of this transmission line and compare it to a simple ideal transmission line model.

8.7   Total L and C in a Transmission Line

In Chapter 2, we showed how the LC circuit model, *in the limit of an infinite number of LC segments*, is used to derive the telegrapher's equation and then the wave equation. One solution to this second-order, linear differential equation is sine waves.

When the properties of the sine waves propagating on this circuit were analyzed, the value of a characteristic impedance and the time delay through the circuit were derived in terms of the total inductance and total capacitance of the transmission line and the length of the line as:

$$Z_0 = \sqrt{\frac{L}{C}} = \sqrt{\frac{L_{total}}{C_{total}}} = \sqrt{\frac{L_{Len}}{C_{Len}}}$$

$$TD = \sqrt{L_{total}C_{total}} = n \times \sqrt{LC} = Len \times \sqrt{L_{Len}C_{Len}} \qquad (8.2)$$

$$v = \frac{Len}{TD} = \frac{Len}{Len \times \sqrt{L_{Len}C_{Len}}} = \frac{1}{\sqrt{L_{Len}C_{Len}}}$$

It should be noted that in the derivation, the assumption is that each L and C element is *infinitesimally small*. In this limit, the n-section LC model is an excellent approximation to the ideal T element and a real lossless transmission line.

This relationship between the characteristic impedance and time delay of the transmission line, and the total L and C in the line offers a very important connection. By combining these relationships, we get

$$C_{total} = \frac{TD}{Z_0}$$

and   (8.3)

$$L_{total} = TD \times Z_0$$

For example, a 50-ohm transmission line that has a time delay of 1 nsec, which could be about 6 inches long, would have a total capacitance of $C_{total} = 1$ nsec/50 ohms = 0.02 nF = 20 pF. The total inductance in the transmission line would be $L_{total} = 1$ nsec $\times$ 50 Ohms = 50 nH.

> *There are no assumptions in this analysis other than the transmission line is uniform. These relationships apply to cables, PCB traces, and all other uniform structures, regardless of the dielectric materials.*

Both relationships show that the total capacitance and total inductance scale with the time delay. The longer the transmission line, the larger the capacitance between the signal and return path. The longer the transmission line, the larger the loop inductance between the signal and return path.

This relationship also points out that the total capacitance in a transmission line scales *inversely* with the characteristic impedance. The *higher* the characteristic impedance, the *lower* the capacitance. For example, one way of decreasing the capacitance would be to pull the signal and return conductors farther apart. This would also result in a higher characteristic impedance.

208

The total loop inductance of a transmission line is *directly* proportional to the characteristic impedance. Increase the characteristic impedance by pulling the signal farther from the return conductor, for example, and the loop inductance will also *increase*.

No matter the details of the construction of the uniform transmission line, the total capacitance between the signal and return path will ALWAYS be the same for the same characteristic impedance and time delay transmission line.

Likewise, every transmission line with the same characteristic impedance and time delay will have the same total loop inductance regardless of its construction.

These are simple and very powerful connections between the "low frequency" electrical properties of total capacitance and loop inductance in the transmission line structure and the transmission line properties of characteristic impedance and time delay.

8.8 The Limit to a Transmission Line as a Lumped C or L

A real transmission line, open at the far end, will look like an ideal capacitor at low frequency. This means at low frequency, the *input impedance* of a real transmission line will start high and decrease with increasing frequency. This is an important property of real transmission lines.

In fact, the ideal capacitor model matches this behavior extremely well. For example, a uniform transmission line with a characteristic impedance of 50 ohms and time delay of 1 nsec has a total capacitance of 1 nsec/50 ohms = 20 pF.

Because the electrical properties of a real transmission line match that of an ideal transmission line from DC to high bandwidth, the input impedance of an ideal transmission line will also behave like an ideal capacitor at low frequency.

At low frequency, the input impedance in the frequency domain of an ideal transmission line open at the far end and the impedance of an ideal capacitor match perfectly.

When the far end of the transmission line is shorted, the transmission line looks like two wires. DC current could travel down the length of the signal wire and then return on the return path. This is a loop. The loop inductance of this signal-return path loop is 50 ohms × 1 nsec = 50 nH.

At low frequency, the input impedance in the frequency domain of an ideal transmission line, shorted at the far end, and this ideal loop inductor match perfectly.

The simulated impedance of an open and shorted transmission line and a capacitor and inductor are shown in **Figure 8.10**.

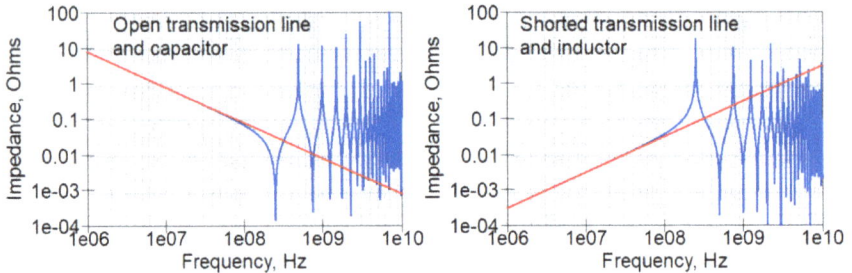

Figure 8.10 Comparing the simulated impedance of an ideal transmission line model and an ideal capacitor or an ideal inductor. Simulated with QUCS.

This says that at sufficiently low frequency, a real transmission line looks like an ideal capacitor or inductor, but only up to a specific frequency.

The dip-frequency, at which the impedance of the ideal transmission line reaches the minimum in impedance, is when the transmission line is one quarter of a wavelength long. This translates to a frequency of:

$$f_{dip} = \frac{1}{4 \times TD} \qquad (8.4)$$

For the case of the time delay of 1 nsec, the frequency of the dip is 250 MHz. This is exactly what is shown in the simulation.

The frequency up to which the single L or C model still matches the ideal transmission line model, the *bandwidth* of the L or C model, is when the length of the transmission line is about one tenth of a wavelength. This corresponds to a frequency of:

$$f_{model-BW} = \frac{1}{10 \times TD} \qquad (8.5)$$

For example, when the time delay is 1 nsec, the bandwidth of the L or C model is $1/(10 \times 1 \text{ nsec}) = 100$ MHz.

Every transmission line will behave like a capacitor or an inductor, depending on the connection at the far end, at low frequency. But there is an upper frequency where this lumped circuit model does not match real interconnects.

The ideal transmission line model, the T element, will have a much higher bandwidth agreement to a real uniform transmission line.

For a real transmission line on an FR4 circuit board that is 12 inches long, its time delay would be about 2 nsec. The highest usable bandwidth up to which the transmission line would behave like an ideal capacitor would be about BW $< 1/(10 \times 2 \text{ nsec}) = 50$ MHz.

The highest usable bandwidth up to which the transmission line would behave like an ideal transmission line could be about BW < 10 GHz, depending on the impact from the losses.

Watch this video and I will walk you through using QUCS to compare the simulated impedance of an ideal transmission line and a single L or C element.

8.9 Capacitance Per Length and Inductance Per Length

Given the characteristic impedance and time delay of an ideal uniform transmission line, we can translate this into the total capacitance and total inductance of the transmission line. Given the length of the transmission line, we can then define the capacitance per length and inductance per length of the uniform transmission line. These are often referred to as the per unit length (PUL) parameters.

The PUL values of the ideal inductance and capacitance are:

$$L_{PUL} = \frac{L_{total}}{Len} = \frac{Z_0 \times TD}{Len} = \frac{Z_0}{v} = \frac{Z_0}{c}\sqrt{Dk} \qquad (8.6)$$

and

$$C_{PUL} = \frac{C_{total}}{Len} = \frac{TD}{Z_0 \times Len} = \frac{1}{Z_0 \times v} = \frac{1}{Z_0 \times c}\sqrt{Dk} \qquad (8.7)$$

where

L_{PUL} = the inductance per unit length of the ideal transmission line

L_{total} = the total loop inductance in the ideal transmission line

Len = the length of the transmission line

Z_0 = the characteristic impedance of the ideal transmission line

TD = the time delay of the ideal transmission line

v = the speed of light in the material

c = the speed of light in air

Dk = the dielectric constant of the laminate

C_{PUL} = the capacitance per unit length of the ideal transmission line

C_{total} = the total capacitance in the ideal transmission line

This is a very important observation. This says that for every transmission line with the same characteristic impedance, made with the same dielectric and with the same Dk value, the capacitance per unit length and inductance per unit length of any transmission line are exactly the same. This is independent of the cross section and the type of uniform transmission line.

For example, in the special case of a 50-ohm transmission line built in FR4, with a Dk = 4, the inductance per inch and capacitance per inch are given by:

$$L_{PUL} = \frac{Z_0}{c} \sqrt{Dk} = \frac{50\,\Omega}{11.8\,in/n\,sec} \sqrt{4} = 8.5\,nH/in$$

and

$$C_{PUL} = \frac{1}{Z_0 \times c}\sqrt{Dk} = \frac{\sqrt{4}}{50\,\Omega \times 11.8\,\frac{in}{n\,sec}} = 0.0034\,\frac{nF}{in} = 3.4\,\frac{pF}{in}$$

(8.8)

This is a remarkable conclusion. All 50-ohm lines in FR4 have a capacitance of 3.4 pF/inch and a loop inductance of 8.5 nH/inch. These relationships can be used to quickly estimate the total capacitance and inductance in a transmission line.

For example, a 50-ohm trace on a board is 1 inch long and its total capacitance is 3.4 pF.

A 50-ohm transmission line is 10 mils wide. The total capacitance in a single square is 3.4 pF/in × 0.01 in = 34 fF. The excess capacitance in a corner is about one half of a square's worth of capacitance. This is 17 fF of excess capacitance.

Do corners affect signals? Yes, of course. They represent an excess capacitance. Using this simple relationship, we can estimate the excess capacitance in a corner. When the trace is 50 ohms and 10 mils wide, the excess capacitance is about 17 fF. Using this estimate we can evaluate the impact on the signal, given its rise time, from the 17 fF lumped capacitor at the corner. Generally, this is such a small capacitance as to not be an issue below 10 Gbps.

But, if using a two-layer board, 62 mils thick, a 50-ohm line will have a width of about 120 mils. The capacitance in a corner is 3.4 pF/inch × 0.12-inch x ½ = 0.2 pF. This amount of capacitance would cause a rise time of about 20 psec, easily noticeable in RF applications or at data rates of 5 Gbps.

The total loop inductance in a 50-ohm line in FR4 is 8.5 nH/inch. A signal-return path loop 2 inches long has a loop inductance of 2 × 8.5 nH = 17 nH and would behave like an ideal inductor element at low frequency.

8.10 Limitations of the n-Section Lumped Circuit Model

An infinite number of infinitesimally small LC elements in series behaves identically to an ideal T element. But it is not practical to create a SPICE simulation with an infinite number of LC elements. The question then is how many LC elements are needed to provide an adequate approximation to a real transmission line?

Given the excellent quality of the ideal T element in matching the behavior of a real transmission line, we can compare the match between the behavior of a circuit model with a finite number of LC elements to that of an ideal T-element.

Given the features of the ideal T-element, we can easily calculate the total L and total C in the ideal transmission line. Given the total number of L and C elements we use in the circuit, we can calculate the value of each L and C element as $L = L_{total}/n$ and $C = C_{total}/n$.

When we construct a collection of n-LC elements in a circuit to approximate the behavior of a real transmission line, we call this an n-section LC model, or an n-section LC lumped circuit model.

It cannot be emphasized enough that an n-section LC circuit model is not a preferred model for a transmission line.

Other than as a conceptual aid in visualizing the infinitesimally small LC model from which we derive the telegrapher's equation, there is no situation where an n-section LC model provides any advantage over a T element model. This is in spite of it being presented in too many textbooks as the recommended approximation to a real transmission line.

The two important behaviors we can evaluate for an n-section lumped circuit model, compared to a T-element are the input

215

impedance of an open circuit and the transmission coefficient through a transmission line with matched source and load.

We set up these two simulations in QUCS, using the circuits shown in **Figure 8.11**.

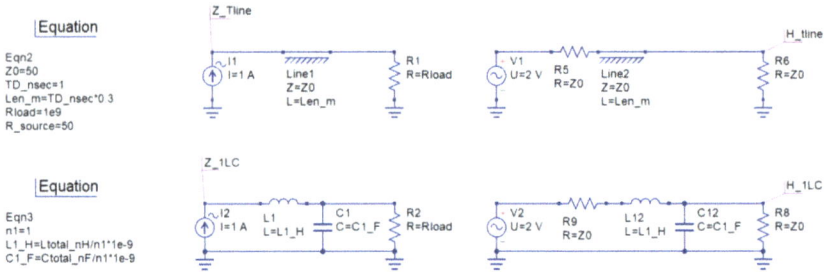

Figure 8.11 Circuits in QUCS to simulate the input impedance and transmission coefficient between an ideal T element and a 1-section LC circuit approximation for a transmission line.

In this circuit, the values of the L and C elements are calculated based on the characteristic impedance and time delay of the ideal T element, given the number of sections to use.

We expect very good agreement between the lumped model and the distributed model at low frequency, but they should differ at a frequency of about $1/(10 \times TD) = 100$ MHz.

The input impedance for each circuit is calculated, and the transmission, or transfer, coefficient. For this special case of 1 element, the simulated performance is shown in **Figure 8.12**.

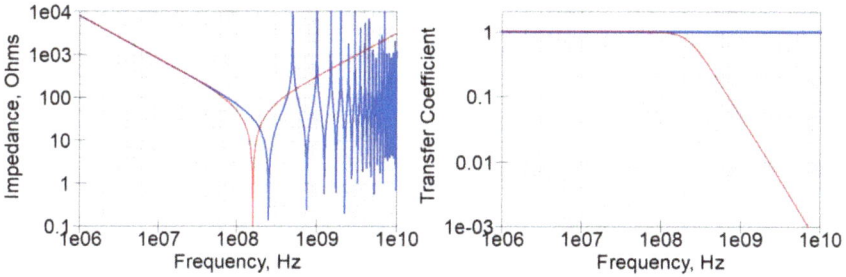

Figure 8.12 Simulated impedance and transfer coefficient between the ideal T element and the 1-section LC model.

As expected, at low frequency, the impedances match really well. Above $1/(10 \times 1 \text{ nsec}) = 100$ MHz, the lumped circuit model does not match the ideal T-element very well.

The LC model is basically a two-pole low-pass filter. The pole frequency for this circuit is

$$f_{\text{pole}} = \frac{1}{2\pi\sqrt{LC}} = \frac{1}{2\pi\sqrt{Z_0 \times TD \times \dfrac{TD}{Z_0}}} = \frac{1}{2\pi \times TD} = 160 \text{ MHz}$$

$$(8.9)$$

The -3-dB frequency in the transfer coefficient matches this estimate really well. If we define the bandwidth of the LC model as the frequency above which it does not match the ideal T element very well, this is approximately the pole frequency, or 160 MHz in this example.

Why does the dip frequency for the LC circuit not match the dip frequency for the ideal transmission line? After all, isn't the total inductance and total capacitance of the ideal transmission line, the exact values used in the lumped circuit model?

This is another example of the fact that the LC circuit model is an approximation to the actual behavior of the distributed ideal T

217

element, which is an excellent match to the behavior of a real transmission line.

The more elements in the n-section LC model, the smaller the value of the L and C elements, since the total L and total C must be constant. This means the pole frequency increases with higher n. This approximates the behavior of an ideal transmission line which has an infinite frequency pole frequency, with higher values of n.

The dip frequency of the ideal T element and the ideal LC model are:

$$f_{dip-T-element} = \frac{1}{4 \times TD} \qquad f_{dip-LC} = \frac{1}{2\pi \times TD} \qquad (8.10)$$

As we add LC elements, keeping their total L and C constant, the match between the LC model and the T elements becomes better. **Figure 8.13** shows the models for an 8-section LC model and a T-element. Note the complexity of the model just using 8 LC sections.

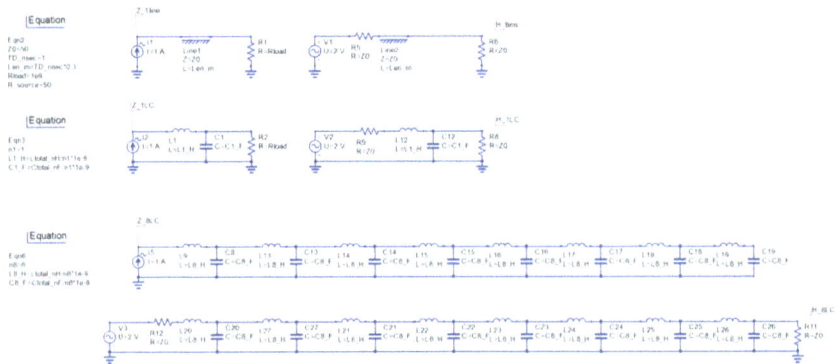

Figure 8.13 Examples in QUCS of an ideal T element and a 1-section and 8-section LC circuit models.

The more elements we add to the LC model, the better we expect the approximation to match the ideal T element. The comparison of the input impedance and transfer coefficient for these three models is shown in **Figure 8.14**.

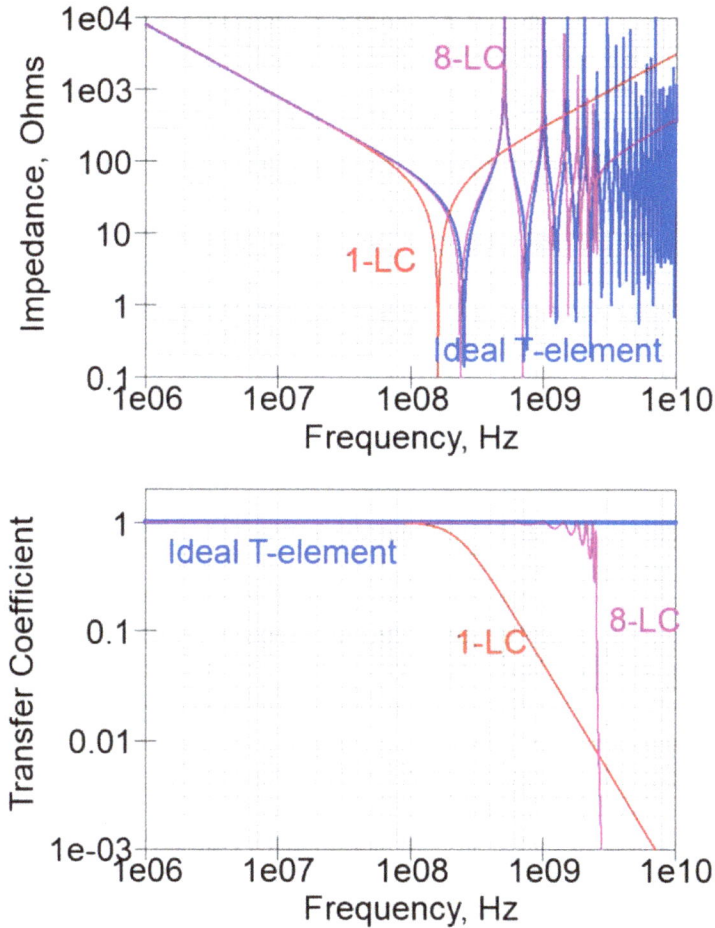

Figure 8.14 Simulated impedance and transfer function of the three ideal models of real transmission lines.

While the 8-section LC model is a better approximation to a real transmission line, for the effort in constructing the model and computation time involved, it is still not as high a bandwidth as the T element all by itself.

When constructing a model for a real, low-loss transmission line, the preferred starting model is always an ideal T element. It matches the behavior of a real transmission line at DC and at very high frequency and is very easy to implement.

This QUCS circuit and its plots in a project file are in the QUCS project folder you downloaded earlier in this chapter.

Watch this video and I will show you how to recreate these simulations.

8.11 Review Questions

1. What is the difference between the real world and the ideal world of electrical models?

2. What are the five ideal circuit element components found in all SPICE simulators?

3. What are three different measurement instruments that can measure properties of transmission lines?

4. What is special about the mathematical world?

5. What is special about the ideal world?

6. What is special about the real world?

7. What is the behavior of an ideal capacitor in contrast a real capacitor?

8. What can be done to the model of a real capacitor to increase its accuracy? Is it still an ideal model?

9. What are the two properties of an ideal lossless transmission line?

10. What features would be added to an ideal lossy transmission line to include the effects of loss?

11. When we describe a real transmission line as a 47-ohm impedance and time delay of 1.3 nsec, what assumption are we really making?

12. Why is the T element a good ideal model for a real transmission line?

13. When we describe a real transmission line as an n-section LC model, what is the assumption we are making about n? What does this make the L and C values?

14. A transmission line has a $Z0 = 45$ohms and a $TD = 1.5$ nsec. How much total capacitance does it have and total loop inductance?

15. A real transmission line is open at the far end. What ideal circuit element will its input impedance look like in the frequency domain? What if it were shorted at the far end?

16. Up to what frequency will a single ideal L or C match the impedance of an ideal transmission line? What frequency is this for a time delay of 1 nsec? 10 nsec?

17. What might be the bandwidth of an ideal T element compared to a real, low-loss transmission line?

18. What is the capacitance per length of all 50-ohm transmission lines in FR4? Suppose a 50-ohm line is 12 inches long? How much total capacitance is in the line? What is the underlying assumption we are making?

19. What is the loop inductance per length of all 50-ohm transmission lines in FR4? Suppose a 50-ohm line is 12

inches long? How much total loop inductance is in the line? What is the underlying assumption we are making?

20.   Why is an n-section LC model not a preferred circuit model for an ideal transmission line?

21.   Why does the dip frequency of the single LC circuit not match the dip frequency for the ideal transmission line when using the total L and total C?

22.   What happens to the pole frequency of an n-section lumped circuit model as we increase the number of LC elements? What is the pole frequency of an ideal transmission line?

Chapter 9   The TDR Principles

Reflections are not always a bad thing. This property of reflections occurring whenever a signal encounters a change in the instantaneous impedance is leveraged in a time domain reflectometer (TDR).

We use the value of the reflected voltage as a measure of the change in instantaneous impedance the signal encounters as it propagates down the line. The time instant the reflections are measured back at the source is a direct measure of their spatial location.

9.1   A Simple Example of a TDR Measurement and What It Can Measure

When the transmission line is uniform, we use the instantaneous impedance as a measure of the characteristic impedance of the line.

Any changes in the instantaneous impedance will be seen as reflections and their time location translates to spatial locations. **Figure 9.1** is an example of the measured TDR response of a transmission line and its physical structure. This is a uniform transmission line with a characteristic impedance of about 48 ohms, with launches on the ends and two strategically placed small discontinuities.

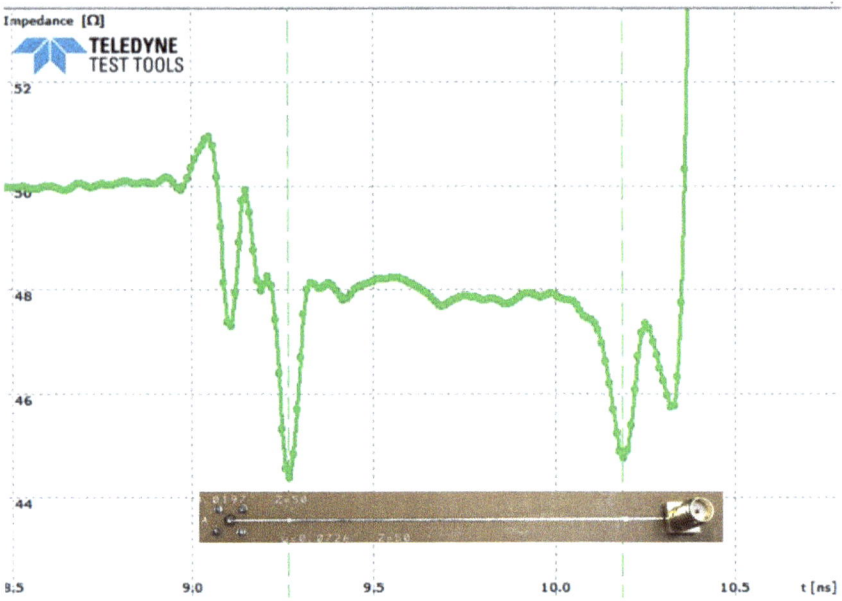

Figure 9.1 The TDR measurement of a uniform PCB trace with small notches producing reflections. The reflected signal has been translated into an instantaneous impedance plot. The TDR in this example is a Teledyne Test Tools T3SP15D.

The vertical scale on a TDR measurement is generally one of three options:

- The measured voltage

- The calculated reflection coefficient

- The calculated first order instantaneous impedance

The horizontal axis is the time the voltage is measured after leaving the source. Generally, this corresponds to the round-trip time delay from leaving the source, to the location of the discontinuity and back again. This round-trip time can also be translated into a distance-to-the-discontinuity scale by taking the one-way time delay and assuming a value for the Dk of the interconnects that gives the speed of the signal.

The TDR is the workhorse instrument for characterizing transmission lines. Each change in the instantaneous impedance as indicated by the reflected voltage is an *impedance discontinuity*.

9.2   Principles of Operation

In a TDR, we launch a well-characterized signal with a fast-rising edge into the interconnect, or device under test (DUT). If this signal encounters a change in the instantaneous impedance, some of it will reflect back to the source.

Inside the TDR, we sit near the source and measure the voltage at an internal node using a very fast sample oscilloscope. A simplified schematic of the structure inside a TDR is shown in **Figure 9.2**.

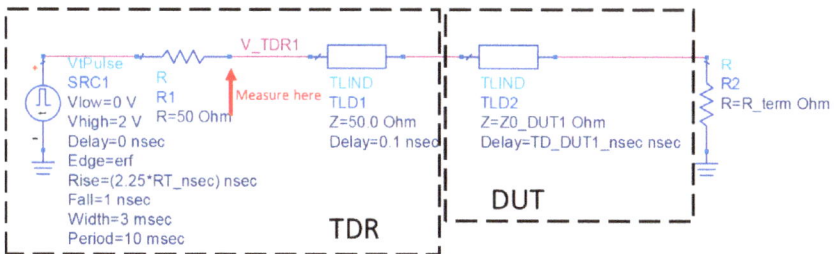

Figure 9.2 The complete circuit to simulate a TDR with DUT and termination connected.

After the initial signal from the pulse generator passes by the measurement point, any additional voltage we see at this internal node can only be due to a reflected voltage having reflected back to the source from an impedance discontinuity.

The time after we launch the fast-rising edge until we see a reflected voltage is the *round-trip* delay to the discontinuity. This

225

time is the round-trip time of how long it takes to get from the pulse generator, to the discontinuity that causes the reflection, and back again.

The round-trip delay depends on the length of the path and the speed of light in the interconnect.

If the interconnect from the pulse generator to the DUT is a circuit board trace, the dielectric constant is about 4, typical of most circuit board laminates, and the speed of a signal on this interconnect is about:

$$v = \frac{c}{\sqrt{Dk}} = \frac{11.8\,in/_{n\,sec}}{\sqrt{4}} = 5.9\,in/_{n\,sec} \sim 6\,in/_{n\,sec} \quad (9.1)$$

If the distance from the pulse generator to the DUT is 3 inch, the time after the edge goes by the scope's measurement point before we see the reflection, is only about

$$T_{round-trip} = 2 \times \frac{Len}{v} = 2 \times \frac{3\,inch}{6\,inches/_{n\,sec}} = 1\,n\,sec \quad (9.2)$$

To measure distances a small fraction of an inch, we need a time resolution of a small fraction of a nsec. This requires a very fast scope.

Since the fast-rising edge signal is repetitive and the reflected signals are very stable and repeatable, we can take advantage of a lower cost instrument called a *sampling scope* to measure the voltage with very short time resolution, typically 10 psec per point. It may take as long as 0.1 second to sample the entire reflected waveform at 10-psec resolution.

We can explore the properties of a TDR in simulation and test out the features of the interconnect that will create specific TDR signatures.

A TDR can be simulated in any version of SPICE in exactly the same way it is constructed, as illustrated in the circuit above.

We use a fast-rising step edge source, from a precision 50-ohm Thevenin source, driving into a short length of 50-ohm transmission line that then connects to the DUT. We sit at an internal point and measure the total voltage.

Of course, any measurement of the voltage cannot tell the direction the signal is propagating on the internal transmission line. It can only measure the net, total voltage at a point relative to the local ground. This voltage will be composed of the incident signal traveling from the source and the reflected signal traveling back to the source.

The way we separate the two signals is by knowing what the incident signal is, which will be constant during the time we do the measurement. Everything other than the incident voltage can only be reflected voltage.

We measure the total, net voltage and subtract off the incident voltage. The difference is the reflected voltage.

To simulate an ideal TDR, we simulate the voltage at the internal node, labeled V_TDR1 in the circuit above. This is the sum of the incident voltage and any reflected voltage.

The fast step edge source is set up as a Thevenin voltage source with a 2 V edge. When it passes through the source resistance of 50 ohms into the 50 ohms of the special reference transmission line at the beginning, it is really a voltage divider circuit. Only half of this source voltage, 1V, appears across the 50 ohms of the input of the transmission line and propagates down the line.

This means the incident voltage at this internal node is really 1 V. Anything other than 1 V at this node must be due to a reflected voltage.

The first quantity we calculate is the reflection coefficient, which is:

$$\text{rho} = \frac{V_{\text{reflected}}}{V_{\text{incident}}} = \frac{V_{\text{TDR}} - V_{\text{incident}}}{V_{\text{incident}}} = \frac{V_{\text{TDR}} - 1\,V}{1\,V} \qquad (9.3)$$

There are three special impedances to consider as the DUT: an open, a short, and a 50-ohm termination. The reflection coefficient we expect to see in each case is summarized here:

Impedance of the end	Reflection coefficient from 50 ohms
Open	+1
Short	-1
50 ohms	0

This is always an important test of any real or simulated TDR.

In this simple simulation, the DUT transmission line characteristic impedance was set to 50 ohms with a TD = 0.4 nsec. The node at which the voltage is simulated is 0.1 nsec from the location of the termination. Using this ideal TDR circuit, each of these three terminations was simulated. **Figure 9.3** shows the simulated reflection coefficients perfectly matching what we expect.

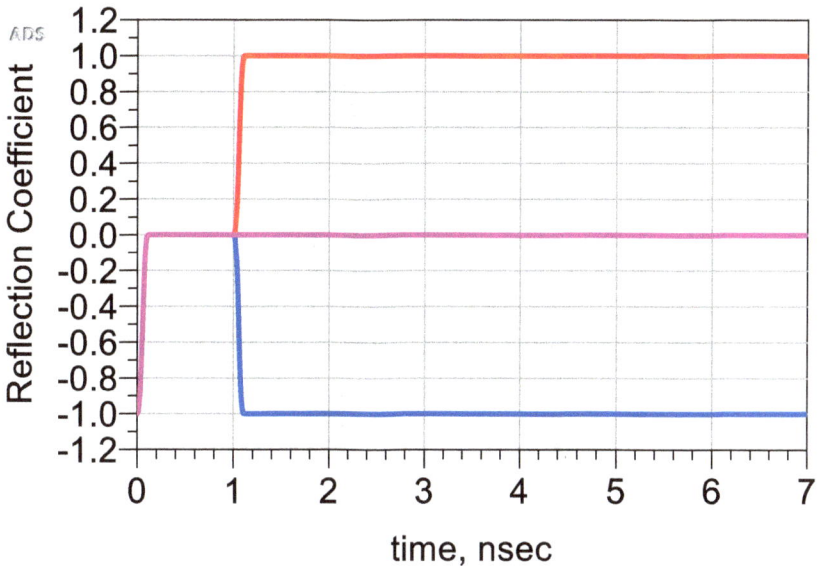

Figure 9.3 Simulated reflection coefficient for the three special cases.

The time at which we see the reflections from the various terminations is the round-trip time for the incident signal to make its way to the termination and the reflection to make its way back to the simulation node, 2×0.5 nsec $= 1$ nsec later, in this circuit.

This is exactly what is measured in a calibrated TDR. **Figure 9.4** shows the measured TDR response for the same three termination conditions.

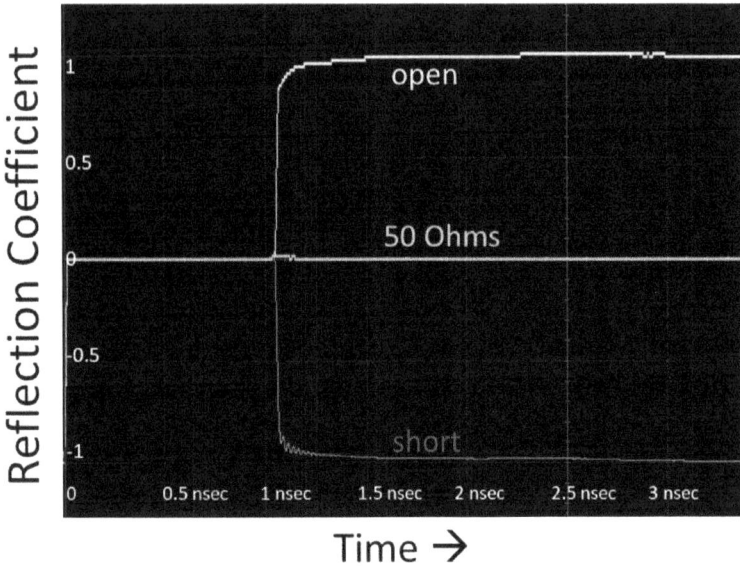

Figure 9.4 Measured reflection coefficients from a calibrated TDR for the three special case terminations. TDR is the Teledyne LeCroy SPARQ Interconnect Analyzer.

Once the reflected signal makes its way to the source, it will see a 50-ohm impedance and no more reflection. The voltage we simulate at the measurement node will be the steady-state, constant voltage after the reflected signal passes by. We could simulate for 10 seconds, or a day, and see the same constant voltage.

9.3   From Reflection Coefficient to Impedance

We next assume that any reflected voltage at an interface is due to an impedance discontinuity from the 50-ohm impedance of the source to the unknown impedance.

This is not strictly correct except at the first reflection and will introduce an artifact we refer to as *masking*. However, if the reflections are small, it can be a very good approximation and will help us get to an acceptable answer faster. If we wait for all the reflections to die out, the impedance on the other side of the discontinuity is the same as would be measured if connected directly.

Only at the first interface, when the signal passes from the 50-ohm reference line to the DUT, can we directly and unambiguously interpret the reflected signal as due to a specific impedance change.

To simplify the analysis, we assume the reflection at every interface is related to the unknown impedance discontinuity by:

$$\text{rho} = \frac{Z_{\text{unknown}} - 50\,\Omega}{Z_{\text{unknown}} + 50\,\Omega} \tag{9.4}$$

After a little algebra, the unknown instantaneous impedance is

$$Z_{\text{unknown}} = 50\,\Omega\frac{1+\text{rho}}{1-\text{rho}} \tag{9.5}$$

We can use simulations to explore how good this simplifying assumption is and when it is a safe assumption. A simulation takes into account all the details of the reflections, including the specific impedance of the two regions and the changes to the signal that continues to propagate.

9.4 TDR Response with a Short Discontinuity Before the Termination

A discontinuity in front of an impedance can mask the value of the impedance as calculated with the first-order estimate from the TDR.

However, the multiple reflections going on inside the discontinuity will die out after about three to five round-trip times, even for a large value discontinuity. When they die out, the discontinuity will be transparent. This means if we wait for all the reflections to die out, we can get an accurate measure of the impedance of the DUT on the other side of the discontinuity.

When the DUT is a short discontinuity, with a uniform transmission line on the other side, the short-time TDR response will be very complicated, but the long-time response will always match the behavior of the impedance on the other side.

If, for example, the discontinuity is a short length transmission line with a very high impedance, there will be multiple reflections between the front of the transmission line and the terminated end of the transmission line. An example circuit is shown in **Figure 9.5**.

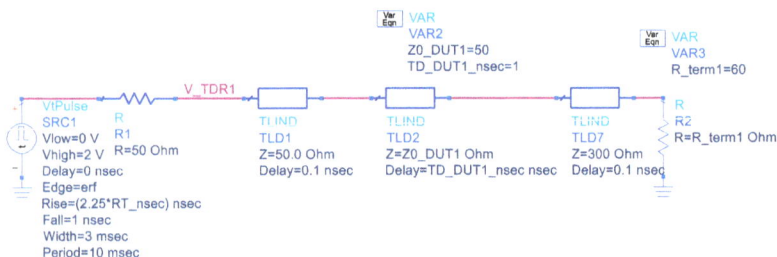

Figure 9.5 Example of a TDR circuit with a large, 300 ohm discontinuity in front of a 60-ohm resistive termination.

The reflections from the front and the back of the discontinuity to the source will add up over time as each reflection makes its way back to the source.

If we wait long enough for the reflections from the discontinuity to die out, roughly a maximum of seven round trip times, we are left with the reflection from the end termination. This is illustrated in **Figure 9.6** of a simulation of this circuit. Included is the case of no discontinuity, the case of a very high-impedance discontinuity and the case of a very low, 5-ohm, discontinuity.

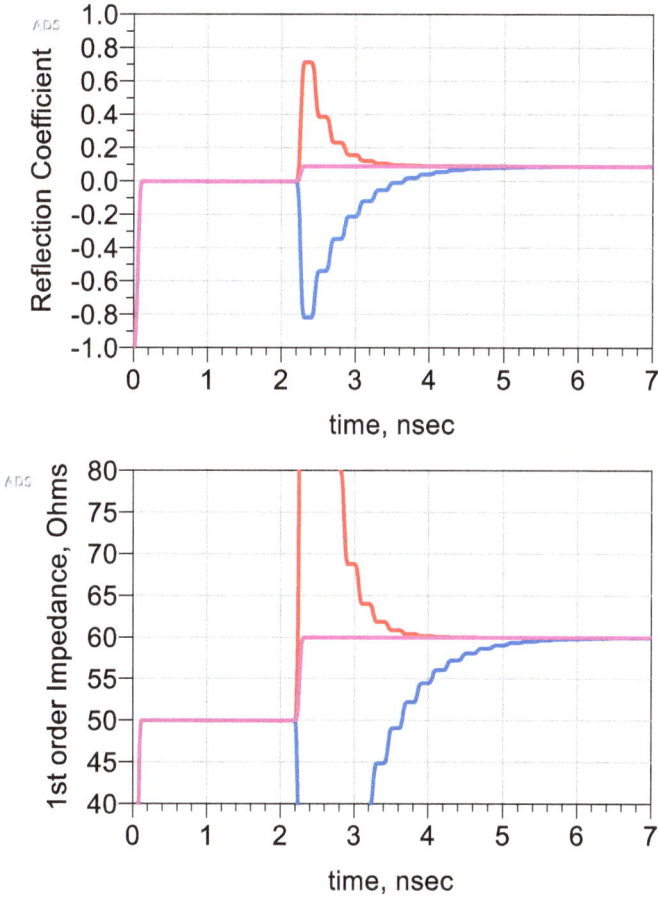

Figure 9.6 The reflection coefficient (top) and the extracted instantaneous impedance (bottom) for a large discontinuity between a 50 ohm line and a 60-ohm resistive termination. Regardless of the discontinuity, we see the impedance of the resistor when all the reflections die out.

We clearly see that if we wait for all the reflections from the discontinuity to die out, the net reflection corresponds to the end termination.

In almost all PCB or cabled structures, there is usually a small discontinuity at the launch. This is due to the geometry

233

transformation from the coaxial geometry of the cable connector to the geometry of the DUT.

When connecting to a cable, sometimes the signal and return path are pulled apart. This happens at the termination of the cable. It is often hidden by the covering on the outside of the cable.

Pulling the signal conductor away from the return conductor at the termination increases the instantaneous impedance at the launch and the impedance looks like a short-duration, high-impedance structure, which we refer to as an *inductive peak*. The TDR response will show a peak in impedance at the launch.

When a coax cable connector is inserted in a circuit board to connect to a circuit board trace, often the connector has an extra capacitance at the pad. This decreases the impedance at the launch and gives a low impedance discontinuity. We refer to this as a *capacitive dip*.

In either case, a few round-trip reflections after the discontinuity, the instantaneous impedance appears uniform.

Figure 9.7 shows examples of the measured TDR responses after the inductive and capacitive launches into otherwise relatively uniform transmission lines.

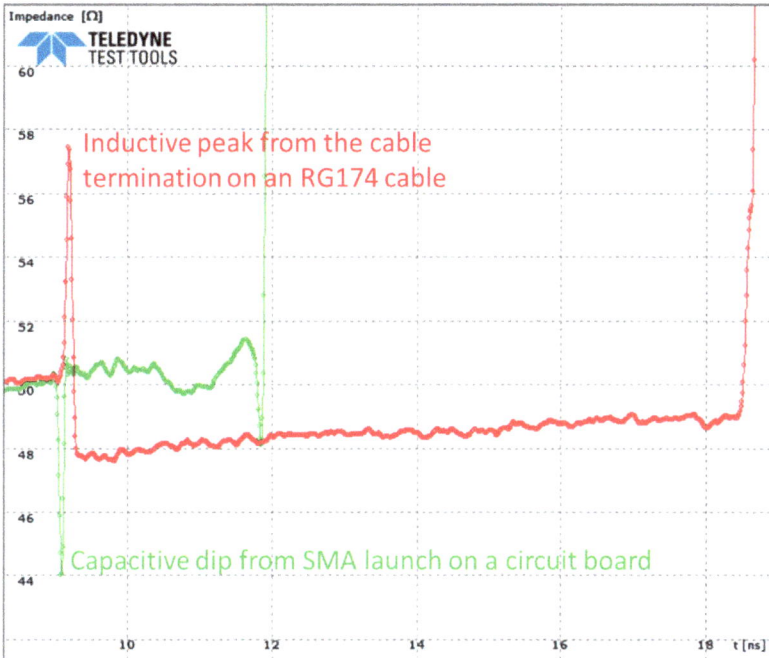

Figure 9.7 The measured launch discontinuity from two different transmission line structures. After the discontinuity the impedance profile of the DUT can be read off the front screen. The TDR rise time was 50 psec.

9.5 The TDR Response from Resistors

In principle, when a TDR step edge reflects from an ideal resistor, the signal sees the instantaneous impedance of the resistor as its resistance. The reflected signal depends on the source impedance and the resistance of the resistor. The impedance we would extract from the TDR response from a resistor would be the same impedance as the resistance of the resistor if measured by a DMM ohmmeter at DC.

Figure 9.8 shows the simulated reflection coefficient and extracted impedance for three different resistors at the far end: 70 ohms, 50 ohm and 30 ohms. The extracted impedance is exactly the values of the resistors.

235

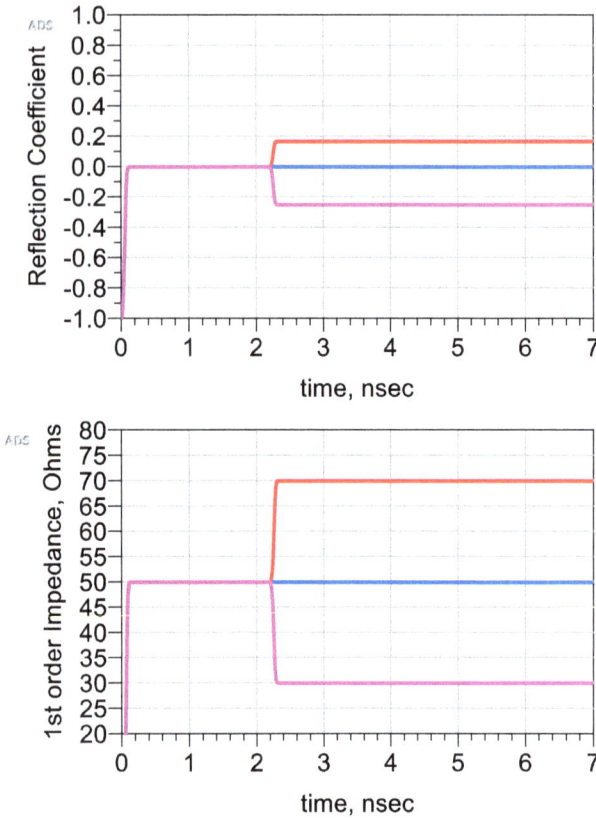

Figure 9.8 Simulated reflection coefficient (top) and extracted instantaneous impedance (bottom) for three different resistor values.

In practice, to measure a resistor with a TDR, we need to attach the resistor to a coaxial connector that connects to the coax connector of the TDR instrument. This will always add a discontinuity, either inductive if it is a long wire loop, or capacitive if there is a large attached pad associated with the resistor.

This discontinuity will affect the initially reflected signal causing a large reflection. If we wait for all the multiple reflections inside the discontinuity to die out, the discontinuity will look transparent and we will see the impedance of the resistor on the other end.

As an example, an axial lead 50-ohm resistor was soldered to an SMA connector with a purposefully large inductive loop. The resistance from the center conductor to the outer shield of the SMA connector was measured with a DMM as 52.0 ohms. The TDR response was measured with a calibrated TDR as 52.2 ohms. The instantaneous impedance extracted from the reflection coefficient is shown in **Figure 9.9**.

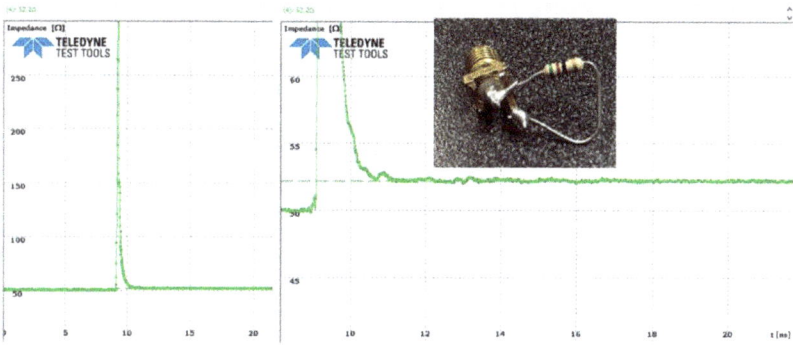

Figure 9.9 Measured instantaneous impedance of an axial lead resistor, shown in the inset, displayed on two different scales. Measured with a Teledyne Test Tools T3SP15D. The TDR rise time was 50 psec.

In this example, the TDR response shows a very large impedance discontinuity from the inductive loop. After the reflections die down, the impedance is constant. On the expanded scale of 5 ohms per division, the impedance is constant to better than 0.2 ohms and reaches a steady state value of 52.2 ohms, as measured by the cursor position.

Measuring a resistor with a DMM and comparing this to the extracted TDR impedance at steady state is a useful method to verify the calibration of a TDR. **Figure 9.10** shows another example of a resistor measured by a DMM at DC as 98.2 ohms and measured by the TDR as 98.4 ohms.

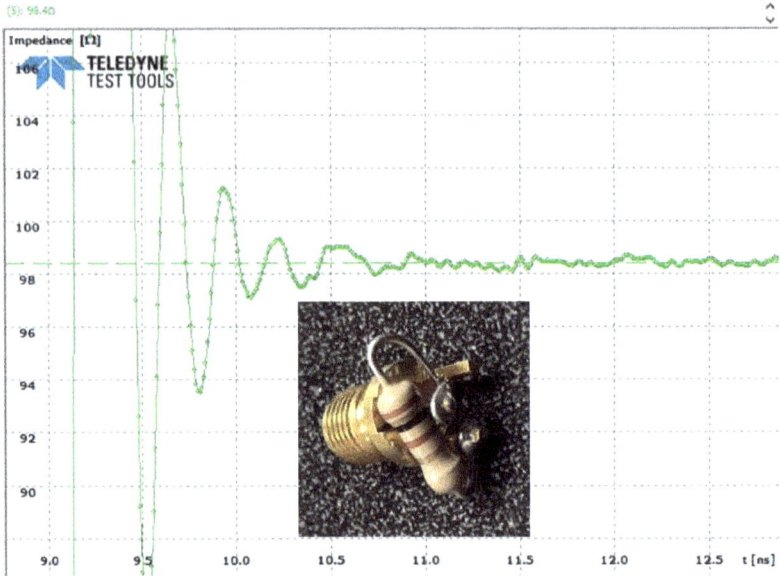

Figure 9.10 TDR measurement of an axial lead resistor soldered to a SMA connector. The resistor was measured by a DMM as 98.2 ohms and by the TDR as 98.4 ohms. Measured with a Teledyne Test Tools T3SP15D. The TDR rise time was 50 psec.

If we place any resistor at the far end of our DUT, the reflected signal will be based on this resistance. After all, the instantaneous impedance the signal sees when it hits a resistor will be its resistance. This will be the instantaneous impedance we measure from the reflected signal at the source.

This suggests an important way of verifying the accuracy of a TDR: we take a resistor, measure its resistance with an ohmmeter at DC and then measure the instantaneous impedance with the TDR. Figure 9.11 shows examples of seven different resistors on SMA fixtures and their TDR measurements.

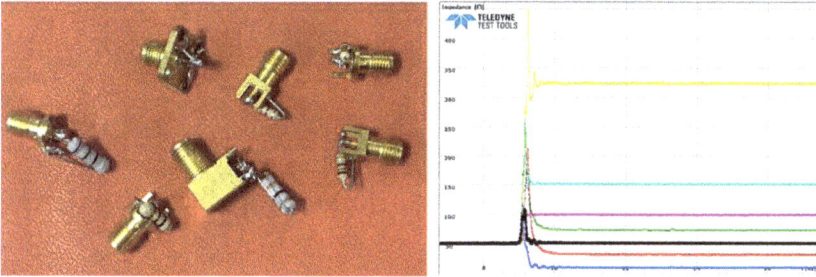

Figure 9.11 Examples of the seven resistor samples measured by TDR.

Depending on the magnitude of the loop created when connecting the resistor to the SMA connector, the inductive peak may be large or small. In every case, a short time after the discontinuity, the TDR reflected response is flat.

The DC resistance of each sample was measured at DC with a DMM Ohmmeter with a NIST traceable accuracy of 0.1%. The measured TDR impedance compared to the DC resistance and their relative errors are summarized here:

DMM Resistance	TDR resistance	Relative error
R = 9.96 ohms	10.1 ohms	1.4%
R = 32.3 ohms	32.5 ohms	0.6%
R = 50.8 ohms	50.9 ohms	0.2%
R = 74.1 ohms	73.8 ohms	0.4%
R = 100.0 ohms	99.7 ohms	0.3%
R = 153.4 ohms	151.8 ohms	1.0%
R = 329 ohms	324 ohms	1.5%

Near 50 ohms, the absolute accuracy is within less than 0.5%.

9.6   A Uniform Transmission Line or a Resistor?

If we were to connect a uniform transmission line with a roundtrip time delay longer than the time we measure, there is no way we could distinguish whether we had a transmission line attached to the end or a resistor.

After all, the only voltage we measure is the reflected signal. When the signal encounters the beginning of the transmission line, we get an initial reflection based on the characteristic impedance of the line. If the signal encounters a resistor, we get the same reflection from the resistor.

In the transmission line, the signal continues to propagate with no additional reflections until it encounters the end of the line. In the case of the resistor, the signal is constant into the resistor, dissipating power in the resistor, with no more reflections. The observed behaviors are the same. **Figure 9.12** shows examples of similar measurements from a coax cable and a resistor on the end of an SMA.

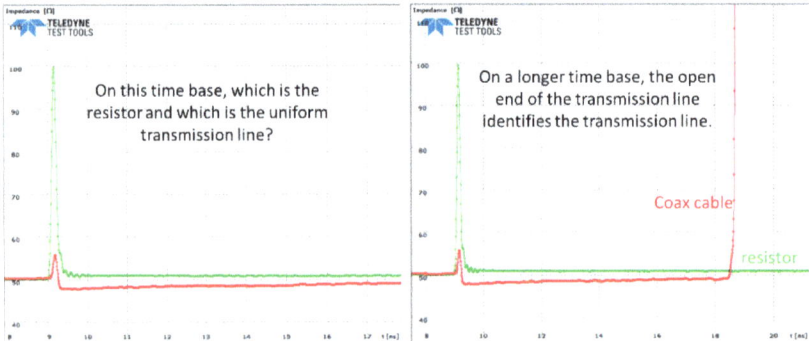

Figure 9.12 TDR response from a coax cable and resistor, nominally the same impedance, measured for a short period of time. Left: In this time base, we cannot tell which is the resistor or which is the transmission line. Right: Seeing the end of the line makes it obvious. The TDR rise time was 50 psec.

240

In practice, there will be a difference between a resistor and a transmission line. Very few transmission lines have a uniform instantaneous impedance over their entire lengths. There will always be small variations of as much as an ohm or more down the length of the line. A resistor will have a TDR response that is mostly flat and constant.

For example, **Figure 9.13** shows the measured instantaneous impedance of two nominally identical transmission line traces on a circuit board, in close proximity. The variation between them is about ± one half a ohm, and their variation down the line, after the launch, is as much as ± 1 ohm.

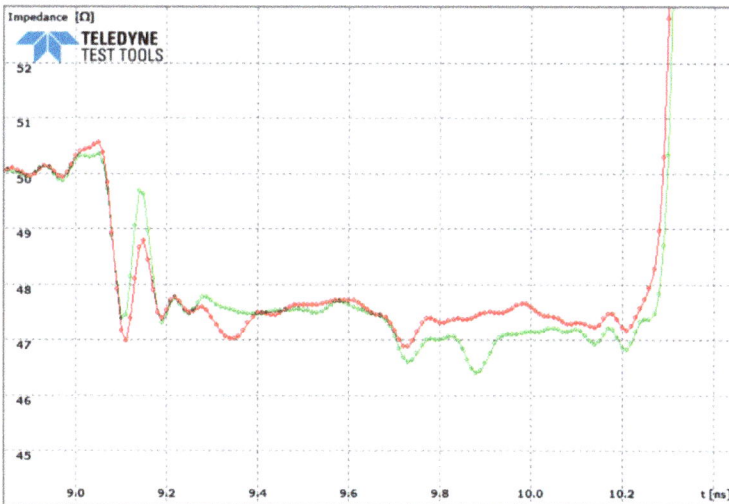

Figure 9.13 In the measured TDR responses of two, uniform circuit board traces. The scale is 1 ohm/div and 0.2 nsec round-trip time delay/div. Measured with a Teledyne Test Tools T3SP15D TDR. The TDR rise time was 50 psec.

Watch this video and I will show you how easy it is to use a TDR to measure the impedance profile of interconnects.

9.7 Electrically Long and Electrically Short Transmission Lines

When looking at the measured impedance profile of a real interconnect, we observe two general behaviors. There are regions where the instantaneous impedance is flat and regions where it is a peak or a dip.

When it is flat, we interpret the impedance profile as from a uniform transmission line. The single value of the instantaneous impedance in this region is the characteristic impedance of the transmission line.

When the instantaneous impedance is not flat but shows a peak or a dip, the transmission line in this region appears to be not uniform. There is no one characteristic impedance that characterizes this structure. Instead, the impedance extracted and displayed by the TDR depends on the rise time of the signal from the TDR source.

An example of the measured TDR response from an interconnect showing these two regions is in **Figure 9.14**. In this example, measured from two sections of uniform coax cable with a connector between them, the cables are seen as uniform transmission lines, but the connector is a peak, a higher impedance.

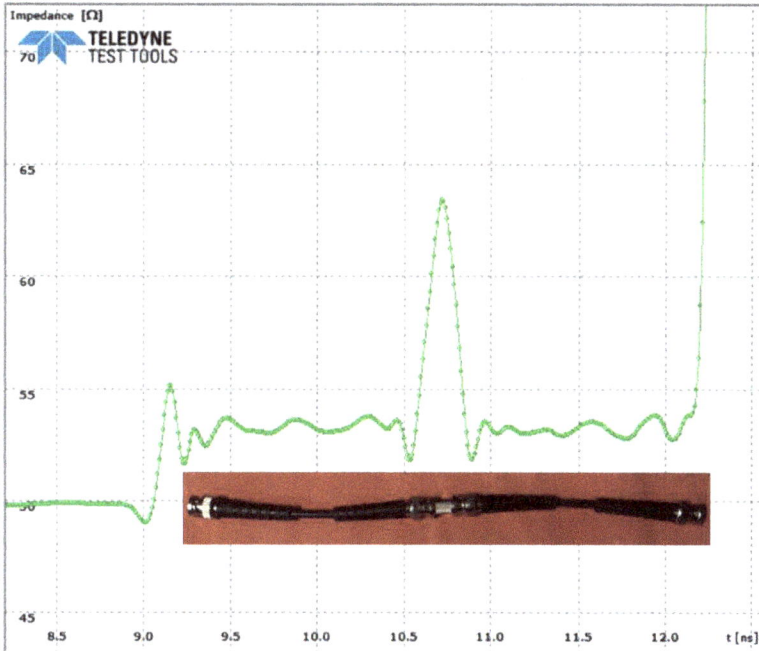

Figure 9.14 Impedance profile of an interconnect with two types of structures, a flat region and a peak or a dip. The TDR rise time was 50 psec.

We refer to the regions of constant, instantaneous impedance as uniform transmission lines. These structures are *electrically long*, long enough to resolve as uniform transmission lines with a characteristic impedance. The regions that are either peaks or dips are discontinuities. They are *electrically short*, too short to be resolved as uniform transmission lines.

The launches, which are usually either peaks or dips, are electrically short. They are discontinuities. The connector, in this example, is a discontinuity.

The discontinuity structure, which is causing the peak or dip reflection, may in fact be a uniform transmission line, with a characteristic impedance different from the transmission line it is embedded in, but its uniform instantaneous impedance cannot be resolved with the rise time of the source.

243

When the signal encounters this discontinuity, which may be a uniform transmission line, the TDR signal will reflect from the front surface. The reflection coefficient will be some value, rho. The reflected signal will have a rise time equal to the rise time of the source.

The transmitted signal will continue on through the discontinuity structure and encounter the end of the structure. There will be a reflection from the backend, with a reflection coefficient of -rho.

This back-end reflection will make its way back to the TDR source right behind the front-end reflection. The delay between these two reflections will be the round-trip delay of the discontinuity, or twice its time delay, $2 \times TD$.

When the round-trip delay is longer than the rise time of the source, the reflections from the front and back will be separate and distinct. Between the rising and falling edge will be a flat top or flat bottom. We will resolve the uniform nature of the transmission line making up the discontinuity. In this case the structure would be just barely discernable as electrically long.

But, if the round-trip delay is shorter than the rise time of the probing signal, the reflection from the front and back ends will overlap and smear out. The net reflection will be just a peak or a dip. We will not see the flat top or bottom. In this case the structure is electrically short.

In principle, the condition for being *electrically long*, and resolving the characteristic impedance of the transmission line is

$$2 \times TD > RT \tag{9.6}$$

or

$$TD > \frac{1}{2} RT \tag{9.7}$$

where

TD = the one-way time delay of the interconnect

RT = the rise time of the TDR source

In practice, to clearly see the flat top or bottom of the short, uniform transmission line requires a slightly longer time delay of the discontinuity compared to the rise time of the probing signal. A better criterion for electrically long is

$$TD > RT \qquad (9.8)$$

This is illustrated in **Figure 9.15**. In this simulation, a discontinuity was created by making a short transmission line with a narrower line width than the transmission line in which it is embedded. It will have a higher characteristic impedance.

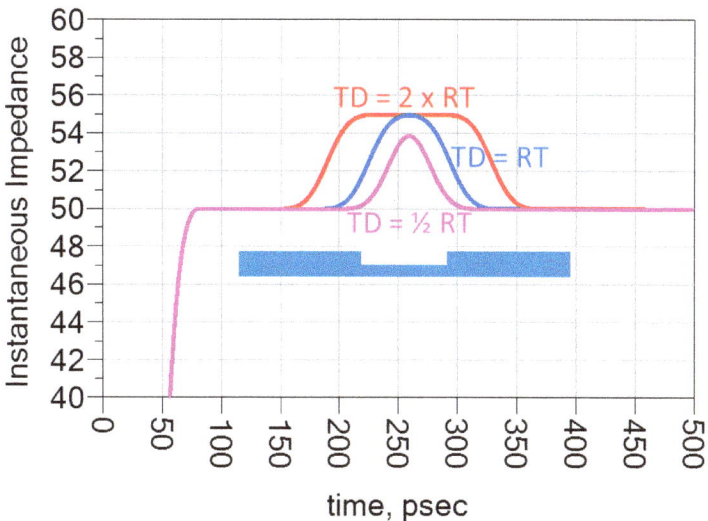

Figure 9.15 The TDR impedance profile of a short discontinuity when changing its time delay compared to the source's rise time.

The round-trip time delay of this discontinuity region is initially longer than the rise time. The interconnect is electrically long, and we resolve the flat top of the impedance.

As the time delay of the discontinuity decreases, the reflections from the front edge and back edge of the discontinuity begin to overlap and smear out.

When the TD = RT, the structure shows an impedance equal to the characteristic impedance of the transmission line, and just barely shows a hint of a flat top. This is at the limit of what could be resolved as electrically long. The reflections from the front and the back are just barely separate.

When the time delay is shorter than RT the reflected signal looks like just a peak. We cannot resolve the flat top as a flat top. As well, the magnitude of the reflected signal is slightly reduced from the characteristic impedance. The structure is *electrically short*.

When the structure is electrically short, we cannot resolve the uniform nature of the discontinuity. If it really is a uniform transmission line, all we can say is its characteristic impedance is higher than the peak value or lower than the dip value of the displayed TDR impedance.

Whether a discontinuity is electrically short or electrically long is not intrinsic to the interconnect but depends on the rise time of the signal interacting with it.

Since its time delay is related to is physical length, we can define the boundary between electrically long or short in terms of its length:

An electrically long interconnect is:

$$TD > RT \tag{9.9}$$

or

$$\frac{Len}{v} = \frac{Len}{c}\sqrt{Dk} > RT \qquad (9.10)$$

or

$$Len > \frac{RT}{\sqrt{Dk}}c \qquad (9.11)$$

In an FR4 laminate transmission line, this translates to an electrically long interconnect being:

$$Len[mils] > \sim 6 \times RT[p\sec] \qquad Len[mm] > \sim 0.15 \times RT[p\sec]$$

For example, if the rise time of the TDR signal is 35 psec, an electrically long interconnect would be longer than about 200 mils or 5 mm.

To turn an electrically short structure into an electrically long structure, we need a shorter rise time TDR source.

The instantaneous impedance measured in an electrically long interconnect will be independent of the rise time of the TDR signal. As long as the rise time is longer than the time delay of the interconnect, the impedance profile will be flat. As the rise time increases, the instantaneous impedance has just one value.

For an electrically short structure, the interaction between the reflections from the front and the back end of the electrical short region will overlap and depending on the rise time and the round-trip time of the interconnect, the peak value reached by the resulting reflected signal will depend on the rise time.

If the electrically short structure is a uniform transmission line, the peak or dip impedance will always be a smaller change than from its characteristic impedance. The peak or dip impedance change always decrease as the rise time increases.

This behavior is illustrated in **Figure 9.16**. This shows the same interconnect structure with different rise times. The electrically long structures show the same instantaneous impedance, and characteristic impedance while the electrically short regions show a peak or dip height that decreases with increasing rise time.

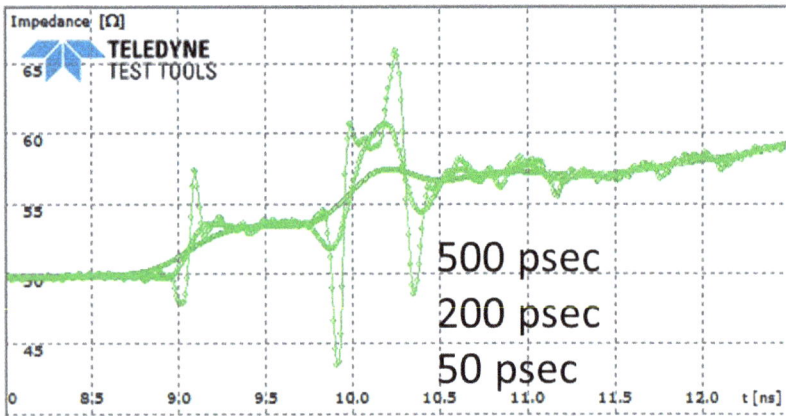

Figure 9.16 The same interconnect measured with three different rise times. Electrically short structures have peak impedances that scale with rise time. Electrically long structures have an instantaneous impedance that is flat and independent of the rise time.

Download this QUCS file and you can explore the transition from electrically long and short interconnects.

Watch this video and I will walk you through distinguishing between electrically long and short interconnects.

9.8 Spatial Resolution of a TDR

Whether we see an interconnect as electrically long or electrically short is not intrinsic to the interconnect. It depends on the rise time of the source. The rise time of the TDR determines how short a structure we can resolve as electrically long. This just defines whether we see the interconnect as a transmission line or a discontinuity.

Just because an interconnect structure, like a via or connector, is electrically short, does not mean the TDR cannot see it and that it cannot be resolved. It just means it cannot be resolved as an electrically long transmission line interconnect. It will appear as a discontinuity.

The noise floor for an impedance change for the T3SP15D TDR is about 0.2 ohms. Any electrically short discontinuity that produces a reflection of 0.2 ohms can be resolved.

This does not mean the TDR with a rise time of 35 psec cannot see a structure that is shorter than 0.15 mm/psec × 35 psec = 5.2 mm. It just cannot resolve a shorter structure as electrically long and see its flat top or flat bottom.

If there are two discontinuities close together, the reflections from each one may smear out if the reflections from one overlap the reflections from the other.

If two discontinuities are adjacent to each other, with a short length of uniform transmission line between them, there will be a reflection from the back edge of the first structure, and a reflection from the front edge of the second structure.

To see that there is a region between them that is different, we do not want these two reflections to overlap completely.

The condition on the round-trip time delay (TD) of their separation is

$$2 \times TD > RT \tag{9.12}$$

or

$$TD > \frac{1}{2} RT \tag{9.13}$$

If the time delay between them is shorter than one half of RT, the reflections from the region between them will overlap and it will be difficult to resolve a separation. The two discontinuities will look like one and we cannot resolve that there are two of them.

To illustrate this, two electrically short discontinuities were separated by an increasing distance and the TDR response simulated.

In the first case, the round-trip time delay between the structures is $4 \times RT$. This is plenty of time to see the reflections from the end of one and beginning of the other as separate peaks. The TDR response of this structure is shown in **Figure 9.17**.

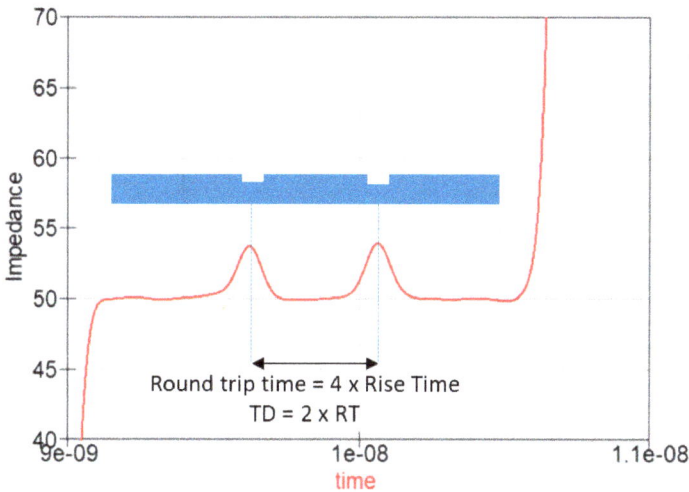

Figure 9.17 When the time delay is > 2 × rise time between two discontinuities, the reflections from the end of one and beginning of another do not overlap.

As the time delay between them, ΔTD, decreases, the reflections from the end of one and the start of the other one begin to overlap and smear out.

When the time delay of the space between them = RT, the reflection peaks begin to overlap. When the time delay = one-half of RT, the peaks overlap but can still be resolved as separate. But when the time delay between them = one quarter of RT, the peaks are so smeared out they cannot really be resolved as two different structures. These conditions of shrinking time delay and the resulting TDR response are shown in **Figure 9.18**.

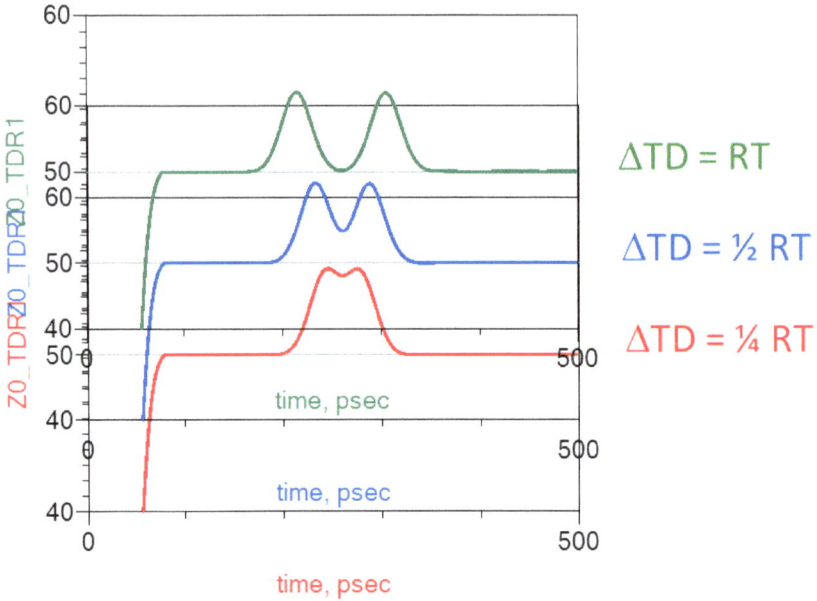

Figure 9.18 Simulated TDR response of two discontinuities as they are brought closer together. When the time delay between them is one half of the rise time, they can be resolved as two separate discontinuities.

The condition to being able to resolve the two discontinuities as two distinct structures is that the time delay between them is TD > one half × RT.

$$TD_{spacing} = Len_{spacing} \frac{\sqrt{Dk}}{c} > \frac{1}{2}RT$$

$$Len_{spacing}[mils] > 12 \frac{\frac{1}{2}RT[p\sec]}{\sqrt{Dk}} \sim 3 \times RT[p\sec] \qquad (9.14)$$

$$Len_{spacing}[mm] > 0.3 \frac{\frac{1}{2}RT[p\sec]}{\sqrt{Dk}} \sim 0.07 \times RT[p\sec]$$

where

Len$_{spacing}$ = the spatial resolution of the TDR to resolve two adjacent discontinuities as two structures

RT = the rise time of the TDR source

Dk = the dielectric constant of the interconnect, assumed to be = 4

c = the speed of light in air, 11.8 mils/psec = 0.3 mm/psec

This is the general criteria that describes the spatial resolution of a TDR.

For example, if the TDR rise time is 50 psec, the spatial resolution for telling two adjacent discontinuities apart is $0.07 \times 50 = 3.5$ mm. If we want to resolve a via structure that might have two ends separated by 60 mils, we would need a TDR rise time shorter than 60 mils/3 = 20 psec.

The resolution of a TDR is the shortest spacing between two structures that can be resolved as two separate structures. The resolution of a TDR does NOT mean how small a discontinuity can be detected.

Watch this video and I will show you how you can explore through simulation this question of spatial resolution.

9.9 Masking: Reflections from Two Uniform Transmission Lines

When the TDR signal encounters an electrically long region where the instantaneous impedance changes, there will be a reflection. If it encounters another electrically long region, there is a second reflection.

At the first interface where the signal transitions from 50 ohm to the next instantaneous impedance, the reflection coefficient is unambiguously:

$$\text{rho} = \frac{Z_{\text{unknown}} - 50\ \Omega}{Z_{\text{unknown}} + 50\ \Omega} \tag{9.15}$$

This means we can interpret the reflection from the first interface and extract an accurate value of the instantaneous impedance as simply:

$$Z_{\text{unknown}} = 50\ \Omega \frac{1 + \text{rho}}{1 - \text{rho}} \tag{9.16}$$

However, once the TDR signal enters this new region, it will not be the same voltage as the original incident signal. In fact, it will be changed to:

$$V_2 = V_{incident} \times T_{1 \to 2} = V_{incident} \times \frac{2 \times Z_2}{Z_1 + Z_2} \qquad (9.17)$$

When it encounters another impedance region, this new incident voltage will suffer a reflection. While the reflection coefficient will be easy to calculate, the voltage that reflects off the interface will suffer another distortion on its way back to the source, and will rattle around between the two ends of the first transmission line.

This means the presence of the first transmission line will distort the reflections from each successive adjacent discontinuity. This makes interpreting the instantaneous impedance of the following region a little difficult from the front screen.

As we increase the impedance of the first section, it will begin to have a larger and larger impact on the extracted first-order impedance of the next section.

Figure 9.19 is an example of the TDR simulation of a 50-ohm transmission line as viewed on the other side of a transmission line with different impedances from 50 ohms to 80 ohms. The presence of the first transmission line *masks* an accurate impedance measurement of the second line using just the first-order interpretation from the front screen of the TDR.

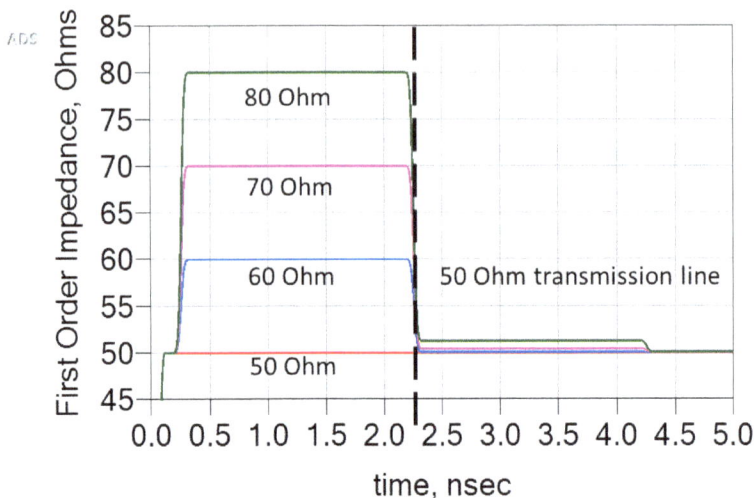

Figure 9.19 TDR simulation of a 50-ohm transmission line on the other side of a different value transmission line.

As the impedance of the first line increases, the value of the 50-ohm transmission line, as read from the TDR front screen, on the other side increases. Its true value is *masked* by the first transmission line.

This behavior suggests that we have to be a little careful interpreting the impedance values after a change in the impedance. When the impedance change is less than 40%, such as 50 ohms to 70 ohms, the impact on the extracted impedance of the following line is less than 1 ohm or 2%. It's when there are large changes that we have to be careful interpreting the impedance from the front screen.

As an example, **Figure 9.20** shows a simulation of two transmission line sections. The second section in each case is 60 ohms. When viewed through a 40-ohm line, the 60-ohm line looks to be 60 ohms to within 1%. When viewed through a 100-ohm line, the second section looks to be slightly higher than 60 Ohms.

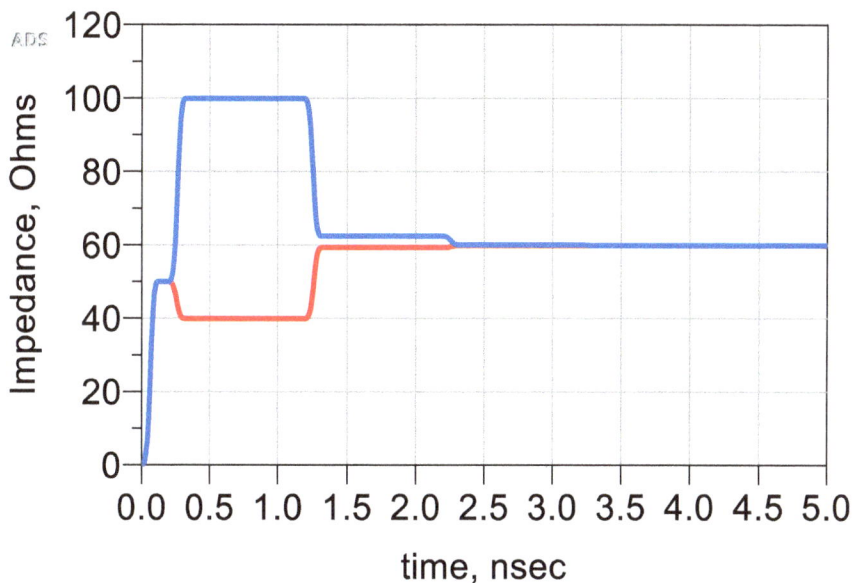

Figure 9.20 The simulated TDR response of a 60-ohm line as viewed through a 40-ohm and 100-ohm line.

The second transmission line is masked by the first structure.

After the reflections from the first structure have died away, roughly two round-trip time delays, the impact from the masking has greatly reduced, and we see the unmasked instantaneous impedance of the second structure.

The shorter the first structure, the sooner into the second structure we can extract its correct instantaneous impedance from the front screen. Or, if the impedance change is small, less than 40%, the impact from the masking on interpreting the impedance of following sections will be less than 1%.

When there are large impedance variations that last for a long time, you cannot interpret the measured reflected signal as an accurate measure of the impedance profile directly from the front screen.

To extract the specific impedance values of the structure beyond the large discontinuities requires a technique called *hacking*. This enables extracting the parameter values from complex interconnects where multiple reflections play a dominant role. This method is covered in detail in the last chapter.

9.10 Review Questions

1. What does a TDR actually measure?

2. What feature of the interconnect corresponds to the time between the start of the signal coming out of the TDR and receiving the reflected signal back in the TDR?

3. What is the shape of the waveform coming out of the TDR and entering the interconnect?

4. If only one node inside the TDR is being measured, how does it separate what is the incident voltage and what is the reflected voltage?

5. What is a discontinuity?

6. What is the impact of a discontinuity at the beginning of a transmission line on the reflected signal and the interpreted impedance?

7. If a discontinuity has a peak in impedance, how else do we refer to this discontinuity?

8. If a discontinuity has a dip in impedance, how else do we refer to this discontinuity?

9. If a resistor is connected to the end of a DUT, what impedance will the TDR measure if we wait long enough?

10. If you measure a resistor or a long transmission line and look for a time short compared to the round-trip time, how could

you distinguish which measurement is the resistor and which is the transmission line?

11. What is the difference between an electrically short and electrically long interconnect?

12. If the rise time of the TDR is 50 psec, how long does a discontinuity have to be in order to see its flat top or flat bottom?

13. What feature would be adjusted in a TDR to resolve a shorter interconnect as a uniform line and not a discontinuity?

14. If two structures are 0.5 inches apart, how short a rise time is needed to resolve them as two different structures?

15. What is the assumption we make when we read the impedance of an interconnect directly from the front screen?

16. When does masking play a role?

17. If there are multiple reflections, how can we extract a more accurate model of the interconnect from the measured TDR response?

Chapter 10 Practical TDR Measurements

It is easy to get a measurement from a TDR. However, it requires some skill and paying attention to best measurement practices to avoid common artifacts.

10.1 Always Use a Torque Wrench

For any TDR instrument, there are two important measurements to do before any serious measurements: an open at the DUT location and a 50-ohm load.

The open measurement defines the end of the TDR cable and the beginning of the DUT. It can sometimes be difficult finding the end of the TDR cable and the beginning of a well-engineered launch to a DUT. Comparing the measurement to an open identifies the transition.

A well calibrated TDR will show an impedance from a 50-ohm load as 50-ohms. Measuring an accurate load is an important verification of the TDR's calibration.

As an example, **Figure 10.1** shows the measured TDR response of an open and a 50-ohm load accurate to 0.1%.

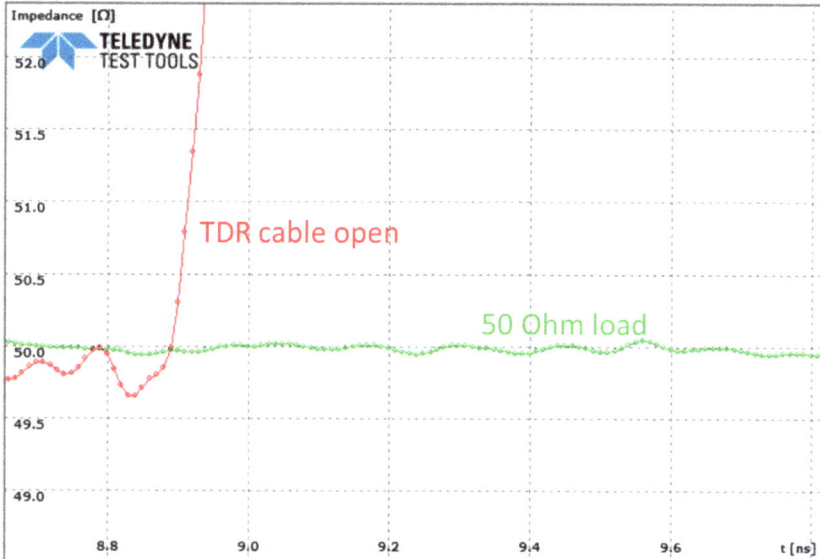

Figure 10.1 First important measurements: The TDR cable open and a 50-ohm load. The TDR rise time is 65 psec. Note the vertical scale is one half ohm/div. The TDR rise time was 50 psec.

In this example, the impedance measured by the TDR is 50 ohms to within ± 0.2 ohms, the limit to the absolute accuracy of the load.

Note also that there is no perceptible launch discontinuity between the end of the TDR cable and the beginning of the 50-ohm load. This is using a Type K, 2.92-mm connector. In this example, the connector was attached on the cable with a torque wrench.

If the DUT is not connected to the TDR cable securely with a torque wrench, there are two artifacts that can result.

First, the seating of the pin in the cup of the connector is offset, leaving an uncontrolled cross section inside the cable. This generally results in a higher impedance discontinuity.

Secondly the poor contact between the signal pin and its cup can have higher contact resistance. This means the impedance looking at the DUT will have an additional resistance in series. The extent of these artifacts is irreproducible. They change each time the connector is screwed onto the TDR's cable and depend randomly on the quality of the connection unless a torque wrench is used.

Examples of the TDR response to the calibrated 50-ohm load, with different levels of finger tightness of the DUT to the cable are show in **Figure 10.2**. The way to avoid these two important artifacts is to always use a torque wrench when connecting a DUT to the TDR cable.

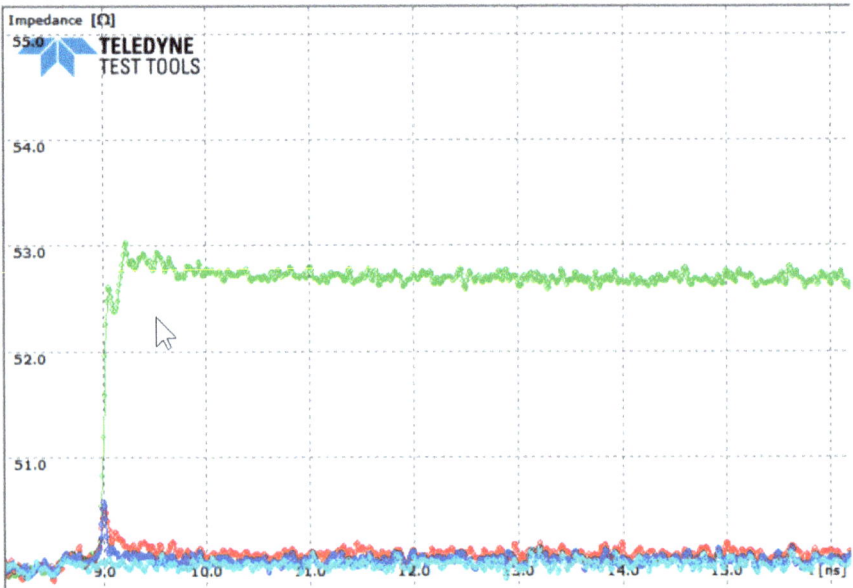

Figure 10.2 Examples of the TDR response of the same 50-ohm load with different, finger-tight connections. Each connection is a different value launch discontinuity and impedance value. The TDR rise time was 50 psec.

The contact resistance appears in the TDR response as an offset impedance. The signal sees the instantaneous impedance of the load through the series resistance of the contact resistance. This means the interconnect looks like a higher impedance. The instantaneous impedance of the load has not increased; we are just seeing another impedance in series.

10.2 Cable Termination Launches

The quality of a coax cable is related to both the cable itself and how the signal conductor and cable shield is terminated to the base of the connector. Generally, more expensive cables have more uniform impedance profiles, but more importantly, more precise cable terminations.

As an example, **Figure 10.3** shows the TDR response of a high-performance coax cable with a 3.5-mm connector and a low-cost RG174 cable with an SMA connector. The high-performance cable has a much smaller launch discontinuity due to a better cable termination.

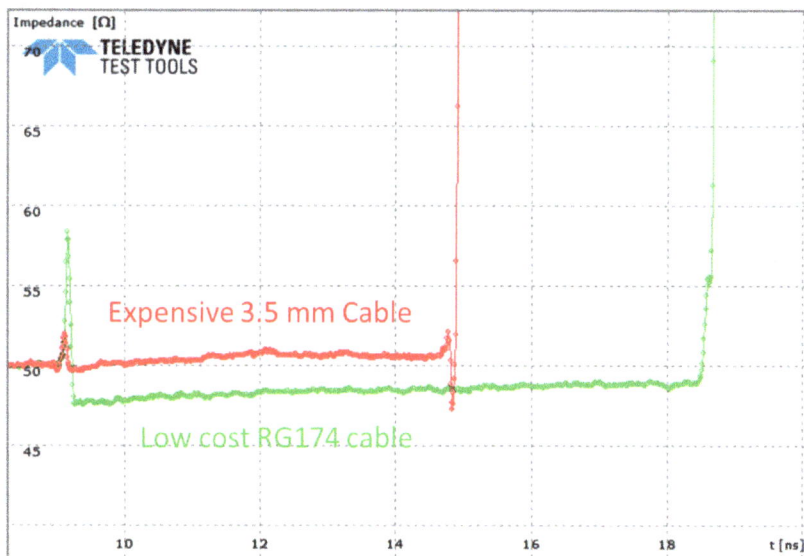

Figure 10.3 Examples of the measured TDR response of an expensive 3.5-mm cable and a low-cost RG174 coax cable. The TDR rise time was 50 psec.

The challenge with using a low-cost cable generally is a poor termination at the launch. This is the trade-off when balancing cost and performance.

10.3 Quality of Cables

Unfortunately, the quality of a cable is not always related to its price. In some cases, seemingly identical, high-performance cables can have very different impedance profiles.

A selection of cables used for routine 10-GHz applications were measured and compared. It was remarkable how much variation was found between these cables. **Figure 10.4** shows just three representative examples. One was very well behaved, showing a uniform impedance, while two others showed impedance discontinuities of as much as 5-ohm variation down the line.

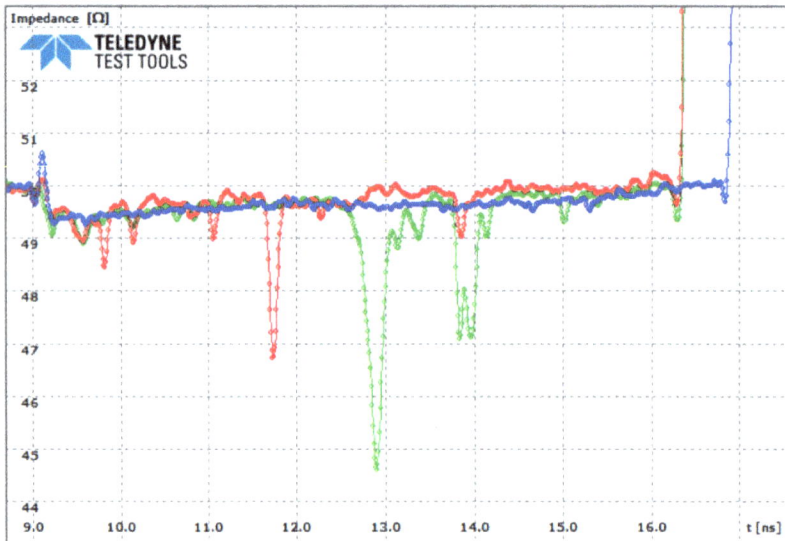

Figure 10.4 Examples of three nominally identical high-performance coax cables, measured at 65 psec rise time.

This is an important reason to always evaluate your cables before assuming they are all in excellent condition. The physical features that give rise to these impedance discontinuities could not be observed by visual inspection or by feel.

10.4 Uniform Transmission Lines and Impedance

All interconnects are transmission lines. When they have a uniform cross section, they will have a constant instantaneous impedance. Even interconnects whose application is not designed for high speed, if they have a uniform cross section, will look like a uniform transmission line.

In **Figure 10.5**, four different examples of common cables not designed for high performance but with a uniform cross section were measured with a TDR. In each case, the two wires at the front end of the cable were pulled apart to solder to an SMA connector. This created a small inductive launch discontinuity.

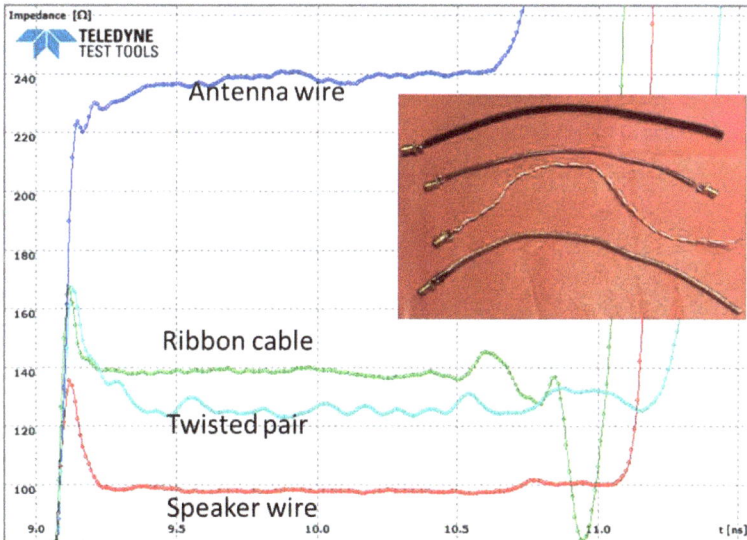

Figure 10.5 Examples of the TDR response of uniform cross section interconnect cables. They all show a very constant and uniform instantaneous impedance, ranging from 240 ohms to 100 ohms in these examples. The TDR rise time was 50 psec.

The characteristic impedances of these cables are all higher than 50 ohms. This is due to the relatively large spacing between the two conductors.

10.5 How Uniform Are Transmission Lines?

The impedance noise of a TDR can be as low as 0.2 ohm peak to peak. It depends on the rise time of the TDR and the averaging time. As an example, **Figure 10.6** shows the measured impedance from a 50-ohm calibrated load with a good- quality launch on a greatly expanded scale.

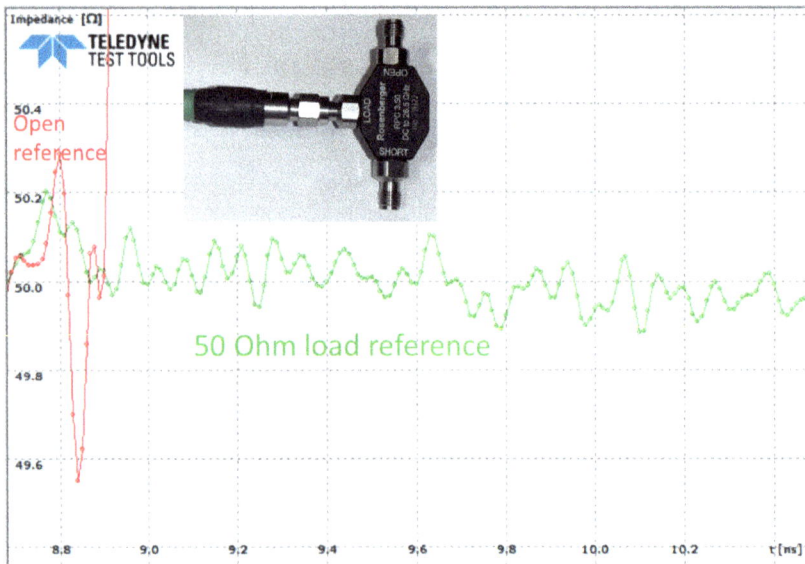

Figure 10.6 Measured TDR response from 50 ohm reference, showing the noise floor of the TDR on the scale of 0.2 ohms/div. The TDR rise time was 50 psec.

In this example using a T3SP15D TDR, the variation in the measured impedance profile is within ±0.1 ohm. This is a measure of the reproducibility and noise of an impedance measurement. Anything more than a 0.2-ohm impedance change is probably a real impedance.

In practice, it is difficult to manufacture a real transmission line very uniform, except in a coax geometry. An airline is constructed from a central solid conductor mounted inside a cylindrical tube. Both can be made very uniform.

The resulting impedance profile of the transmission line can be very constant. **Figure 10.7** shows an example of the measured TDR response of an airline. The absolute impedance is 48.5 ohms, and it is uniform to better than 0.28 ohms peak to peak.

Figure 10.7 Measured impedance profile of an airline. In the region of the airline section, the impedance is constant to better than 0.3 ohms. The TDR rise time was 50 psec.

However, this is not really the case when constructing a real transmission line on a circuit board. However carefully we design a transmission line on a circuit board to have a constant instantaneous impedance, the impedance will vary when it is manufactured.

This is due to variations in the plated thickness of surface traces, the etching variation of line widths, the thickness variation in laminates, and even the local variation in the Dk of the laminates due to glass weave or resin cure variations.

267

In the best case, with a wide conductor that is less sensitive to line width and trace thickness variations, and a two-layer board with a uniform dielectric thickness, the impedance variation on a trace will still typically be about ±0.5 ohm. An example of a reference trace on a circuit board is shown in **Figure 10.8**.

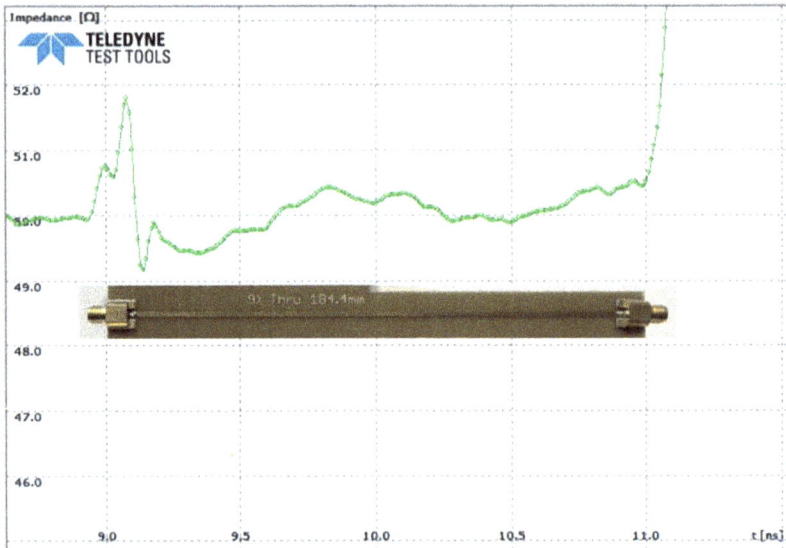

Figure 10.8 An example of a well-designed and precision circuit board trace. The impedance variation is about 2% peak to peak. The TDR rise time was 50 psec.

This impedance variation is exceptional rather than typical.

The typical variation in impedance across a PCB transmission line fabricated with FR4 with a large capacitive launch is shown in **Figure 10.9**. This is the measured TDR profile for an 8-inch-long microstrip fabricated on a four-layer board.

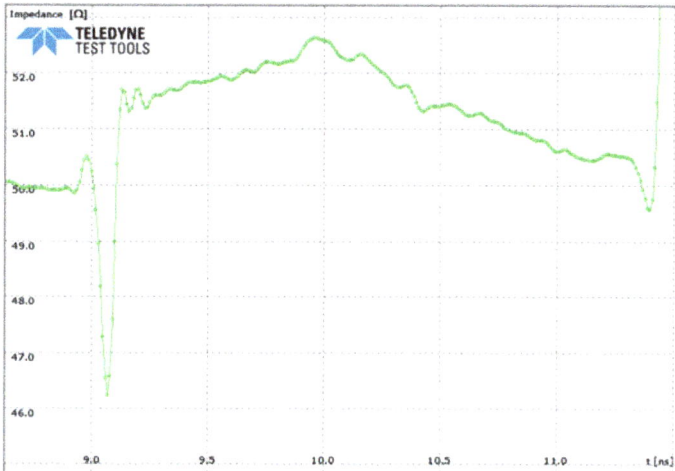

Figure 10.9 Measured TDR impedance profile for an 8-inch microstrip on a four-layer board. The TDR rise time was 50 psec.

The variation in this example is over a 2-ohm peak to peak range out of a 50-ohm line. This is only a 5% peak-to-peak variation. Due to the sensitivity of the TDR, this small variation is easily measured.

Is this impedance variation real, or due to some sort of artifact?

An important consistency test is to measure the TDR profile from both ends. If this impedance variation is real, the impedance profile from the other end should be the mirror response.

The TDR impedance profile measured from both ends of the interconnect are shown in **Figure 10.10**. In this example, the measurement from the left end is the mirror impedance of the impedance profile measured from the right end. This demonstrates that the measured impedance profiles are consistent with a real variation in impedance across the trace length.

269

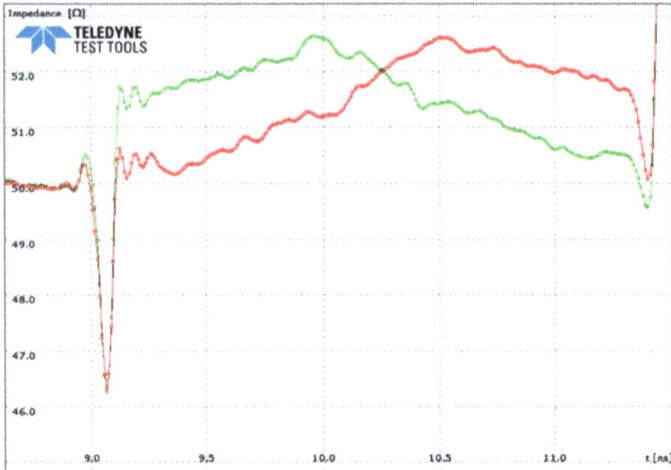

Figure 10.10 Measured impedance profile from the left and right edge of the same PCB trace showing a real impedance variation. The TDR rise time was 50 psec.

Unfortunately, the price of a PCB is also no guarantee of the uniformity of the impedance profile to expect of the traces.

Another board fabricated as a high-performance (more expensive) board with microstrip and stripline traces was measured with a Type K high-performance connector. **Figure 10.11** shows the impedance profile of two transmission lines on this board, with variations of about 2.5-ohms peak-to-peak variation after the very clean launch.

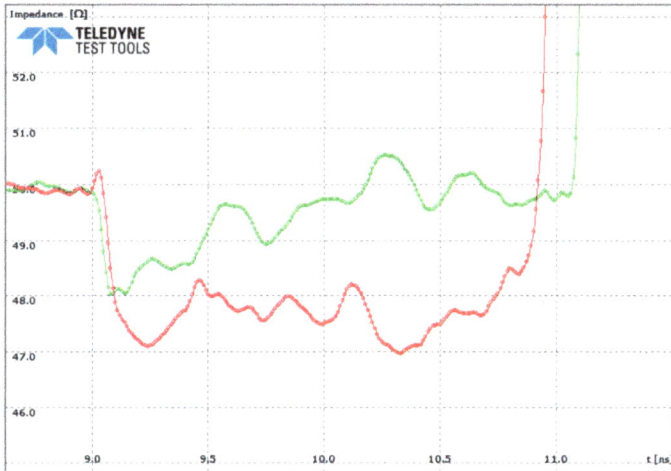

Figure 10.11 Example of the measured impedance profiles for two transmission lines with a clean launch on a circuit board. The TDR rise time was 50 psec.

When a circuit board transmission line is designed as a uniform transmission line, the best case you can expect is to see an impedance variation that is constant to within ±1 ohm. This is still very constant, roughly ±2% variation down its length.

10.6 Increasing Impedance Down the Line

An important consistency test to verify if an impedance variation is real, is to measure the impedance from both ends of the transmission line.

When the impedance variation is a real variation in the instantaneous impedance, the impedance from one side is the mirror image from the other side.

However, there is another common cause of apparent impedance variation down the length of a transmission line: high series resistance.

271

As we noticed with some contact resistance, a series resistance will appear as an additional impedance resulting in a higher displayed instantaneous impedance. Does this mean the instantaneous impedance of the transmission line is really higher?

When there is a large series resistance along the transmission line, as from a thin or narrow conductor, the observed TDR impedance profile increases down the length. But this does not mean the *instantaneous impedance* increases.

The impedance that causes the reflections at each step the TDR signal takes is the series resistance *and* the instantaneous impedance.

As an example, a simple transmission line was simulated with a 0.5-ohm resistor in series with a 50-ohm segment. Ten of these segments were connected in series and the TDR response simulated.

The total resistance was 5 ohms broken into ten segments. The simulated and extracted TDR impedance profile is shown in **Figure 10.12**.

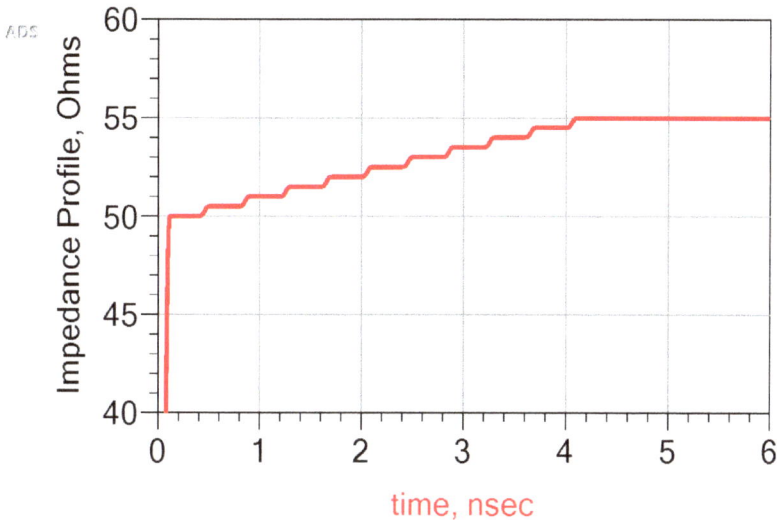

Figure 10.12 Simulated TDR impedance profile of a series of resistor and transmission line segments.

The total resistance in the series path is 5 ohms, broken into 10 segments. Each segment is clearly seen as steps with a total change of 5 ohms. If each segment were a smaller length and a smaller resistance, the staircase rise would look smoother and a more continuous ramp up in impedance.

This is exactly what a high series resistance transmission line looks like. The resistance is distributed down the length of the transmission line. **Figure 10.13** shows the measured TDR response of a thin, narrow conductor flex transmission line with an impedance that ramps up.

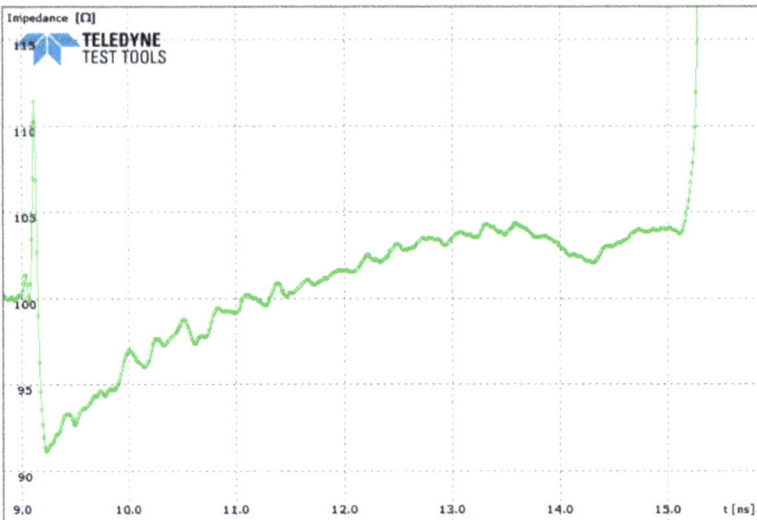

Figure 10.13 The measured impedance of a high-resistance transmission line showing the TDR impedance profile ramping up. The TDR rise time was 50 psec.

Is this a real impedance variation down the length of the transmission line, or due to the distributed series resistance introducing the artifact of the apparent increase in impedance?

One consistency test is to look at the impedance from the other end of the transmission line. If this is a real impedance variation, the TDR profile from the other end will start high and drop lower down the length. If it is due to a high distributed series resistance, the impedance will be uphill in both directions. **Figure 10.14** shows the measurements from both ends.

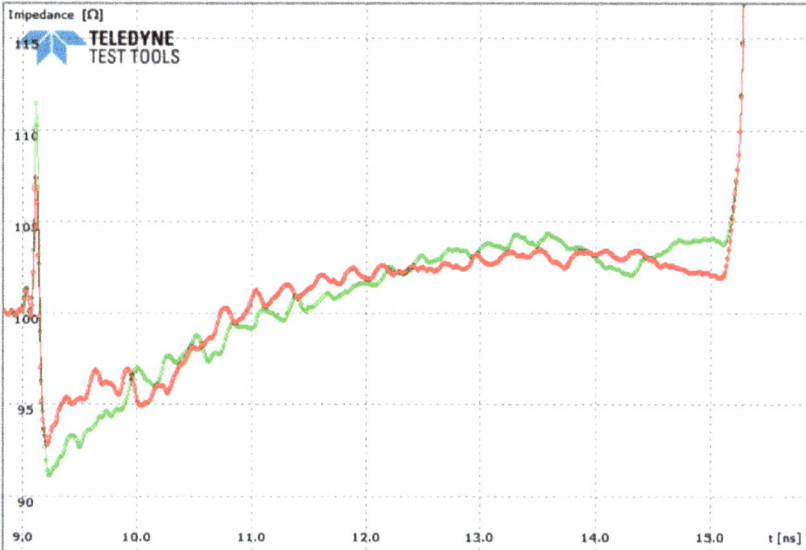

Figure 10.14 Measured TDR impedance from both ends of the transmission line showing the impedance profile is uphill in both directions. The TDR rise time was 50 psec.

The measurement from both ends is consistent with a large series resistance. From the beginning to the end of the line, the series resistance has increased by 12 ohms. This would be the series resistance of the interconnect.

It is difficult to get a good measure of the instantaneous impedance of the transmission line, or how uniform it might be, due to this series resistance dominating the impedance profile. Using the initial impedance, we can estimate the instantaneous impedance at one end as 93 ohms and 92 ohms from the other end.

10.7 Discontinuities from Different Ends

A discontinuity is a structure that is electrically short. This means its TDR response will depend on the rise time of the TDR signal. The longer the rise time, the lower the peak or less the dip.

The shorter the rise time, the less overlapping of the reflections from the front and back of the structure and the less smearing out of the reflected waveform at the TDR. The larger the reflected signal, the larger the apparent impedance change.

The magnitude of the reflected signal and the apparent instantaneous impedance will depend on the rise time of the TDR signal. What is important is the rise time of the signal, *when it encounters the discontinuity*.

As a signal travels down a real transmission line, there are always losses in the line. These are due to conductor loss and dielectric loss. Both effects are frequency-dependent, getting larger at higher frequency.

This means high frequencies are attenuated more when traveling down a real transmission line than low frequencies. If you take a signal with a very fast rise time, like 50 psec, and remove some of the high-frequency components by attenuation, the rise time of the signal will increase.

As the TDR signal travels down a real transmission line, the rise time of the signal increases. This means that the same discontinuity feature if located near the TDR source will show a larger reflected signal than the same discontinuity located farther down the transmission line, where the rise time would be longer.

As an example, **Figure 10.15** shows the measured TDR response from a capacitive pad of an SMA launch located at both ends of an otherwise uniform transmission line. The launch on each side of the transmission line are identical, yet, when viewed from one end, they look very different.

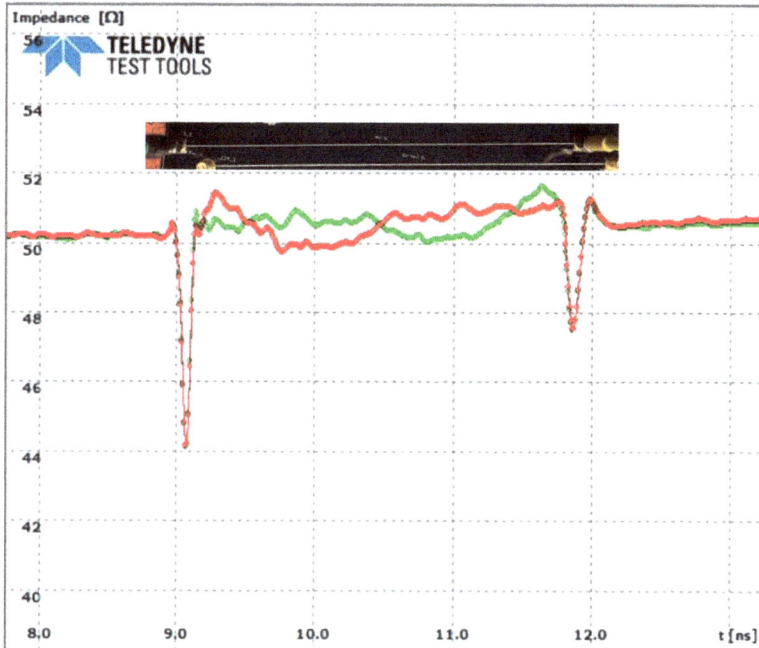

Figure 10.15 The measured TDR profile of a launch on either side of the transmission line viewed from both ends. The TDR rise time was 50 psec.

When measured from the left side, the left launch has a peak value of 44 ohms. When measured from the left-side, the right launch at the far end is 48 ohms. Are the launches on the two ends of the transmission line really different?

The important consistency test is to measure the impedance profile from each end. When viewed from the right end, the right launch has an identical TDR response as the left launch when viewed from the left side.

The reason the two launches look different when viewed from one end of the transmission line is because the TDR signal's rise time has increased by the time it reaches the far end and smears out when reflected by the electrically short launch.

277

10.8 TDR Response from a Transmission Line with an Open or Short Far End

The initial reflection from the interface between the TDR instrument's 50-ohm cable and the DUT tells us the initial instantaneous impedance of the transmission line.

When it is higher than 50 ohms, the initial reflection is positive. When it is lower than 50 ohms, the initial reflection is negative.

The initial impedance is completely independent of the far end termination of the transmission line. In fact, we cannot see the far end of the transmission line unless we measure for at least a round-trip time delay.

If the far end is open, we should see a reflection of 100% if we wait long enough.

If the far end is shorted, we should see a reflection of negative 100% if we wait long enough. This will appear as a short.

And if the far end is terminated with 50 ohms, eventually there will be no reflected signals. These three examples for the same interconnect are shown in **Figure 10.16**.

Figure 10.16 Measured TDR impedance profile of the same interconnect with three terminations at the far end. The initial impedance is independent of the refection from the far end.

The secret is to pay attention to the source impedance as well as the impedance of the line. It is the initial impedance that influences the first reflection. The interactions of the incident source signal is time separated from the following part of the interconnect. The initial reflection at each interface is independent of the rest of the interconnect.

If the reflections are large enough to continue to bounce around in the system, then the impedance profile of the entire system may influence the long-term TDR response of the interconnect. But this is after all the initial reflected signal makes its way back to the source.

10.9 TDR Response from a Very Low Impedance Transmission Line

A low impedance transmission line is easily measured with a TDR. The reflection coefficient from the 50-ohm source impedance or 50-ohm cable impedance to a 10 ohm transmission line, for example, is

$$\text{rho} = \frac{10-50}{10+50} = -67\% \tag{10.1}$$

This is a large, negative value, but easily measured and converted into a first-order impedance. An example of a very wide transmission line is shown in **Figure 10.17**. This particular transmission line has two small added capacitive loads that act as precision markers. The SMA launch is connected only at one end, inserted from the back surface of the board where the return plane is, not visible in this view.

Figure 10.17 A very wide microstrip transmission line with an SMA launch at one end.

The measured TDR impedance profile, shown in **Figure 10.18**, shows a very low impedance at the first reflection, exactly as expected. Since it is a uniform transmission line, except for the two small extra pads, the instantaneous impedance is mostly constant, exactly as expected.

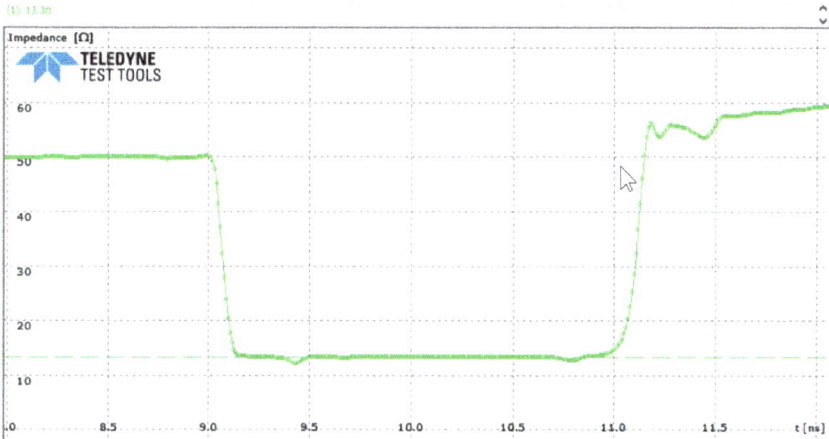

Figure 10.18 Measured impedance profile from this very wide, very low impedance microstrip transmission line. The cursor measures 13.3 ohms, noted in the upper left corner of the screen.

The impedance, which we can read off the front screen, is about 13.3 ohms. The far end of this transmission line is open. We expect to see a reflection coefficient of 100% and all of the signal coming back to us. But we don't.

The reflected signal we get back corresponds to seeing a 55 ohm impedance at the far end. This is completely wrong. There is an open at the far end. Yet, because of the very large impedance discontinuity of the 50 ohms encountering the 13.3 ohm trace, the transmitted signal is distorted and the reflection from the far end, traveling back to the source, encountering the change in impedance from the 13.3-ohm line into the 50-ohm cable causes another reflection, which bounces back to the open at the far end.

All of these distortions from the very low impedance of the DUT results in a distorted series of reflections due to the signal bouncing back and forth between the high impedance open at the far end and the 13.3-ohms to 50-ohms interface at the source ends. This is an example of masking.

These reflections are difficult to keep track of without either a bounce diagram or a circuit simulation. We can easily analyze the extreme cases from the front screen.

Right at the first interface, the initially measured instantaneous impedance of the DUT is accurately measured as 13.3.-ohms. After all the bounces have settled down, which might be 10 round-trip delays, we expect to see an open at the far end.

The time delay of this interconnect is about 1 nsec, resulting in a round-trip delay of about 2 nsec. After about 20 nsec, the reflections should have finished, and we should see the reflections from an open. **Figure 10.19** shows the reflection coefficient for this initial step edge encountering the low impedance line, bouncing back and forth internally, and eventually reaching a reflection coefficient of 1. How many bounces are required depends on just how low the line impedance is.

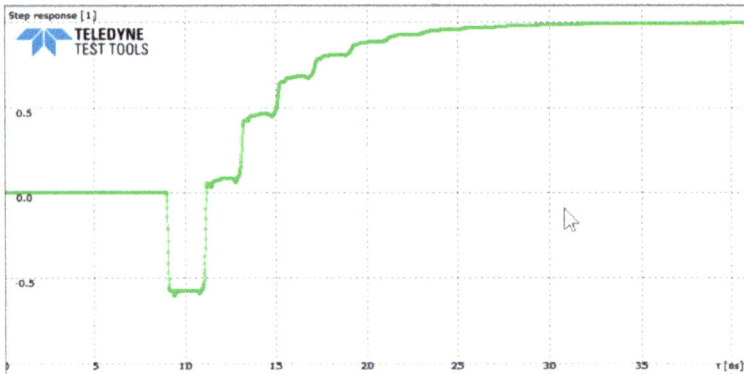

Figure 10.19 The measured reflection coefficient from the front of a 13.3-ohm transmission line viewed over a long period of time.

The low impedance masks the open at the far end. We have to wait for the multiple reflections to die out before we can read the impedance of the line and the other side of this very low impedance.

10.10 When There is Coupling to a High Q Resonator

So far, we have considered the behavior of a single transmission line isolated from others. If there is any adjacent metal, the TDR signal traveling down the transmission line can couple, through normal electric and magnetic field coupling, to the adjacent conductor. This is called cross talk.

The driven transmission line is the aggressor and the floating conductor is the victim.

If the ends of the victim conductor are open, or shorted at just a few points, the victim conductor will have natural resonant frequencies where the coupled voltage wave will bounce back and forth reflecting from the ends, or boundary conditions.

The resonant frequencies will depend on the length of the floating conductor and the dielectric constant of the material between the signal and return path of the floating conductor.

The impact of this behavior will be to cause additional reflected signal in a TDR measurement. The coupling to the floating victim line may not affect the initial TDR impedance profile from the DUT, but after energy has coupled over to the victim conductor and is trapped at the resonant frequency, it may couple back to the driven line and appear as though there is additional structure in the DUT. This is an artifact.

For example, if there is an adjacent trace that is unconnected to a driver and receiver, the TDR response of the aggressor line will show additional reflected signal long after the signal has reflected from the far end of the test line. This is an artifact of energy being coupling back to the test line from the adjacent victim line. **Figure 10.20** is an example of two close-spaced microstrip transmission lines with the second line floating.

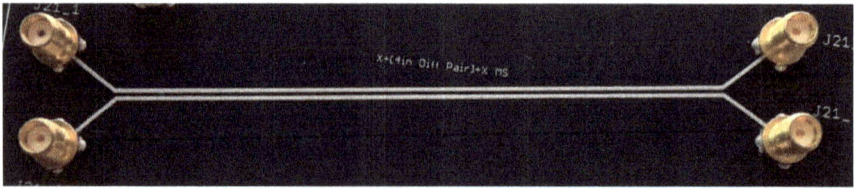

Figure 10.20 Two closely spaced microstrip transmission lines. One end of one line is driven by the TDR. The other ends of the other line are open.

The TDR impedance profile of the interconnect looks perfectly well behaved. **Figure 10.21** shows the impedance profile with the low impedance of the launch, the uniform impedance of about 57 ohms, the capacitive dip at the far end from the other launch, and the open at the end of the line.

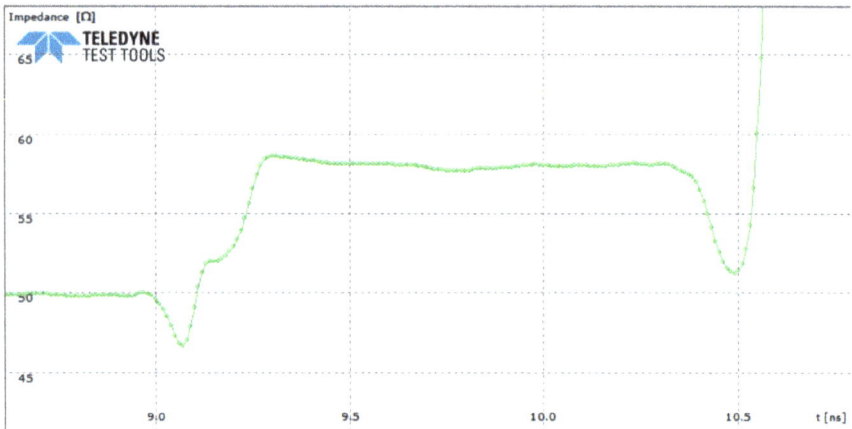

Figure 10.21 Measured impedance profile of a uniform transmission line that is adjacent to a floating line open at its ends.

There is no impact on the initial impedance measured by the TDR from the adjacent floating conductor.

However, if we look at the reflected signal after the signal has reflected from the far end, the reflection coefficient tells a different story. **Figure 10.22** shows the measured reflection coefficient.

There is energy coupled back into the driven line long after the initial signal has reflected from the far end open.

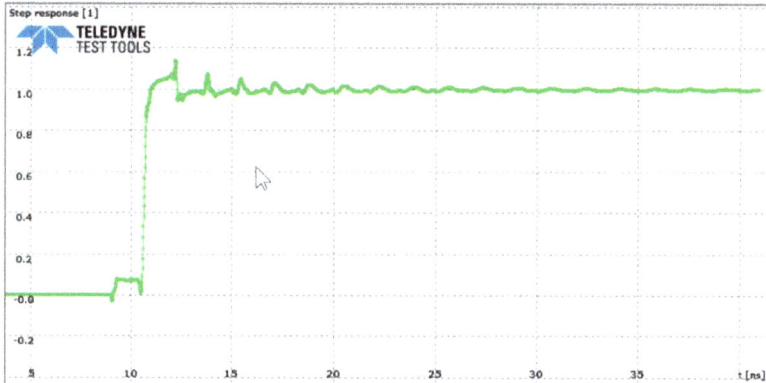

Figure 10.22 Measured reflection coefficient from this same driven line, but on an expanded time base showing the energy coupling back from the stored energy that is in the adjacent floating conductor.

If we spoil the Q of the resonator by damping out the reflections at the ends using 50-ohm resistors, we can prevent the energy coupling into the victim line from building up and slowly leaking back. The terminating resistors on the ends of the victim line do not change the energy coupled into the victim line. They just dissipate this energy and prevent it from building up.

Figure 10.23 shows the measured reflection coefficient from the first aggressor line, when the second adjacent line is floating and when its ends are terminated with 50-ohm resistors to damp out the reflections and spoil the Q of the resonance.

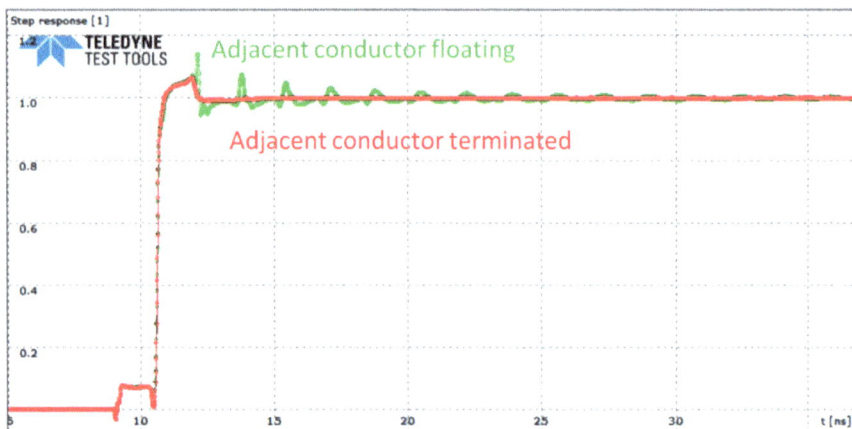

Figure 10.23 Comparing the measured reflection coefficient of one line when the adjacent conductor is floating, acting as a high Q resonator, leaking coupled energy back, and when the ends of the victim line are terminated, damping out the resonance.

This is why using a copper pour on a board is a really bad idea. This situation can easily arise with isolated islands of copper that remain floating on a board. In such a case, if care is not taken to add enough shorting vias, any floating metal acts as a high Q resonator.

Energy from aggressor signal lines will coupling to the floating metal and recouple to other nearby conductors. This pollution is a form of cross talk. Sometimes engineers think that just throwing copper on a design will reduce cross talk. In reality, there is a high risk of a copper pour increasing the cross talk between aggressors and victim line.

Sometimes an engineer will try to add a shorting via at each of the floating metal in the hopes this will somehow direct the coupled noise voltage to ground. In reality, when the coupled noise that is propagating toward the ends of the victim conductor encounters a short at the far ends, it will just reflect back. Adding shorting vias at the ends of the copper pour will not change the Q of the resonator.

An example of the reflection coefficient measured on the driven line of the coupled pair of transmission lines, with the second line floating, terminated, or shorted on the ends is shown on **Figure 10.24**.

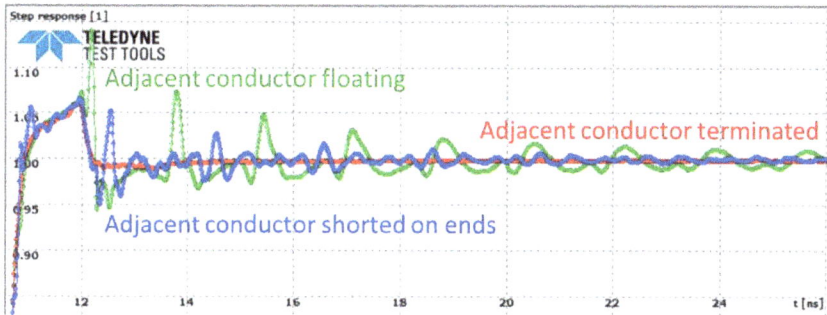

Figure 10.24 The measured reflection coefficient of the driven line with the adjacent conductor floating, terminated, and shorted on the ends.

There is almost as much noise coupling back from the adjacent conductor when its ends are grounded as when they are open. Any adjacent floating metal or shorted at the ends will resonate and be a source of enhanced noise.

There is rarely a good reason to use a copper pour. As a good habit you should never use a copper pour on a layer other than ground unless you have a strong compelling reason to do so, and then be very careful to avoid inadvertently creating high Q resonant structures that can add to the noise.

Watch this video and I will demonstrate these measurements with a TDR.

10.11 Review Questions

1. What are the two most important measurements to take with a TDR before doing any other measurements?

2. Why is it important to always use a torque wrench?

3. What is the shape of the reflection from a connection that is not tight?

4. Why is the launch into a coax cable usually inductive?

5. What is the impedance noise floor of a typical TDR?

6. What is the typical impedance variation in an airline?

7. What is the typical impedance variation in a PCB trace?

8. What is the most important consistency test to determine if a measured impedance variation is real?

9. If there is a large series resistance distributed down a transmission line, what will the impedance profile look like?

10. How would you verify that an impedance variation is due to a high series resistance or a real spatial variation in impedance?

11. Why do identical connectors look different when measured from different ends of an interconnect?

12. How would you confirm that the two ends of the transmission line have identical connectors?

13. If the far end of a transmission line is open, what will be the reflection coefficient we see if we wait long enough?

14. Why is a copper pour on a signal layer a bad practice?

Chapter 11 Measuring Dk with a TDR

The dielectric constant (Dk) of a laminate is an important electrical property needed to accurately design a target characteristic impedance. It will affect the characteristic impedance of a transmission line and the time delay of the line.

We refer to the bulk dielectric constant as the dielectric constant of a homogenous, uniform material. But laminate dielectrics used in circuit boards generally are not homogenous, uniform materials.

Instead, we use the Dk value representative of the composite combination of resin and glass yarn. It will depend on the specific construction — the type of resin, the type of glass yarn, the number of layers of glass yarn, and the total thickness. This makes it a little difficult describing the Dk of a laminate in general and why being able to measure the Dk of an as-manufactured layer on a board is important.

11.1 The Dk and the Speed of a Signal

The dielectric constant is a measure of the speed of a signal in the material compared to the speed of light in air. The larger the dielectric constant, the slower the speed of the signal. But the speed varies with the square root of the dielectric constant:

$$v = \frac{c}{\sqrt{Dk}} = \frac{Len}{TD} \tag{11.1}$$

where

v = the speed of the signal in the material

c = the speed of light in air

Dk = the dielectric constant of the laminate

Len = the length of a section of interconnect

TD = the time delay for a signal to travel down the length

Fundamentally, the TDR can measure the round-trip time delay down and back of a precision length. From the round-trip delay, the one-way time delay can be calculated and then the velocity, and from the velocity of the signal, the dielectric constant.

From the one-way time delay and length of the interconnect the Dk is:

$$Dk = \left(\frac{TD \times c}{Len}\right)^2 = \left(\frac{TD[n\sec] \times 11.8}{Len[in]}\right)^2 \qquad (11.2)$$

The fundamental method that is used to measure the Dk of a material is to fabricate a uniform transmission line using the dielectric and measure the time delay of a known length interconnect. From the time delay and the length of the transmission line, the Dk is calculated.

The challenge is measuring the time delay of a known length of a transmission line when launches are involved. Where do you decide the launch ends and the uniform section of transmission line begins?

There are three simple approaches to this problem:

- Guess — assume the uniform transmission line begins at the end of the launch, which is at the dip in the TDR response.

- Measure two different length transmission lines with similar launches and take the difference in length and difference in time delay.

- Measure the time delay between two small discontinuities a precision length apart.

Figure 11.1 shows an example of a test board with three stripline transmission lines that demonstrate these three approaches. One line is 6 inches long, one line is 2 inches long, and one line is 6 inches long with precision pads in the middle.

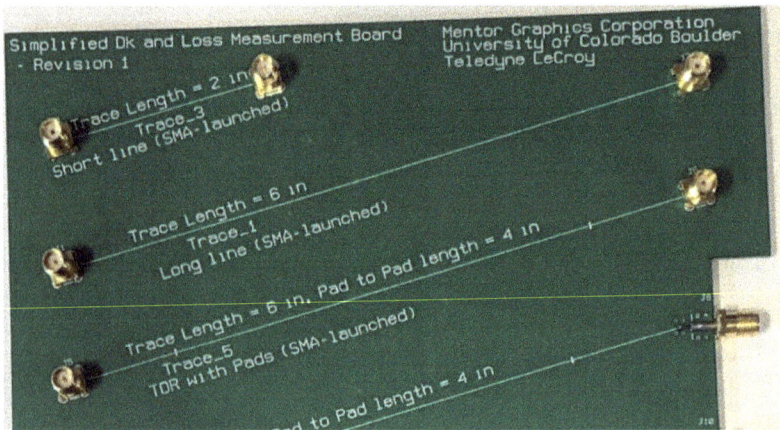

Figure 11.1 Example of a test board with three stripline transmission lines demonstrate the three methods. Note they are rotated at 15 degrees to the glass weave axis to average the glass weave distribution.

This board was designed and developed by Neeti Sonth and Priya Guruswamy of Mentor Graphics and the author as part of a joint Mentor Graphics, Teledyne LeCroy, and University of Colorado, Boulder, collaboration. It will be used in the following sections to compare the three different techniques to measure the Dk.

11.2 Dk$_{eff}$ of Microstrip and Stripline

The dielectric constant is an intrinsic property of an insulating dielectric material. In most circuit board structures, the dielectric material is not homogenous, uniform, and the same above and below the signal trace. Instead, there may be different materials above and below the signal line.

Most printed circuit board laminates are composed of combinations of a resin and a glass yarn. This is the fundamental structure of fiberglass. **Figure 11.2** shows a cross section of a signal line between two layers of fiber glass embedded in a resin composite.

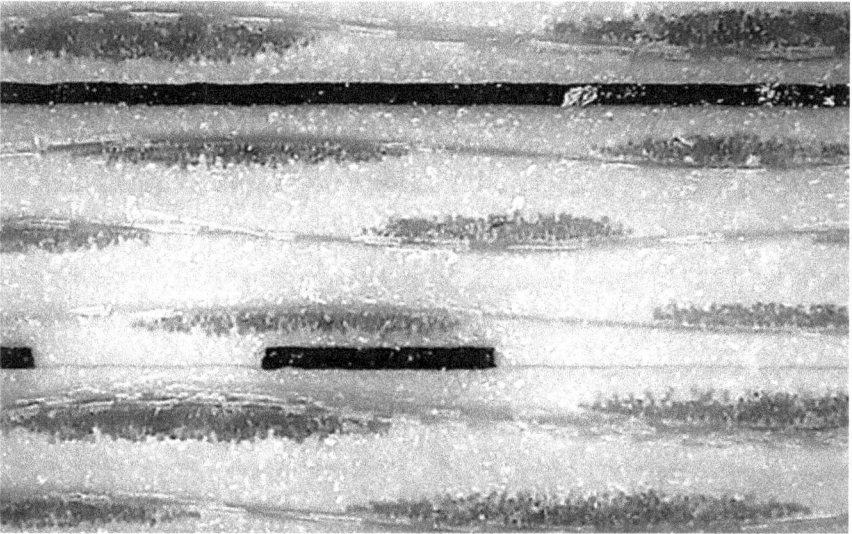

Figure 11.2 Close-up of a circuit board trace on one layer and resin and glass yarn above and below.

The resin typically has a low dielectric constant on the order of 3.2. The glass has a higher dielectric constant on the order of 6.7. This means the composite Dk of a fiberglass system depends on the relative amount of glass fiber compared to the resin. As the signal line moves around the microstrip of the glass yarn, its local Dk may vary. This makes characterizing the Dk of the laminate a little ambiguous. The best that can be done is, measure an average composite value of Dk.

In a stripline, there are generally two dielectric layers, one above and one below the signal line. One is generally a core layer that is fully cured when assembled, and one is a prepreg. That is cured during the lamination process. The measured Dk in a stripline is a composite value of the contributions from the top and the bottom layers. It is sometimes difficult to extract the Dk of each layer from the single, average, composite Dk.

In microstrip, there is always a dielectric layer below the signal line to the return plane, but air above the trace. In many cases, there may also be a thin layer of solder mask on top of the trace. In each case, the electric field lines, which are sensitive to the Dk value, will see a mixture of materials. This is illustrated in **Figure 11.3**.

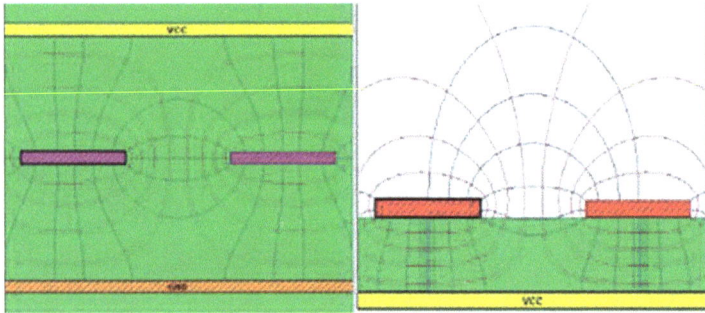

Figure 11.3 An example of the electric field lines around stripline and microstrip traces. Signals on these interconnects will see a Dk value that is some combination of the various materials.

What the TDR signal sees as it propagates down the transmission line is an effective Dk based on some combination of the values of the different materials and their geometrical distribution.

From a measurement of the effective Dk based on the time delay of a known length, it is difficult to back out the bulk Dk of each of the layers, except in the special case of a microstrip. In this case, the top Dk value of air, is known to be 1.

In general, all we measure is the effective Dk of the transmission line. We have to carefully interpret how we use this number to characterize the interconnect and as an input to a simulator.

In the case of a stripline, if we assume the Dk of each layer is the same, then the effective Dk is the same as the bulk Dk of either layer.

11.3 Measuring Dk, Method 1: Guess the Location of the Ends of the Line

The TDR has a time resolution to measure the round-trip time delay to a specific impedance feature on the order of a small fraction of the rise time. This can be a resolution on the order of 10 psec, depending on the TDR.

When we guess the time delay, there is some uncertainty about where the end of the launch is and the beginning of the line. This can be as much as 0.05 nsec. When the total time delay of the line is 1 nsec, this is a 5% impact on the time delay.

For example, **Figure 11.4** shows the measured TDR response of a nominally uniform *stripline* transmission line with two through-hole SMA launches on the ends, separated by 6 inches. It is measured from both ends. The launches are nearly identical and there is a small impedance variation down the line. The characteristic impedance of this line is about 65 ohms.

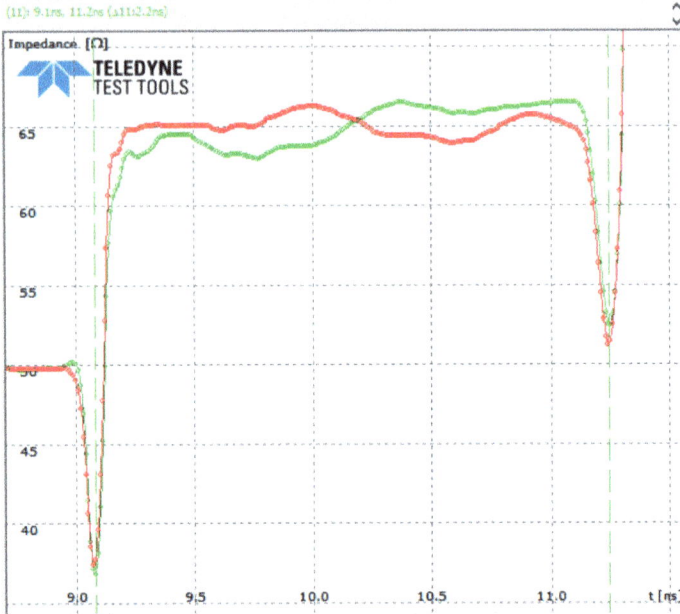

Figure 11.4 Measured TDR response of a 6-inch-long stripline transmission line with SMA launches on the ends, measured from both ends. The TDR rise time was 65 psec.

Using cursors, the round-trip time delay between the dips of the connector launch is measured as 2.16 nsec. This makes the one-way time delay 1.08 nsec. There is some uncertainty of where the beginning of the trace is, where the launch ends, and how much additional delay the launch contributions. This is inherent in this method.

The extracted Dk is

$$\text{Dk} = \left(\frac{\text{TD}[\text{n sec}] \times 11.8}{\text{Len}[\text{in}]} \right)^2 = \left(\frac{1.08 \times 11.8}{6} \right)^2 = 4.51 \quad (11.3)$$

This is the composite bulk dielectric constant of the two layers that make up the laminate material surrounding the stripline trace and between the two planes. If we assume the Dk of each layer is the same, then the bulk Dk for this laminate, composed of glass yarn and resin, is 4.51.

11.4 Measuring Dk, Method 2: Two Different Line Lengths

The second method to measure the Dk is to use two different length, but otherwise identical, transmission lines. If they have the same launches, then the difference in time delay between them should be due only to their difference in length.

It is only necessary to use the same features on each line to determine a round-trip delay time. The difference between the delay in these two lines for the same structures is what is important.

The measured TDR response from a 2-inch-long and 6-inch-long transmission line is shown in **Figure 11.5**. The launches from the ends of the lines are seen to be very similar. This is an important consistency test.

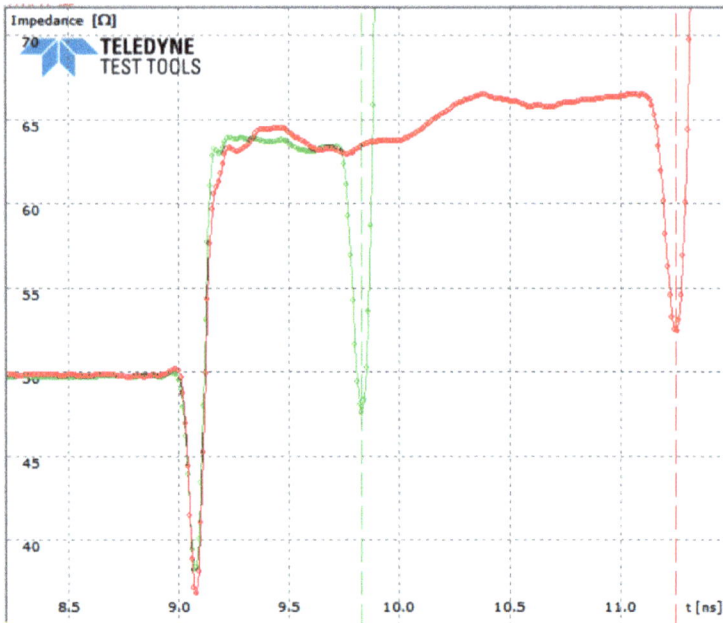

Figure 11.5 Measured TDR response from two identical stripline transmission lines with different lengths. The TDR rise time was 65 psec.

The round-trip time delay difference measured by the cursors between the ends of the launch discontinuity, which is a very sharp feature, is 1.42 nsec. The one-way time delay value is 0.71 nsec, for the 4-inch difference in length.

The effective Dk extracted is:

$$Dk = \left(\frac{TD[n\sec] \times 11.8}{Len[in]} \right)^2 = \left(\frac{0.71 \times 11.8}{4} \right)^2 = 4.39 \quad (11.4)$$

This is the effective Dk for the two layers that make up the stripline structure.

This same method can be applied to any interconnect structure, such as a coax cable.

In this example, a low-cost RG174 coax cable was cut with one length 5.125 inches long, from the beginning of the BNC connector to the cut end of the cable, and the second one was 35.375 inches long.

The length difference was 30.25 inches. The TDR responses of both cable fragments is shown in **Figure 11.6**.

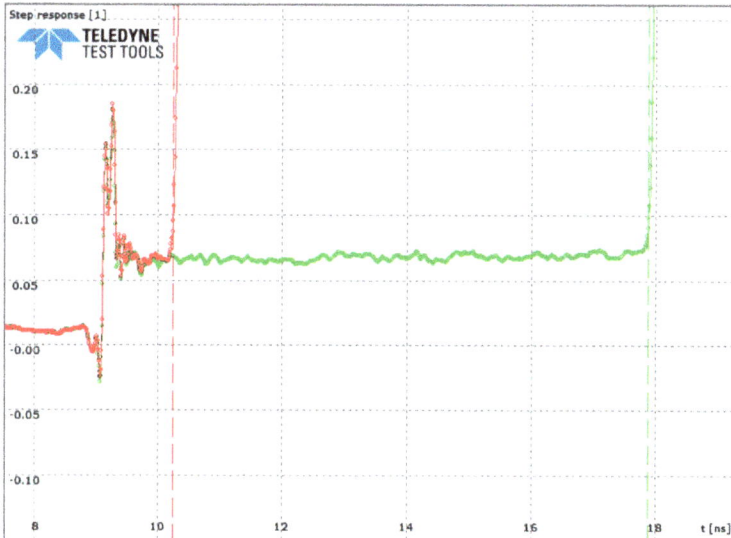

Figure 11.6 Measured TDR step responses from the two cable fragments. The round-trip time delay difference is 7.64 nsec.

From the round-trip delay of 7.64 nsec, the one-way time delay is 3.82 nsec and the length difference of 30.25 inches, the Dk of the cable dielectric is calculated as:

$$Dk = \left(\frac{TD[nsec] \times 11.8}{Len[in]} \right)^2 = \left(\frac{3.82 \times 11.8}{30.25} \right)^2 = 2.22 \qquad (11.5)$$

299

This value is close to the published value of low-density polyethylene, 2.26. The 2% lower value measured is probably due to the slight air gap between the insulating layer and the braid over the dielectric.

With a dielectric constant of 2.22, the wiring delay of a cable would be 1.5 feet/nsec, the commonly used rule of thumb.

11.5 Measuring Dk, Method 3: Special Test Structure with Small Pads

The third approach is to build a special transmission line consisting of a uniform transmission line with two small discontinuities strategically placed a precision distance apart.

They should be far enough away from the launches to clearly see the peaks or dips. This should be at least $2 \times RT$. In this specific TDR with a 65-psec rise time, this is a 130-psec delay, or 0.8 inches.

The distance between the discontinuity features should be at least 1 nsec round trip time, so that the 10-psec resolution can provide a round-trip time delay, or a one-way time delay, accurate to 1%. A reasonable distance is 4 inches.

The discontinuities can be narrower regions, creating a high impedance, or wider regions, making a lower impedance.

It is only necessary that they be visible as a peak or a dip. They should have a length significantly shorter than the TDR resolution so that the peak width is based on the TDR rise time, not the length of the discontinuity.

The resolution of the TDR, with a 65-psec rise time is

$$\text{Len}_{resolution}[\text{in}] = 2.95 \times RT[\text{n sec}] = 2.95 \times 0.065 = 0.2 \text{inches}$$

A smaller structure with large impedance change is preferred. It will result in a well-defined dip or peak.

In the test structure fabricated, the length of the extra pad was chosen as 0.05 inches which is one quarter of the spatial resolution of 0.2 inches. The width of the pad was chosen as 18 mils, 3 × the nominal line width. **Figure 11.7** shows the artwork for the signal layer with the through-hole SMA launch and the first small discontinuity.

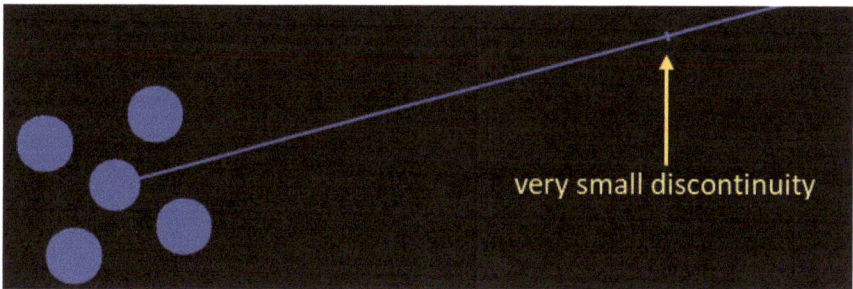

Figure 11.7 Artwork for the signal line in a stripline with a through-hole SMA launch and a small discontinuity reference pad exactly 1 inch from the launch.

With these conditions, the dips are very clear and easily measured. **Figure 11.8** shows the measured TDR response of this structure.

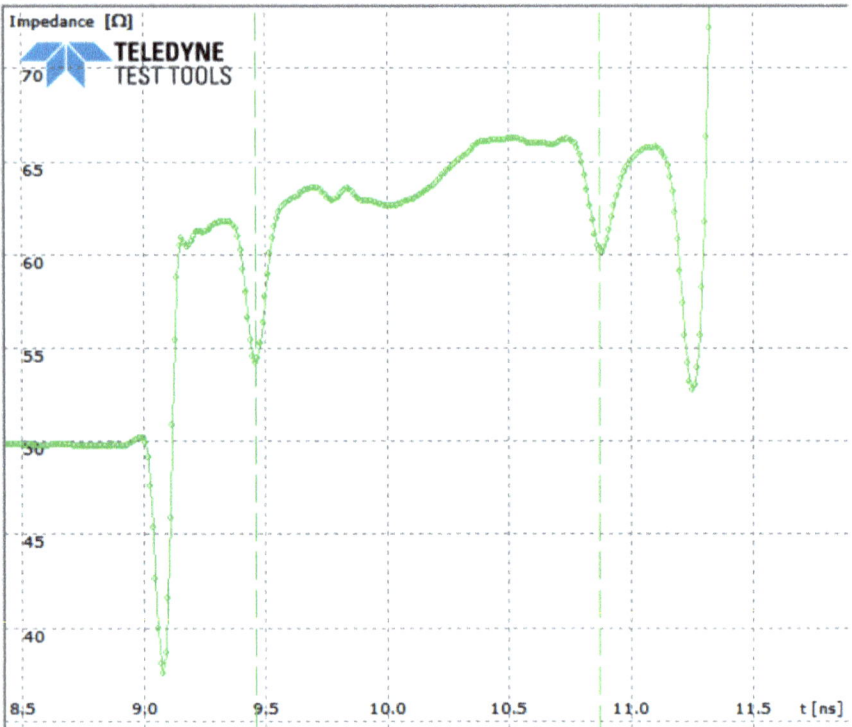

Figure 11.8 Measured TDR response of one transmission line with two small pads located 4 inches apart. The TDR rise time is 65 psec.

The measured round-trip delay between the centers of these dips is 1.42 nsec. The one-way delay time is 0.71 nsec. This is identical to within the ±10-psec resolution of the TDR, as measured by the two different line length method. The extracted Dk is the same value:

$$Dk = \left(\frac{TD[n\sec] \times 11.8}{Len[in]} \right)^2 = \left(\frac{0.71 \times 11.8}{4} \right)^2 = 4.39 \quad (11.6)$$

It should be noted that the two-line method and the single line with precision discontinuities both give the same value of the effective Dk, 4.39.

302

But the guess method results in a Dk of 4.51. This is higher by 2.7%. In the measurement of the single-line method, to get the lower value of Dk as measured in the other methods, would have required a shorter measured time delay.

There is inherent ambiguity of where to place the markers to measure the time delay for the 6-inch section. Where on the TDR response does the 6-inch section begin and end?

If better accuracy than 5% is required, the two-line or the one-line-two-discontinuity method should be used.

11.6 From Effective Dk to Bulk Dk in Microstrip

Microstrip is a special case. Generally, there will be one layer of dielectric between the signal trace and the return plane. When there is no solder mask over the signal trace, the effective DK just depends on the bulk Dk of the single layer and the cross-section geometry of the line.

We still only measure an effective Dk value based on the time delay of a precision length, but we can use simple analytical approximations to convert the bulk dielectric constant from the effective dielectric constant.

The starting place is to measure the effective Dk using either the two-line method or the two-discontinuity method. **Figure 11.9** is an example of the measured TDR response using the two-discontinuity method, and a close-up of the microstrip pattern with the small pads.

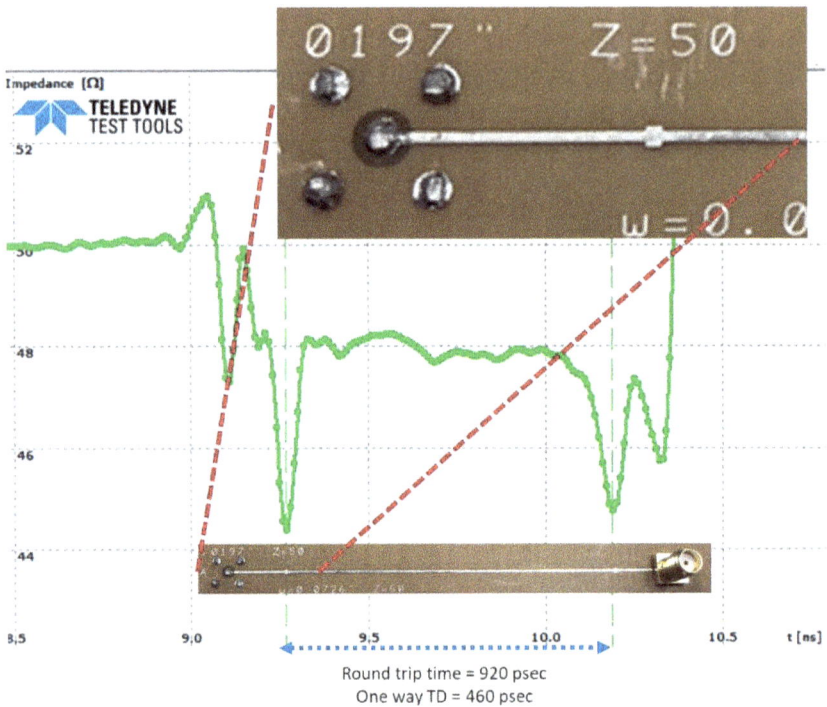

Figure 11.9 Example of a microstrip transmission line with two discontinuities and the measured TDR response using a rise time of 50 psec.

In this transmission line, the one-way delay between the two pads, located 3 inches apart, was 0.46 nsec. The extracted effective Dk is

$$Dk = \left(\frac{TD[\text{nsec}] \times 11.8}{Len[\text{in}]}\right)^2 = \left(\frac{0.46 \times 11.8}{3}\right)^2 = 3.27 \quad (11.7)$$

An effective Dk value of 3.27 sounds like a very low value, but this is the effective value of the combination of the bulk dielectric constant and the contribution from the field lines in air, with a Dk = 1.

There are a variety of approximations that relate the cross section and the bulk dielectric constant to the effective dielectric constant. A simple one is the following:

$$a = \frac{1}{\sqrt{1 + 12\dfrac{h}{w}}}$$

and (11.8)

$$Dk = \frac{2 \times Dk_{eff} - 1 + a}{1 + a}$$

where

Dk = the bulk dielectric constant

Dk_{eff} = the effective dielectric constant of the microstrip transmission line based on the speed of the signal

a = a correction factor

w = the line width

h = the dielectric thickness

There are many on line calculators that implement this simple calculation, such as this one.

We can verify the quality of this analytical approximation by comparing its calculations to that of a 2-D field solver for the same conditions.

In the Polar SI9000 field solver, the effective Dk calculated by the field solver is not directly displayed on the results screen. To view the calculated effective Dk value for a specific set of geometry and bulk dielectric constant values, you have to click on the more button, located just under the calculate impedance button, highlighted with the red arrow in **Figure 11.10**.

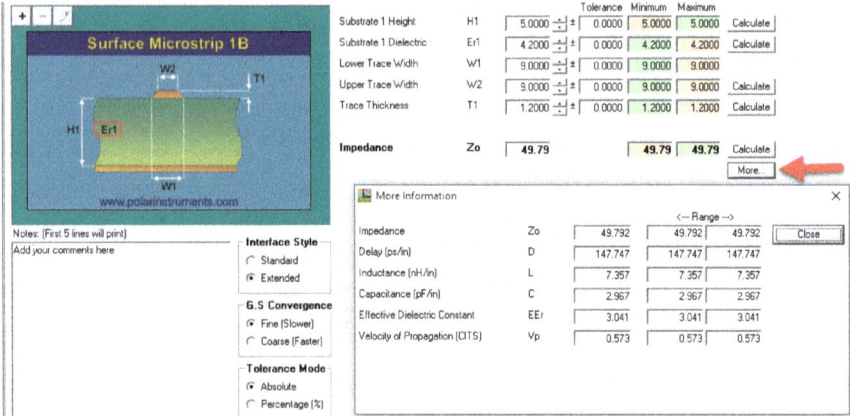

Figure 11.10 To display the calculated effective DK, press the more button under the calculate button and the effective Dk is displayed along with other information.

For the specific case of:

h = 57 mils

w = 12 mils

we change Dk$_{eff}$ and calculate Dk$_{bulk}$. The comparison of the analytical approximation and the results of the Polar Instruments SI9000 2-D field solver, for the special case of h = 57 mils and w = 12 mils, is shown in **Figure 11.11**. The agreement is excellent. This adds a level of confidence to using this simple relationship.

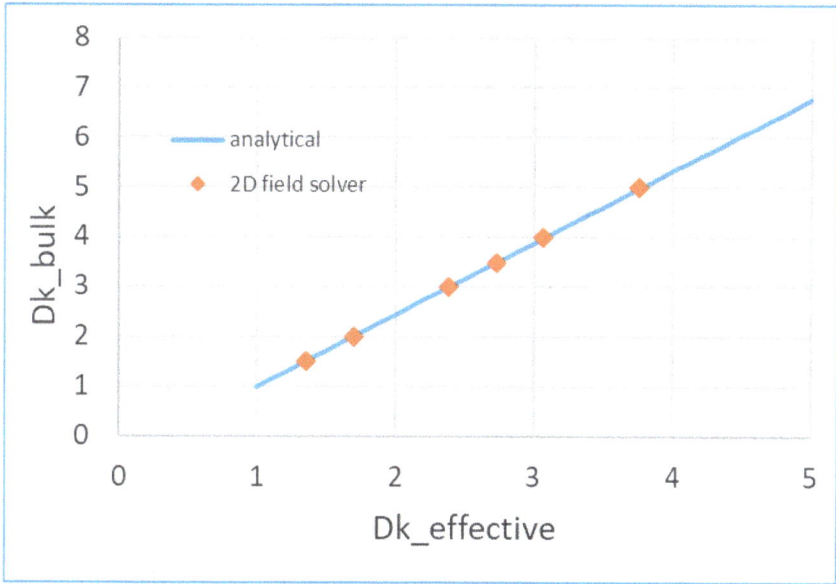

Figure 11.11 Comparing the analytical approximation and the predictions of a 2-D field solver. The agreement is excellent.

For example, in this microstrip transmission line, w = 20 mils, h = 10 mils and Dkeff = 3.27.

The bulk dielectric constant is

$$a = \frac{1}{\sqrt{1+12\dfrac{h}{w}}} = \frac{1}{\sqrt{1+12\dfrac{10}{20}}} = 0.378$$

and

$$Dk = \frac{2 \times Dk_{eff} - 1 + a}{1+a} = \frac{2 \times 3.27 - 1 + 0.378}{1 + 0.378} = 4.29$$

(11.9)

The bulk dielectric constant of this laminate used with this microstrip is 4.29.

11.7 At What Frequency Does the TDR Measure the Impedance or Dielectric Constant?

The dielectric constant of most materials is slightly frequency-dependent. An example of the measured Dk of an FR4 stripline transmission line is shown in **Figure 11.12**. In this sample, the Dk varied from 4.4 at 1 GHz to 4.2 at 20 GHz. This is a 5% change. It decreases roughly linearly on a log frequency scale.

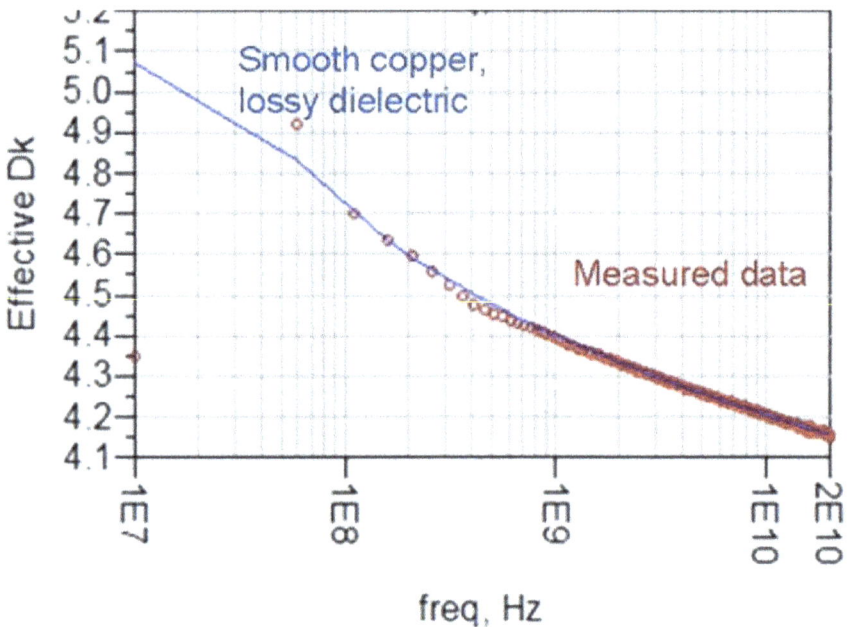

Figure 11.12 Measured Dk over a wide frequency range for an FR4 type material. This was measured with a network analyzer in the frequency domain. The line through the data is based on a Svensson-Djordjevic model.

This is an example of one of the most frequency-dependent interconnect materials. In most other materials, the Dk varies less with frequency. This property of a material to have a Dk and a speed of propagation varying with frequency is called *dispersion*.

The Dk value of the laminate, in addition to the geometry, influences the characteristic impedance of a transmission line. But it affects the characteristic impedance by the square root of Dk.

This means that with a variation of 5% from 1 GHz to 20 GHz, the impact on the characteristic impedance is about 2.5%. While the impact of the frequency-dependent Dk is easily measured, it is a smaller contribution to the characteristic impedance than the manufacturing variation from the combination of line width etching, trace thickness variation, and dielectric thickness control.

If the Dk of a material varies with frequency, even only 5%, which value do you use to characterize the material? At what frequency do you measure the Dk, and to what frequency does the TDR measurement correspond?

When a value of the Dk is used in calculating the characteristic impedance of a transmission line or the time delay of a signal, the value used is typically at 1 GHz. This is for two reasons.

From 1 GHz to 20 GHz, the Dk will vary, at worst case by 5%, and will always be higher at low frequency. If you are going to use one value, 1 GHz is the lowest frequency value you should use as it gives a rough value for the range of frequency components around 1 GHz.

If the properties of dispersion are important in an application, the frequency-dependent properties of the dielectric should be included in the model of the dielectric used in a simulator.

Many simulators use a frequency-dependent model for the dielectric constant. These models are called *causal models*, as their predictions generally predict the time domain, causal properties, of the signal in real systems more accurately than models using a constant dielectric constant with frequency.

A popular causal model is the Svensson-Djordjevic model. It assumes the dielectric constant varies with the log of frequency:

$$Dk(f) = Dk_2 + \left\{ \frac{\Delta Dk}{\log(f_2) - \log(f_1)} \right\} (\log(f_2) - \log(f)) \qquad (11.10)$$

where:

$Dk(f)$ = the dielectric constant at the frequency, f

Dk_2 = the dielectric constant at frequency values f_2

ΔDk = the change in the Dk value between the frequency points f_2 and f_1

f_2 = a higher frequency point

f_1 = a lower frequency point

This model uses a frequency dependence of Dk that varies with the log of frequency. The slope inside the brackets is a constant for the material.

The causal nature of this model means there is a connection between the frequency dependence of the Dk and the value of the dissipation factor, Df. The dissipation factor is a measure of the dielectric loss in the material.

The larger the dissipation factor, the larger the loss and the larger the slope of the frequency dependence of Dk. The lower the dielectric loss in a laminate, the less dispersion and the more constant the Dk with frequency.

In the Svensson-Djordjevic model, the slope of Dk, on a log frequency scale, the term in the brackets, is related to the Df by

$$\left\{ \frac{\Delta Dk}{\log(f_2) - \log(f_1)} \right\} = 1.47 \times \left(Df(f_2) \times Dk(f_2) \right) \qquad (11.11)$$

This says, measure the Dk and Df at one frequency, f_2. This gives the slope of how Dk varies with frequency. From this one frequency value and the model, we can calculate the Dk(f) at any other frequency.

While in principle it does not matter what the f_2 frequency is, most simulators use a value of 1 GHz as the frequency at which the Dk and Df are specified. From this single frequency value, the frequency dependence of Dk can be calculated.

However, to measure the frequency dependence of Dk and include it in a Svensson-Djordjevic type model, we need BOTH the Dk and the Df at one frequency.

A TDR measurement ONLY provides the Dk value. When you use the Dk from a TDR measurement you cannot build a frequency dependent casual model. You will only get the Dk value.

When an electrical property is strongly frequency dependent, the TDR response strongly depends on the TDR's rise time. For example, the reflection from an electrically short structure is strongly rise time dependent and the impedance of an electrically short structure is frequency dependent.

The dielectric constant of the laminate in a transmission line is not strongly frequency dependent. It varies at the extreme, with the log of the frequency.

When we use the position of the peak as a measure of the time delay of an interconnect length, the highest frequency component in the edge corresponds to the bandwidth of the TDR signal. This is roughly

$$\mathrm{BW} = \frac{0.35}{\mathrm{RT}} = \frac{0.35}{0.035\,\mathrm{n\,sec}} = 10\,\mathrm{GHz} \qquad (11.12)$$

To first order, the Dk measured by the TDR signal is related to the speed-of-the-high frequency components at roughly 10 GHz for a 35-psec edge.

For shorter or longer TDR rise times, the frequency components in the peak will change, but the Dk will generally change only slightly, with the log of the frequency, and will be insensitive to the exact frequency.

As an example of the lack of sensitivity of the extracted Dk to the rise time of the TDR signal is shown in **Figure 11.13**. In this example, the rise time of the TDR signal is changed from the shortest rise time, 35 psec, to 200 psec. The effective highest bandwidth changes from 0.35/0.035 nsec = 10 GHz to 0.35/0.2 nsec = 1.7 GHz over this range of rise times.

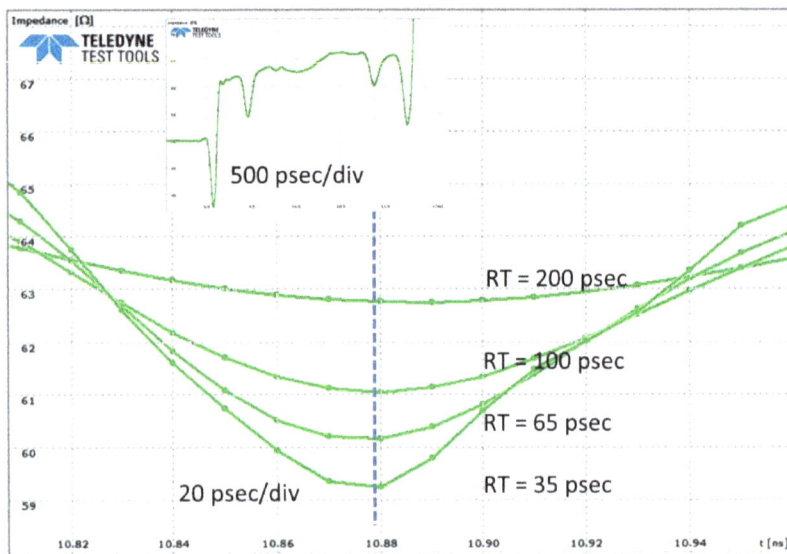

Figure 11.13 The measured time location of the second dip at different rise times. Inset is the overall TDR response showing the location of the second dip.

The time position of the dip from the second discontinuity, a measure of the round-trip delay to the second discontinuity, is slightly rise time dependent.

In this example, the position of the dip varies slightly with the rise time. The longer the rise time, the more the dip shifts to longer time, corresponding to a longer delay and a higher Dk value. This is an example of the dispersion in the Dk for this specific circuit board example.

On this scale of 20 psec/div, and a with a sample interval of 10 psec per point, the shift in delay is about 20 psec from a 35-psec rise time to 200 psec rise time. This is out of 1.5-nsec total round-trip delay, or about $0.02/1.5 = 1.4\%$.

This example illustrates that to first order, it does not matter what rise time is used to measure the Dk. A shorter rise time makes it easier to measure the dip or peak position. This would correspond to a frequency value of about 5 GHz.

If the difference in Dk value from 5 GHz to 1 GHz is important in your application, don't use a TDR to measure the Dk. But, if you want a good starting place for a Dk that is at roughly 5 GHz, a TDR can provide a quick, simple estimate.

If the frequency dependence of the Dk is important in an application, do not use the time domain display of the impedance profile. Instead, plan to get the S-parameters of the interconnect and use the phase of the insertion loss to extract the frequency dependent time delay and from this the frequency-dependent speed and Dk.

A TDR is not a good tool to use to measure an electrical property that is strongly frequency dependent.

11.8 Review Questions

1. What is the Dk of a dielectric a measure of?

2. What combination of materials contribute to the Dk of a laminate layer?

3. Why would the measured Dk of a stripline structure not be the Dk of each layer?

4. What are three ways of measuring the Dk of a transmission line?

5. Why is the measured Dk based on the time delay from the beginning to the end of the line inaccurate?

6. What is the advantage of measuring the Dk using two different length transmission lines?

7. What is the assumption we are making when using two different transmission lines to measure the Dk?

8. What is the advantage of using one line with two small strategically placed discontinuities?

9. What is a typical value measured for the composite Dk of a stripline composite?

10. How does this compare with the bulk Dk of the microstrip layer?

11. Why is the Dk measured for a microstrip not the bulk Dk?

12. What do you have to know about the microstrip to convert the effective Dk into the bulk Dk?

13. What do we call the property of the material if the Dk varies with frequency?

14. If the dispersion of a material is large, what must also be true of the electrical properties of the material?

15. What does a causal model mean?

16. How does the Dk vary with frequency in most materials?

17. If the small variation in Dk with frequency is important in your application, what should you do?

18. When we use a TDR to measure interconnect properties, what are we assuming about the frequency dependence of the figures of merit?

Practical Transmission Lines

Chapter 12 Calculating the Characteristic Impedance from Geometry and Material Properties

A uniform transmission line with a constant cross section throughout its length has a constant instantaneous impedance and a single characteristic impedance. Since the geometry has exactly the same cross section down the length of the transmission line, only the cross section, the 2D geometry, and the dielectric constant of the materials influence the value of the characteristic impedance.

Knowing how the geometry and materials affect the characteristic impedance will allow us to design a transmission line for a target impedance value.

We will use this connection between how the geometry affects the electrical properties of a transmission line in two different ways:

In *analysis*, we use as input the geometry of the cross section and the material properties to *calculate* as output, the characteristic impedance and time delay:

Analysis: geometry + Dk → Z0 and TD

In *design*, we start with a target Z0 and a specific Dk and *synthesize* the geometry that will result in the required Z0:

Design: Z0 and Dk → geometry

When possible, we try to limit the input variables to 2 or 3 and explore design space consisting of the combination of all values of the variables that results in the target impedance. This is the space from which we make design trade-offs.

For both processes, it is essential to have an analytical connection between how the geometry and Dk affects the Z0. We will use this for both processes.

12.1 Characteristic Impedance and Geometry

In Chapter 2, the characteristic impedance of a uniform transmission line was shown to be:

$$Z_0 = \frac{1}{vC_{Len}} = \frac{\sqrt{Dk}}{c \times Dk \times C_0} = \frac{1}{cC_0\sqrt{Dk}} \qquad (12.1)$$

where

Z_0 = the characteristic impedance

c = speed of light in air

C_{Len} = the capacitance per length

C_0 = the empty space capacitance per length

Dk = the laminate dielectric constant

This simple relationship is independent of the specific geometry. The only assumption is that the interconnect is uniform. It separates the factors that influence the characteristic impedance into three pieces:

- Fundamental constant (c)
- Material properties (Dk)
- Geometry terms (C_0)

The characteristic impedance is inversely related to the empty space capacitance per length and inversely to the square root of the dielectric constant.

For example, if the line width increases, the area of overlap between the signal and return conductors increases and the empty space capacitance would increase. This would cause the characteristic impedance of the line to decrease.

Any geometrical feature that increases the capacitance per length of the transmission line will decrease the characteristic impedance. Likewise, any geometrical feature that decreases the capacitance per length will increase the characteristic impedance.

This relationship is key to understanding the design of a transmission line. As young engineering students, we learn the connection between geometry and capacitance. By inverting this relationship, we now understand the connection between geometry and characteristic impedance.

If the dielectric constant of the material between the signal and return path increases, the characteristic impedance would decrease, but with the square root of the dielectric constant.

The challenge is calculating exactly how much the geometry and material properties affect the characteristic impedance, and based on a target value, a strategy to explore design space and find an acceptable design.

12.2 Analytical Exact Examples

Fundamentally, it is the precise shape of the electric and magnetic fields around the signal and return paths that result in the characteristic impedance. There are only three specific geometries where the fields can be calculated exactly, which are based on a solution of Maxwell's equations in cylindrical geometry.

These geometries, shown in **Figure 12.1**, are:

- Coax

- Twin rod
- Rod over infinite plane

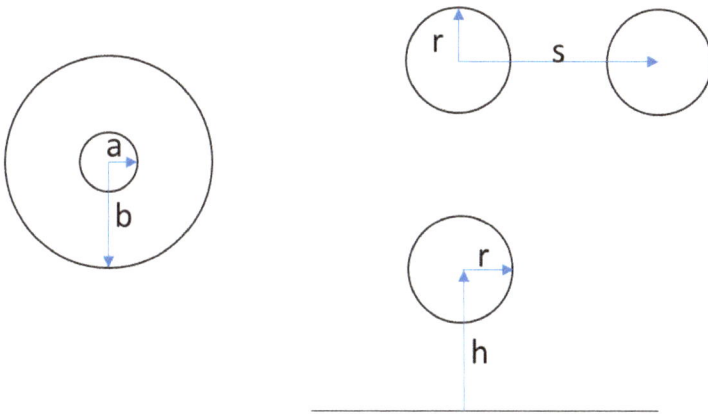

Figure 12.1 The three specific geometries where Maxwell's equations can be solved exactly as a closed-form equation.

For these specific geometries, Maxwell's equations can be solved analytically, and the characteristic impedance calculated.

The resulting values are:

Coax:

$$Z_0 = \frac{377\Omega}{2\pi\sqrt{Dk}} \ln\left(\frac{b}{a}\right) \qquad (12.2)$$

Twin rod:

$$Z_0 = \frac{120\Omega}{\sqrt{Dk}} \ln\left(\frac{s}{2r} + \sqrt{\left(\frac{s}{2r}\right)^2 - 1}\right) \qquad (12.3)$$

Rod over infinite plane:

$$Z_0 = \frac{60\Omega}{\sqrt{Dk}} \ln\left(\frac{h}{r} + \sqrt{\left(\frac{h}{r}\right)^2 - 1} \right) \qquad (12.4)$$

For any other cross section, a field solver is required to accurately calculate the characteristic impedance from the cross-section geometry.

Any equation you might see published describing the characteristic impedance and geometry of any other cross section, such as microstrip or stripline, is an approximation. Without comparison to a field solver, the accuracy of analytical approximations is unknown. Unfortunately, how complicated the equation may look is not an indication of its accuracy.

12.3 Using a 2D Field Solver to Calculate the Characteristic Impedance

The cross section of a uniform transmission line is constant down the length of the transmission line. If we assume the propagation of the signal is as transverse electric and magnetic (TEM) fields, then the field lines will all be in the plane of the cross section and constant amplitude down the length of the transmission line.

This is a good assumption only if the separation between the signal and return path is much smaller than a wavelength. This condition for which the TEM approximation would apply is:

$$h < \frac{1}{10}\lambda = \frac{1}{10}\frac{c}{f\sqrt{Dk}} \qquad (12.5)$$

where

h = the dielectric thickness between the signal and return path c = the speed of light in air

λ = the wavelength of the frequency component

f = the frequency of the highest frequency

For the case of FR4 with Dk = 4 and the speed of light in the dielectric is 6 inches/nsec, and the frequency below which propagation is TEM is roughly:

$$f[\text{GHz}] < \frac{600}{h[\text{mils}]} \qquad (12.6)$$

where

f is the upper frequency limit in GHz

h is the dielectric thickness between the signal and the return path in mils

For example, if the dielectric thickness is 10 mils, the frequency up to which the TEM mode approximation would apply is f < 60 GHz.

This simple analysis suggests that the approximation of TEM propagation and the use of a 2D field solver is a very good approximation for most digital applications.

For higher frequencies, a full wave solver is required that takes into account the non-TEM modes of the propagating electric and magnetic fields. A 2D field solver will not give an accurate result.

In the case of TEM propagation, the electric and magnetic fields in one cross section of the interconnect will be identical to any other cross section down the interconnect. There are no fields in the direction down the length of the transmission line. When we calculate the field lines, it is just the cross-section information that influences the field lines. This is a two-dimensional problem, which we refer to as a 2D problem.

To calculate the electric and magnetic fields in the 2D geometry, we use a 2D field solver.

While we could take a three-dimensional section of a transmission line with some length, and use a 3D field solver to calculate the electric and magnetic fields, just because it is a 3D solver does not mean that the result will be more accurate than a 2D solver.

We would only use a 3D solver when the cross section is not uniform. For example, if the signal line has a bend where its width changes at the corner of where the signal line passes over a gap in its return path, a 3D solver is the right tool to use. For these structures, only a 3D field solver will give an accurate result.

In the case of a uniform cross section, a 3D field solver does not provide a more accurate answer than a 2D field solver. But, a 2D solver can analyze the characteristic impedance and time delay more than $100 \times$ faster than a 3D solver.

There is an important limitation of every field solver. A field solver can only solve a specific problem. The precise geometry and distribution of materials with the specific Dk values must be defined. For this specific case, the distribution of conductors and dielectrics represent the boundary conditions for which the electric, E, and magnetic, H, fields are solved.

From the resulting E and H fields, the important electrical properties such as characteristic impedance and time delay are calculated. This process is illustrated in **Figure 12.2**.

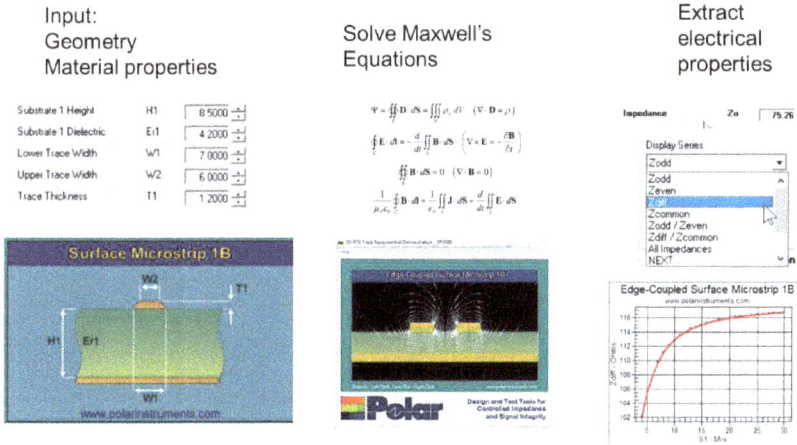

Figure 12.2 The steps using a 2D field solver.

The result of a field solver is a point solution. The calculation is for one specific set of conditions.

In some tools, a geometrical or material property can be automatically varied or swept through different specific values, and the electrical properties calculated at each changed value. This makes for a much more versatile tool, as we can visualize the trends and use this changing feature to describe a design space. But each point is still a calculation for a specific set of input conditions.

There are a number of commercially available 2D field solvers. They differ in the range of cross sections they deal with, the ease of exploring design space, and the simplicity or complexity of the user interface.

The most common, easiest to use, and suitably versatile commercial 2D field solver is the SI9000 from Polar Instruments.

Any reader interested in a free, 2-week evaluation license for this software can contact Polar at this link.

323

While this tool will calculate the properties of lossy lines and export S-parameters, we will use it primarily to calculate only the characteristic impedance of lossless transmission lines.

A key feature of this tool is the capability to both *analyze* a characteristic impedance and time delay for a specific set of input features, and to *explore design space* to find a range of geometries that result in a target impedance.

The SI9000 tool is designed specifically to analyze printed circuit board structures. These are planar geometries with one or more signal conductors and at least one return plane that is infinite in extent. There are dozens of predefined and parameterized structures to choose from in this tool.

12.4 Finite Width of the Return Path

One very important limitation with this tool is the assumption that the return conductor, usually one or more planes, is infinite in lateral extent. This means the impact from a narrower width of the return path cannot be analyzed with this 2D tool.

As we show later, the fringe fields from the edge of the signal line extend about 5 × h on either side of the signal line. If we keep the width of the return path equal to the linewidth + 10 × h, all the field lines will be contained in this region and making the return path wider should not affect the characteristic impedance.

In a 50 ohm line, the ratio of linewidth to dielectric thickness is about 2:1. This means that h ~ ½ × w. The minimum width for the return path so it does not affect the characteristic impedance is return-path-width > w + 10 × ½ w = 6w. For the width of the return path to not affect the characteristic impedance of a roughly 50-ohm line, it should span about 6w centered underneath the signal line.

Using another field solver tool, we can calculate the characteristic impedance as the width of the return path is changed. **Figure 12.3** shows that providing the total width of the return path under the signal trace is > 3w, the impedance is independent of the return path width to within 1%. Our estimate of 6w is a conservative estimate.

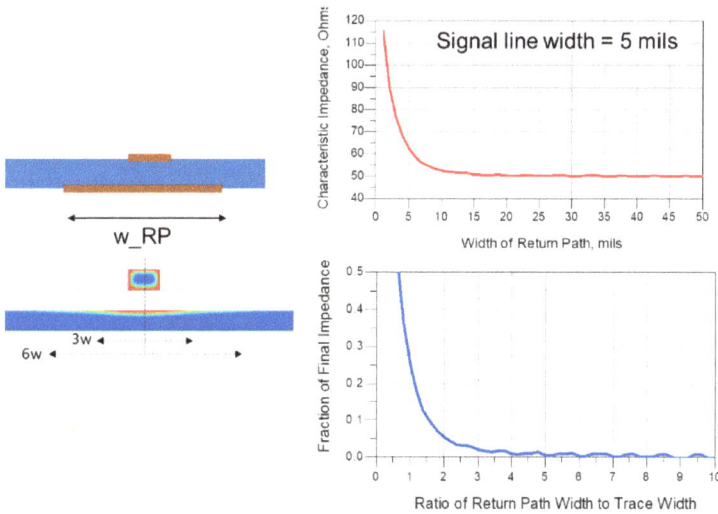

Figure 12.3 Changing the total width of the return path and calculating the characteristic impedance. Note the current distribution and the size of the return path distribution.

While there is still some return current outside the 3w region in the return plane, it is not enough to affect the impedance more than 1%. All of the return current is contained within a total width of 6w for a 50-ohm transmission line.

The return current distribution shows that most of the return current is directly underneath the signal line with a little leaking out on either side. If the return path is made wide enough to contain all the return current, making it wider has no more impact.

325

Using other field solvers, such as the Polar Instruments SI9000 that assume the return path is infinite, is a very good approximation providing the return path is wider than about 3 linewidths.

12.5 Practice Safe Simulation and Rule #9

Regardless of the tool we use, it is usually very easy to type in a few parameters, push the run the button, and get a result. However, it is sometimes a challenge to get a result that is meaningful for the problem for which we are interested.

While it is impossible to ever prove your answer is correct, it is possible to test the consistency of your answer. The most important consistency test to always perform is the comparison to what you expect to see.

This is such an important rule, it has become known as Bogatin's Rule #9:

"Before you do a measurement or simulation, first anticipate what you expect to see. If you are wrong, there is always a reason why it's not what you expected. Don't proceed until you can identify the inconsistency.

Sometimes it is because you made a mistake in the setup, sometimes it's because a cable is bad, sometimes it is because the DUT is not working, and sometimes it is because your understanding is off, and this is a chance to accelerate up the learning curve.

If you are right, and you see the behavior you expected, you gain a little confidence that maybe you do understand the system."

It is such an important rule it has become popular on hats, as shown in Figure 12.4.

Figure 12.4 This rule is so important, it even appears on hats.

Rule #9 is the first and most important step in practicing safe simulation.

Take every opportunity to think of other consistency tests to compare to your result. This is how you will often catch errors in the values typed in as initial conditions, or errors in how the field solver tool was set up.

The most important consistency test is based on the central connection between geometry and material properties and characteristic impedance:

$$Z_0 = \frac{1}{cC_0\sqrt{Dk}} \tag{12.7}$$

The better you understand this relationship, the more confident you will be about the results from a 2D field solver.

12.6 Review Questions

1. What is the difference between analysis and design?

2. Why are we able to calculate the characteristic impedance of a coax exactly?

3. Why is TEM mode propagation important when considering using a 2D field solver to calculate the characteristic impedance of a transmission line?

4. How high a frequency can we go and have TEM field lines in a microstrip if the dielectric thickness is 20 mils? 5 mils?

5. What is the hidden assumption in most 2D field solvers about the return plane?

6. At a minimum, how wide should the return plane be to not affect the characteristic impedance of the line to less than 1%?

7. What is rule #9 and why is it important?

8. What is an example of an error you could make using a 2D field solver that you could catch if the result was way off from what you expected?

Chapter 13 A Microstrip Transmission Line

There are two very important, common transmission line structures to analyze: a microstrip and a stripline.

A microstrip is a signal line over a dielectric layer, a bottom return plane, and air above the signal line. There are many variations of the simple microstrip:

- With coplanar ground
- With solder mask
- With two or more dielectric layers below the signal line
- More than one coating over the signal line

Each of these have a specific name and are a different type of microstrip. A few examples in the SI9000 tool are shown in **Figure 13.1**.

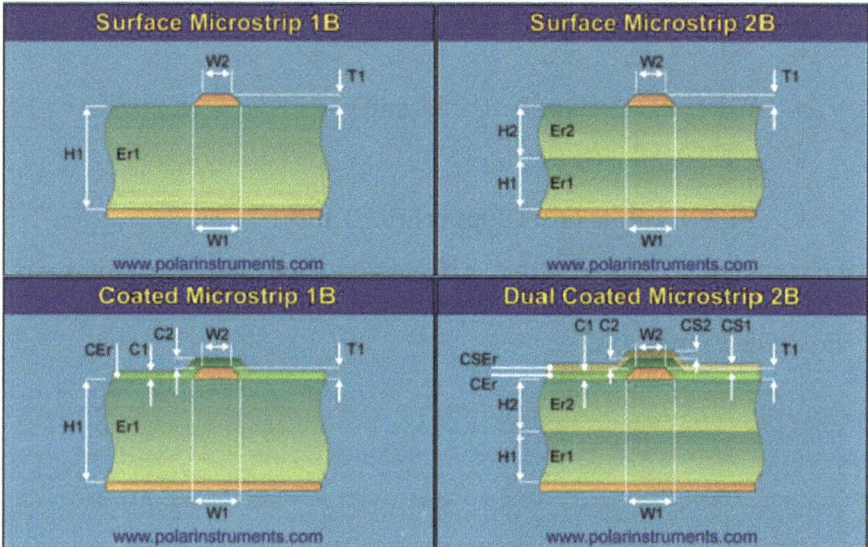

Figure 13.1 Example of four different microstrip structures with different parameters.

13.1 The Surface Microstrip

The simplest structure is the surface microstrip 1B. There are only five parameters that define this structure:

- The top-line width, W2
- The bottom-line width, W1
- The dielectric thickness, H1
- The dielectric constant, Er1
- The trace thickness, T1

To calculate the characteristic impedance, a specific value for each term is entered and the characteristic impedance for this specific case is calculated. Applying Rule #9 as a rough rule of thumb, a microstrip in FR4 will be about 50 ohms when the ratio of the line width to the dielectric thickness is 2:1.

For example, for the specific case of a:

331

- Er1 = 4.2

- W1 = W2 = 10 mils (0.254 mm)

- H1 = 5 mils (0.126 mm)

- T1 = 1.2 mils (1 oz copper) (0.31 mm)

the calculated Z0 is 47 ohms. This is very close to the 50 ohms we expected to see. The setup screen for this specific problem is shown in **Figure 13.2**.

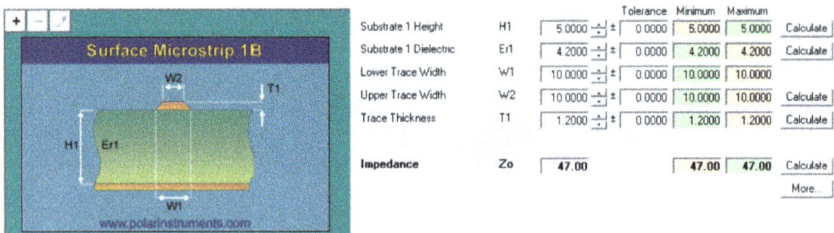

Figure 13.2 The basic calculation of the Z0 in the SI9000 tool.

This is the basic operation of the tool: specific parameter values are input and the characteristic impedance for that specific case is calculated.

In addition, one or more parameter values can be swept through a range and the characteristic impedance calculated for each parameter.

For example, if the dielectric thickness, H1, were to change from 1 mil to 50 mils, we would expect to see a low impedance for a thin dielectric and a high impedance for a thick layer.

Using the sensitivity tab found on the bottom of the screen, we can set this problem up. The resulting plot of Zo as the dielectric thickness changes is shown in **Figure 13.3**.

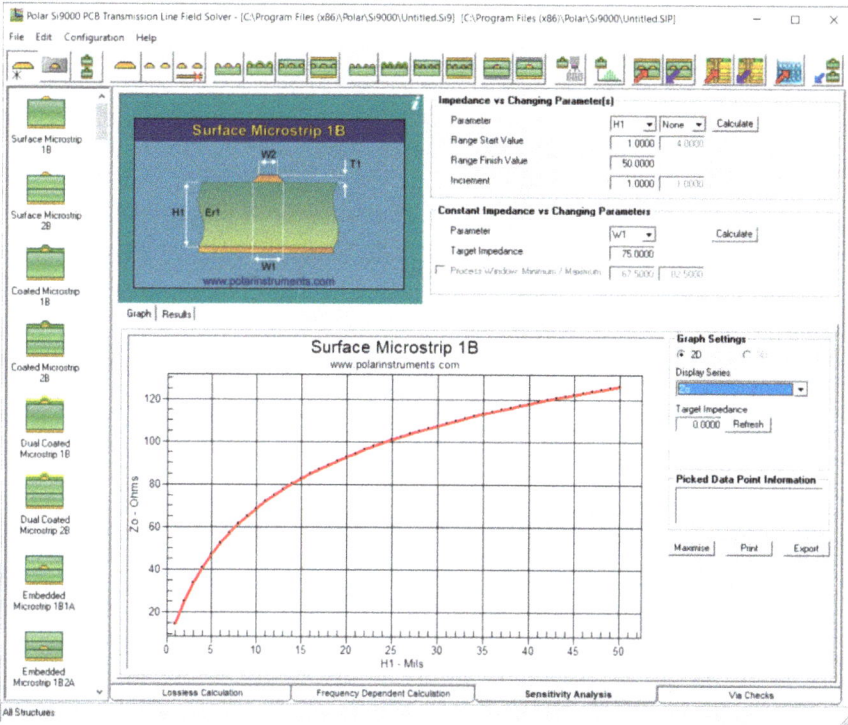

Figure 13.3 Setting up a sensitivity analysis by sweeping the dielectric thickness.

The difference in width between the top of a trace and its bottom is the etchback factor. A value of 0 means the trace has a rectangular cross section.

If you want to keep the top and bottom conductor widths the same, or a fixed difference, for example, you can set the etchback factor in the menu under configuration/parameters. In this screen, shown in **Figure 13.4**, select the etchback value of 0 in the right-hand screen identified by the arrow for a rectangular cross section.

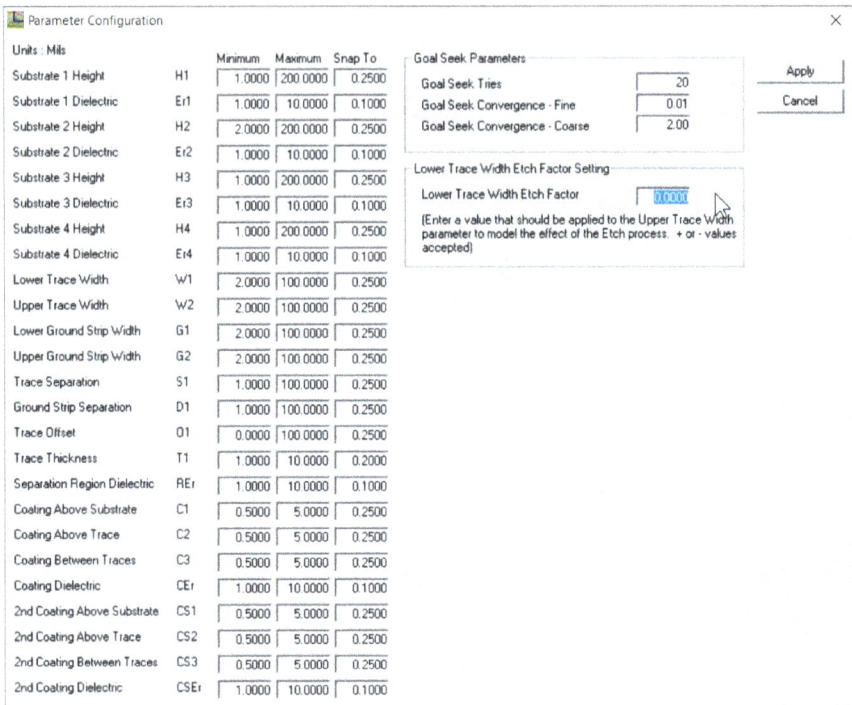

Figure 13.4 Select the etchback value from the configuration/parameters menu.

When the linewidth, W1 or W2 is selected as the parameter to swept, this etch back factor is used as one of the width terms and swept to calculate the other width term. To calculate the characteristic impedance for a rectangular trace, use an etchback factor of 0.

Terms that affect the characteristic impedance the most are considered *first-order* terms. The terms that are not as important in affecting the characteristic impedance are *second-order* terms.

The trace thickness and etchback factor are second-order terms. Changes in these parameters have a small impact on the characteristic impedance of the line.

Watch this video and I will show you how to get started using the Polar Instruments SI9000 2D field solver.

13.2 A Simple Rule of Thumb

The characteristic impedance of a transmission line is usually constant if the linewidth and dielectric thickness are scaled together. This means, to first order, the characteristic impedance depends on the ratio of the linewidth to dielectric thickness.

This is a very useful observation to create a simple rule of thumb for the specific case of:

W1 = W2

Dk = 4.0

T1 = 1.2 mils

As we sweep the dielectric thickness, H1, we can calculate the line width needed to synthesize a 50-ohm line. This maps out design space for a 50-ohm microstrip in FR4.

If the dielectric thickness increases, the capacitance would decrease and the impedance would increase. To bring the impedance back down to 50 ohms, the linewidth would have to increase.

As the dielectric thickness increases, we can calculate the linewidth to maintain 50 ohms. This is shown in **Figure 13.5**.

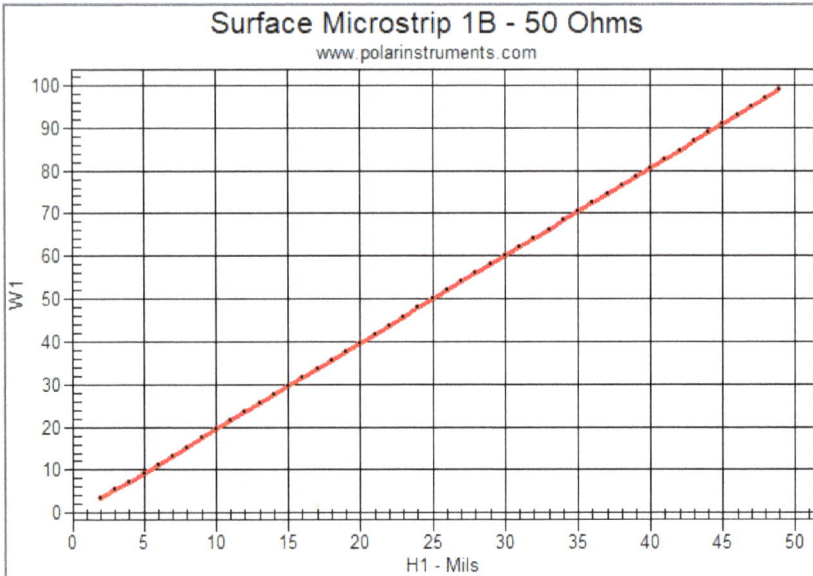

Figure 13.5 Mapping design space for a 50-ohm line in FR4. As the dielectric thickness increases, the line width must change to compensate. This is a line of constant 50-ohm impedance.

This straight line suggests a constant ratio for a 50-ohm line. In **Figure 13.6**, we plot the ratio of linewidth to dielectric thickness for various dielectric thicknesses to achieve a 50-ohm line. It is nearly 2 across the entire design space.

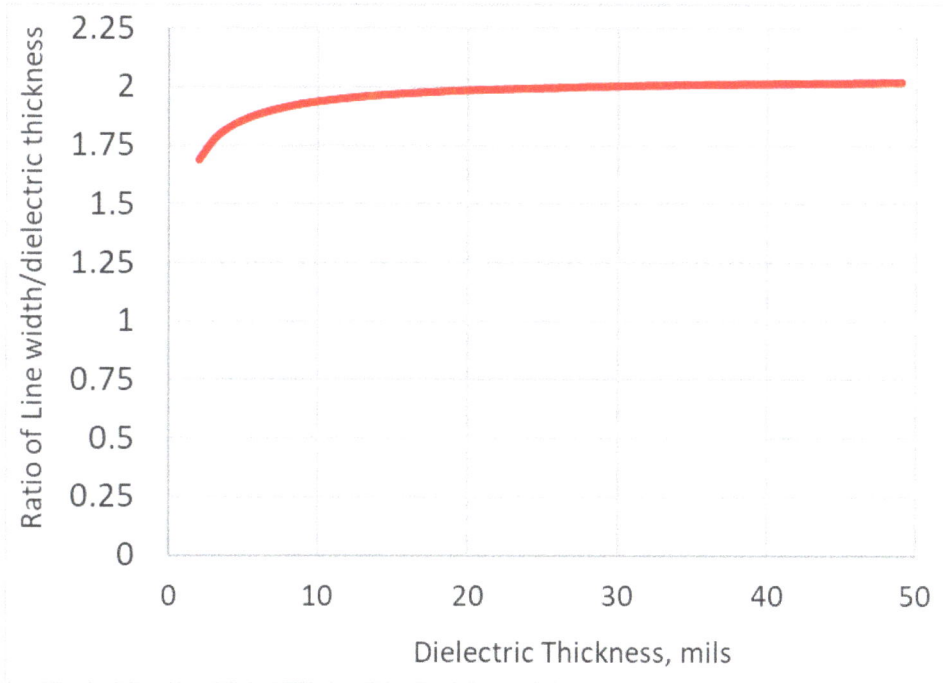

Figure 13.6 Ratio of the linewidth to dielectric thickness for a 50-ohm line in FR4 with 1-oz copper.

This is the origin of the very simple rule of thumb that in microstrip, for 1-oz copper and FR4, the linewidth to dielectric thickness is 2:1.

This should be the starting place for estimating the characteristic impedance of a microstrip. If the Dk is higher than 4, the linewidth would have to be slightly narrower.

Watch this video and I will show you how to map design space using SI9000.

13.3 Sensitivity Analysis: First- and Second-Order Factors

To determine the impact of changing a geometrical feature on the impedance of the line, we look at how much the characteristic impedance changes for a 10% change in the input parameter from the nominal conditions.

When changes are small, the impact on the characteristic impedance is linear and we can define the sensitivity as the leading term in the Taylor series expansion. We can normalize the sensitivity as the ratio of the relative change in each term. For example, the relative sensitivity of an impedance change to a linewidth change is

$$\text{relative sensitivity} = \frac{\dfrac{\Delta Z_0}{Z_0}}{\dfrac{\Delta W}{W}} \qquad (13.1)$$

This is the relative sensitivity. For a percentage change in an input term, the percentage change in the characteristic impedance is the percentage change in the input term $\times$ relative sensitivity. The larger the relative sensitivity, the more sensitive the impedance is to the selected parameter.

For small changes in an input term, the output term will be linear. A negative value means the impedance decreases for an increase in the term.

338

We start with the nominal values of a 50-ohm line:

- W1 = 10 mils

- W2 = 9 mils (etchback factor of 1 mil)

- H1 = 5 mils

- Dk = 4.0

- T1 = 1.2 mils

This results in a characteristic impedance of 48.65 ohms. The setup for this condition is shown in **Figure 13.7**.

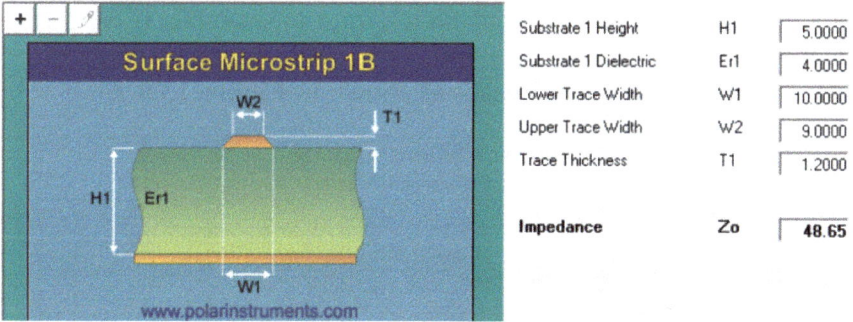

Substrate 1 Height	H1	5.0000
Substrate 1 Dielectric	Er1	4.0000
Lower Trace Width	W1	10.0000
Upper Trace Width	W2	9.0000
Trace Thickness	T1	1.2000
Impedance	Zo	48.65

Figure 13.7 The nominal conditions calculated in the Polar Instruments SI9000 tool.

We change each parameter by 10% and look at the impact on the Z0 and calculate the relative sensitivity. As part of this analysis, we should always apply Rule #9 and anticipate what to expect.

If increasing a term increases the capacitance per length, the characteristic impedance should decrease. The linewidth, dielectric thickness, and dielectric constant should be first-order terms and changing the trace thickness should be a second-order term.

Increasing the linewidth will *increase* the capacitance and *decrease* the characteristic impedance.

Increasing the dielectric thickness will *decrease* the capacitance and *increase* the characteristic impedance.

339

Increasing the Dk will *increase* the capacitance and *decrease* the characteristic impedance.

Increasing the trace thickness should *increase* the capacitance very slightly. This is because the fringe electric field lines would have a slightly *larger* surface to start from and *decrease* the characteristic impedance.

Increasing the etchback should *decrease* the capacitance very slightly, and *increase* the characteristic impedance.

Using the SI9000 tool we can calculate the sensitivity of the characteristic impedance from each term. For example, if we change the linewidth from 10 mils to 11 mils, the field solver-calculated impedance decreases as shown in **Figure 13.8**.

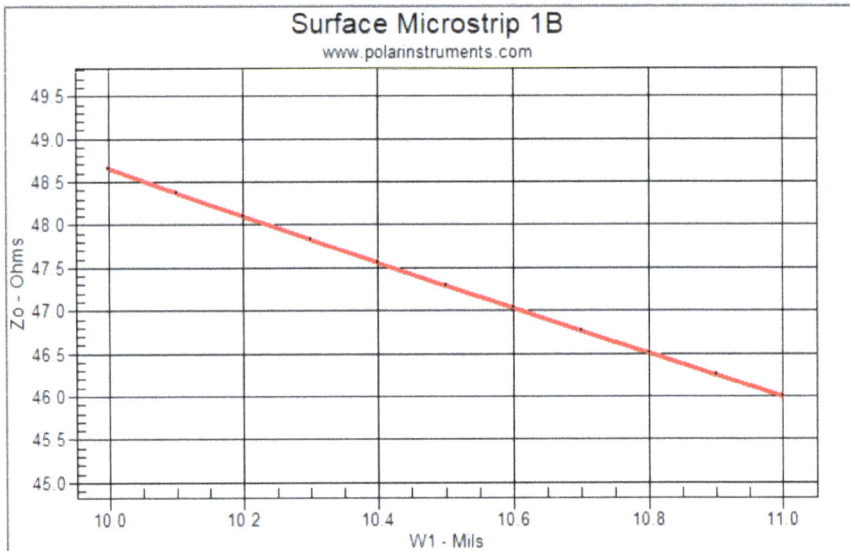

Figure 13.8 For the nominal conditions, just increasing the line width by 10% decreases the impedance from 48.65 ohms to 46.01 ohms. This is a -5.4% change.

The sensitivity in Z0 to a line width change is -5.4%/10% = -0.54.

This means for a 10% change in the linewidth, the characteristic impedance would change by 10% × -0.54 = -5.4%.

We can evaluate the impact from the etchback. In these examples, we assumed the etchback is 1 mil. A 10% increase in the etchback is a 0.1-mil decrease in the width of the top of the trace.

By decreasing the top of the trace width, W2, by 0.1 mils but keeping the bottom linewidth constant, we introduce a 10% increase in etchback. This decrease in the top of the trace width should decrease the capacitance per length slightly and increase the characteristic impedance slightly.

The relative sensitivity on the characteristic impedance is calculated as 0.01.

Continuing this analysis through each of the other terms, using the SI9000 tool for the sensitivity analysis, we find the relative sensitives to be:

W1:   -0.54

H1:   0.59

Dk :   -0.41

T1:   -0.038

Etchback (W1 – W2) : 0.01

This clearly identifies the first-order terms as W1, H1, and Dk. Changes in the trace thickness and the etchback have a second-order impact on the impedance. *Increasing* all the terms except the dielectric thickness will *decrease* the characteristic impedance.

This behavior is important when analyzing the impact on impedance variations from manufacturing variations. If each input term can be controlled to the same relative amount, then the relative sensitivity shows directly the impact on the impedance variation.

In order to reduce the variation in the characteristic impedance, the most important terms to control in the manufacturing process are the dielectric thickness, the linewidth, and the dielectric constant, in that order.

13.4 A Comparison to Analytical Approximations

The IPC-2251 specification offers a few simple analytical approximations to the characteristic impedance of microstrip and stripline structures.

For example, the approximation for the characteristic impedance of a microstrip with air above, and rectangular cross section, is

$$Z_0 = \frac{87}{\sqrt{Dk + 1.41}} \ln\left(\frac{5.98H1}{0.8W1 + T1}\right) \qquad (13.2)$$

We can use the 2D field solver to compare this approximation to a more accurate calculation. This comparison is shown in **Figure 13.9** for the nominal case of Dk = 4, T1 = 1.2 mils, and H1 = 10 mils, while sweeping W1.

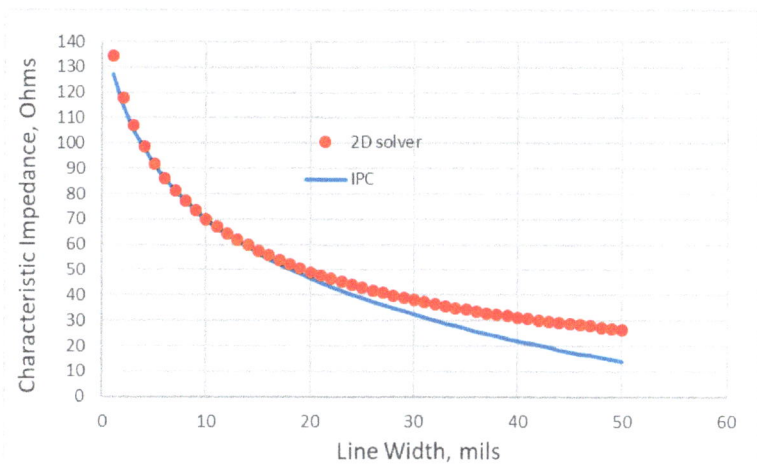

Figure 13.9 Comparing the IPC approximation with the results from the SI9000 field solver.

The approximation is very good for impedances of 50 ohms and above. However, for a 30-ohm impedance line, the approximation is off by 30%, predicting a 20-ohm impedance, when the field solver predicts a 30-ohm impedance line.

13.5 Comparison to a Measurements of a Simple Test Board

While the accuracy of the Polar SI9000 field solver has been compared to other field solvers and matched to better than 1%, it's always of value to test how accurately the measured impedance of real boards can be predicted with this or any tool.

A simple test board was fabricated on a double-sided 1-oz copper substrate. Individual traces were machined out of the solid copper plane on the top layer using an LPKF circuit board prototype machine. This work was done by Mohammed Al Hasani, a senior at the University of Colorado, Boulder. **Figure 13.10** shows a top image of the test board.

Figure 13.10 Simple test board with different linewidths and SMA connectors on one end to measure the characteristic impedance for different linewidths. Note the bottom trace is designed to measure the Dk of the board for input to the 2D solver, which was measured as 4.49.

The TDR response of each trace was measured. An example of the composite of each measurement is shown in **Figure 13.11**.

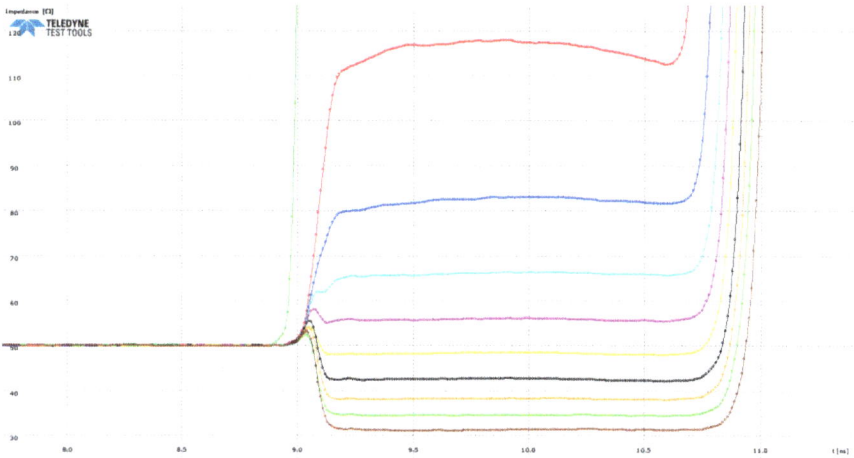

Figure 13.11 Measured TDR response for each of the uniform lines in the test board. Note also the increasing delay time, and Dk value for wider lines.

Using the copper thickness from the spec sheet as 1-oz copper, the measured dielectric thickness of the board as 56 mils, and the measured bulk dielectric constant of 4.49 for this specific board, the predicted impedance for various linewidths was simulated using the SI9000 tool.

On the same plot, the measured impedance and linewidth was added using the error bars for the 3% uncertainty in-line width and variation in impedance of the traces. **Figure 13.12** compares the measured impedance and the simulated impedance. The agreement across the range is within the 3% uncertainty in linewidth.

345

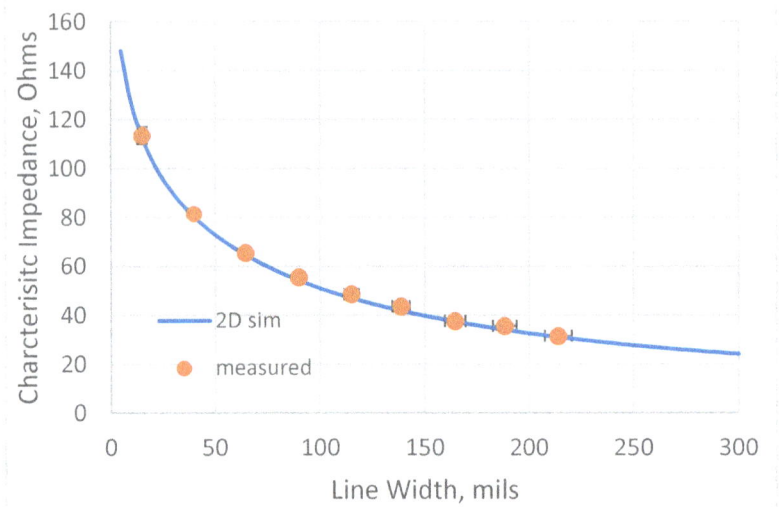

Figure 13.12 Comparing measured and simulated characteristic impedance of microstrip traces as the linewidth is varied.

This simple experiment indicates that if the geometry and material properties are well known, and extracted directly from the same board in which other measurements are performed, a 2D field solver can accurately predict the electrical performance to within the accuracy of the measurement.

13.6 Second-Order Factors: Solder Mask, Trace Thickness, Etchback, and Adjacent Conductors

In order to analyze the impact these additional features have on the impedance of a microstrip, we have to use a model that contains the features so we can adjust them.

A good candidate structure to use that contains all of these elements are the coated coplanar strips with ground 1B. This cross section is shown in **Figure 13.13**.

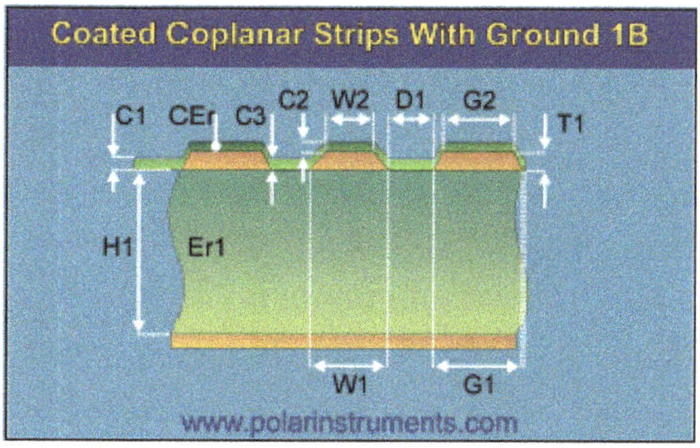

Figure 13.13 This structure has all the features we need to analyze these second-order effects.

This is a model for a microstrip trace, with a solder mask coating and two adjacent traces that are assumed to be grounded. The adjacent traces have exactly the same linewidth and etchback as the central trace and are spaced the same distance from the signal trace on either side of it.

New parameters are introduced in this model to describe the new features. These are:

- G1: lower width of the adjacent ground strip

- G2: upper width of the adjacent ground strip

- D1: separation of the adjacent ground strip

- C1: thickness of the solder mask on the substrate region

- C2: thickness of the solder mask over the trace

- C3: thickness of the solder mask between the traces

- CEr: dielectric constant of the solder mask

We can set up an initial set of parameters that are nominal values and then look at the sensitivity of each term on the characteristic impedance. These are defined in **Figure 13.14** with the calculated value of the impedance as 50.45 ohms.

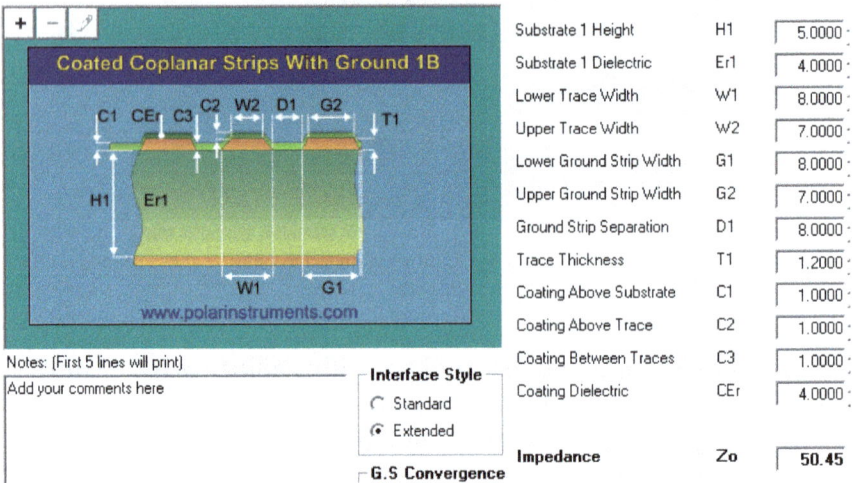

Substrate 1 Height	H1	5.0000
Substrate 1 Dielectric	Er1	4.0000
Lower Trace Width	W1	8.0000
Upper Trace Width	W2	7.0000
Lower Ground Strip Width	G1	8.0000
Upper Ground Strip Width	G2	7.0000
Ground Strip Separation	D1	8.0000
Trace Thickness	T1	1.2000
Coating Above Substrate	C1	1.0000
Coating Above Trace	C2	1.0000
Coating Between Traces	C3	1.0000
Coating Dielectric	CEr	4.0000
Impedance	Zo	50.45

Figure 13.14 The problem setup with the calculated impedance of 50.45 ohms.

In this case, the nominal conditions are:

- 1 mil etchback
- Linewidths of 8 mils, spacings of 8 mils
- Dielectric thickness of 5 mils
- Solder mask Dk same as the laminate, 4
- Solder mask thickness 1 mil everywhere

With each term, we can evaluate the relative sensitivity by varying it 10% and looking at the impact on the impedance.

As part of the process, we should apply Rule #9 and anticipate what to expect.

If increasing a feature will increase the capacitance between the signal and return, then it should decrease the characteristic impedance. Other than the first-order terms of the linewidth, dielectric thickness, and dielectric constant of the laminate, each of the other terms should be second-order terms.

Increasing G1, the linewidth of the adjacent trace will slightly *increase* the capacitance of the signal trace. This should *decrease* the impedance.

Using the SI9000 field solver, we find the relative sensitives are:

Substrate height, H1:	0.47
Substrate 1 DK, Er1:	-0.36
Line width W1:	-0.58
Etchback: W1-W2	0.02
Adjacent trace width, G1:	-0.001
Gap spacing, D1:	0.07
Trace thickness, T1:	-0.06
Solder mask thickness, C1:	-0.04
Solder mask Dk, CEr:	-0.04

This analysis points out that the most important terms influencing the characteristic impedance are also the substrate height, the line width, and the dielectric constant.

The other factors are all second-order.

In addition to evaluating the relative sensitivities, we can evaluate the absolute impact on the characteristic impedance. This is the impact on the impedance if this feature were not present.

From the nominal conditions, if there was no etchback, but the linewidth was rectangular, the impedance would change from: 50.45 → 49.17, a 2.5% decrease. The slight decrease in impedance is due to the very slight increase in fringe capacitance from the wider top section of the trace.

The etchback of the line increases the impedance by 2.5%, compared to no etchback.

The presence of the adjacent conductor being ground, or floating, is estimated by comparing the impedance with it present or moved very far away. The impedance changes: 50.45 → 52.19, 3.4%. The presence of the adjacent grounded conductors on either side, with a space equal to the linewidth, decreases the impedance by 3.4% compared to them not being present. This is when the spacing is equal to the linewidth.

To get a better idea of the sensitivity of the spacing to the adjacent conductors, we can sweep the edge-to-edge spacing, D1, and see the impact on the characteristic impedance. This is shown in **Figure 13.15**.

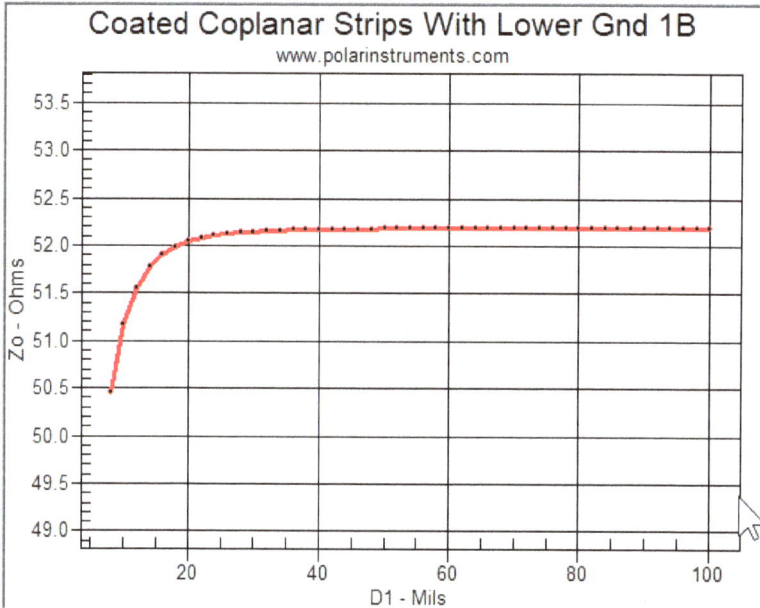

Figure 13.15 Moving the two adjacent grounded traces farther away (D1) increases the characteristic impedance of the trace. Note the scale is 0.5 ohms/div or 1% per division. This is for an 8-mil-wide line.

In this geometry, the line width is 8 mils and the dielectric thickness is 5 mils. When the two grounded traces, one on either side of the signal line, are moved about 25 mils away, 5 × H1, the impedance is not affected by the presence of the other traces.

> *This is the origin of the common rule of thumb*
> *that the fringe electric fields from the edge of a*
> *conductor extend about 5 dielectric thicknesses*
> *from the edge of the conductor.*

In this example, the dielectric thickness of the substrate is 5 mils and when the edge-to-edge spacing is 25 mils, there is no impact from the other conductors.

351

In the nominal case, we assumed the trace thickness was 1-oz copper, T1 = 1.2 mils. If this were to decrease to ½-oz copper, T1 = 0.6 mils, the characteristic impedance would increase from 50.45 → 52.24, a 3.5% increase.

The last term that is important is the presence of the solder mask. We can make it disappear by changing its dielectric constant to 1. This effectively makes the coating air.

The characteristic impedance goes from 50.45 → 53.53. This is an increase of 6.1% on the characteristic impedance.

This analysis suggests four important observations about these second-order factors:

1. A nominal 1-mil etchback on a 1-oz trace can have a 2.5% impact on the characteristic impedance.

2. The presence of closely spaced traces on either side can reduce the impedance by 3.4%.

3. Using 1-oz copper rather than ½-oz copper can affect the characteristic impedance by 3.5%.

4. Not taking into account the solder mask can affect the impedance by 6.1%.

While these factors are all second order, in the worst case, if they add in the wrong direction, they can contribute a total impact on the characteristic impedance of more than 15%.

This is why a 2D field solver is so important. With little effort, the impact from these second-order factors can be included in the calculation of the characteristic impedance and taken into account in the design phase to increase the chance of hitting a target impedance.

13.7 Review Questions

1. What are the two most commonly used transmission line topologies found in all circuit boards?

2. For a 50 ohm FR4 microstrip, if the linewidth is 10 mils, what should the dielectric thickness be approximately?

3. As the dielectric thickness increases, what should happen to the characteristic impedance?

4. What is the difference between a first-order and second-order term? What are examples of a first-order term and a second-order term in the design of a microstrip?

5. What is a good rule of thumb for the ratio of the linewidth to dielectric thickness for a 50-ohm microstrip line?

6. If the laminate layer is 5 mils thick, what linewidth would give about 50 ohms microstrip?

7. What design terms influence the characteristic impedance of a microstrip the most?

8. What is the limitation to using online calculators for the characteristic impedance of transmission lines compared to a 2D field solver?

9. What are three examples of second-order terms in the design of a microstrip that influence the characteristic impedance only slightly?

10. How many dielectric thicknesses do the fringe fields extend laterally from the edge of a trace?

11. If you do not consider the various second-order design features in the design of a microstrip, how far off can the fabricated impedance be from what was estimated?

Chapter 14 Stripline Analysis

A stripline topology is a signal layer between a top and bottom return plane. The DC voltage on either plane does not influence the characteristic impedance of the line. What is affected by the choice of the DC voltage is the switching noise that may arise when a signal line changes its return planes.

This is why, in all the analysis for characteristic impedance, the DC voltage on the planes is never an input criterion. What is important is proximity of the planes.

There are a number of variants of a stripline transmission line with different names:

Triplate: an early name for stripline

Symmetrical stripline: the dielectric layer above and below the signal layer is the same thickness

Asymmetric stripline: the dielectric layer above the signal line has a different thickness than below the signal line

Dual stripline: two signal layers between the two planes. The two signal layers are usually routed at 90 degrees to each other to reduce the interlayer cross talk

14.1 Simple Stripline Analysis

There are many different stackup configurations for stripline structures. This makes it difficult to generalize with rules of thumb how geometrical features affect characteristic impedance.

However, there are two common configurations for which some insight can be gained from analysis: a symmetrical stackup with a top and bottom dielectric layer with the same thickness, and an asymmetrical stackup with three dielectric layers, each with the same dielectric thickness.

An example of a symmetrical stripline with a 10-mil-wide line and 10-mil-thick dielectric layers, resulting in a 46.8-ohm characteristic impedance is shown in **Figure 14.1**.

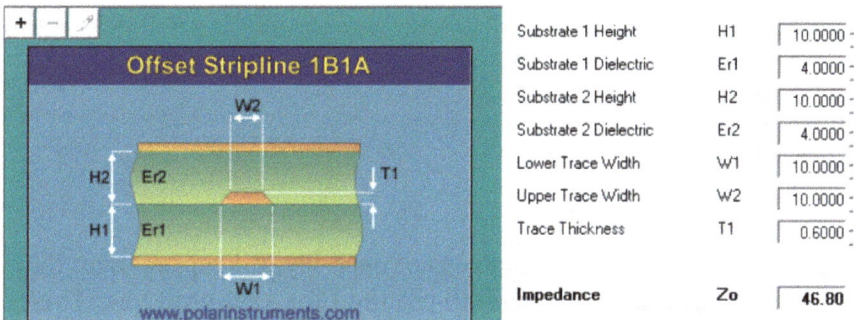

Substrate 1 Height	H1	10.0000	
Substrate 1 Dielectric	Er1	4.0000	
Substrate 2 Height	H2	10.0000	
Substrate 2 Dielectric	Er2	4.0000	
Lower Trace Width	W1	10.0000	
Upper Trace Width	W2	10.0000	
Trace Thickness	T1	0.6000	
Impedance	**Zo**	46.80	

Figure 14.1 A typical symmetrical stripline structure with nominal conditions.

The case of an asymmetrical, dual stripline with a 10-mil-wide trace and all dielectric thicknesses also 10 mils results in a characteristic impedance of 54.6 Ohms.

The asymmetrical stackup analysis using the Polar SI9000 tool is shown in **Figure 14.2**.

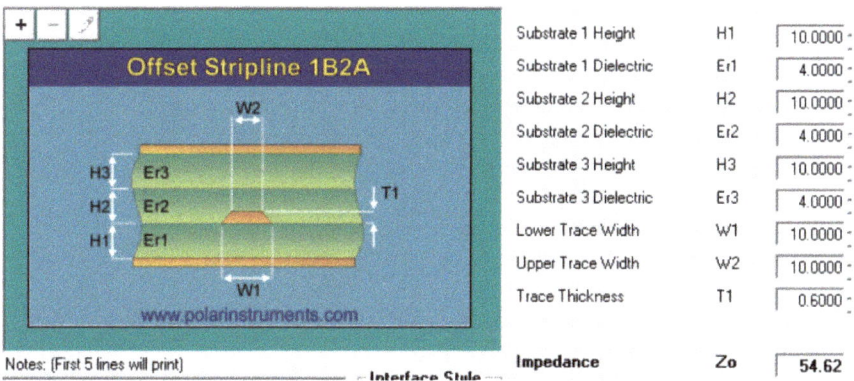

Substrate 1 Height	H1	10.0000	
Substrate 1 Dielectric	Er1	4.0000	
Substrate 2 Height	H2	10.0000	
Substrate 2 Dielectric	Er2	4.0000	
Substrate 3 Height	H3	10.0000	
Substrate 3 Dielectric	Er3	4.0000	
Lower Trace Width	W1	10.0000	
Upper Trace Width	W2	10.0000	
Trace Thickness	T1	0.6000	
Impedance	**Zo**	54.62	

Notes: (First 5 lines will print)

Figure 14.2 The nominal stackup for a dual stripline with all dielectric layers the same thickness.

We can generalize the case of a symmetrical stripline and what ratio of linewidth to dielectric thickness per layer results in a 50-ohm line. Using the SI9000 tool, we start with nominal conditions of:

- T1 = 1.2 mils (0.031 mm)
- Er1 = 4.2
- Er2 = 4.2
- H1 = H2
- W1 = W2

We sweep the dielectric thickness of each layer, keeping them the same and calculate the linewidth required for a 50-ohm line. This analysis maps out the design space for a 50-ohm symmetrical stripline.

We expect that as the dielectric thickness of each layer increases, the line width to achieve 50 ohms would also have to increase. **Figure 14.3** is the result from the Polar SI9000 tool showing exactly this behavior.

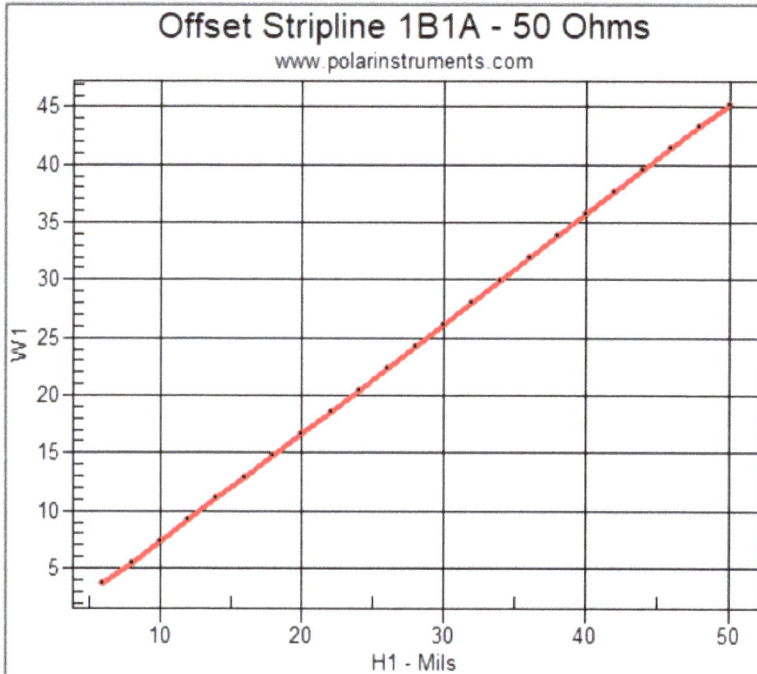

Figure 14.3 Design space for a 50-ohm symmetrical stripline. The line is the value of linewidth and dielectric thickness of each layer for a 50-ohm line.

However, when we plot the ratio of the linewidth to the dielectric thickness to achieve 50 ohms, it is not quite a constant value. The ratio of linewidth to dielectric thickness varies from about 0.6 to 0.9, as shown in **Figure 14.4**.

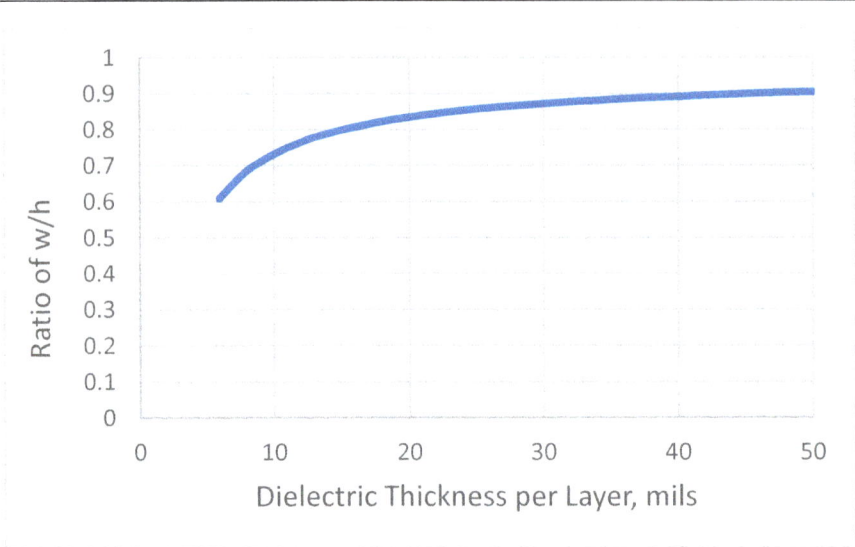

Figure 14.4 Ratio of linewidth to dielectric thickness for FR4 type laminates in a symmetrical stripline. It is about 0.75.

As a rough rule of thumb, for a symmetrical stripline, the ratio of line width to dielectric thickness of one layer is:

$$\frac{w}{h} = 0.75 \qquad (14.1)$$

Since linewidths will generally be on the narrow side, using a value of 0.75 is a great starting place.

This means that for a dielectric thickness of 10 mils per layer, the line width for 1-oz copper to achieve 50 ohms is about w = 0.75 × 10 mils = 7.5 mils (0.19 mm)

Recall that for a microstrip, the ratio was 2:1 linewidth to dielectric thickness. The presence of the top plane requires a much thicker dielectric spacing from the signal line to either plane to achieve the same impedance.

In an asymmetrical stripline structure, it is difficult to generalize a simple rule of thumb for a stackup. There are just too many parameters. This is why a 2D field solver is an essential tool to explore design space.

14.2 Comparison to Analytical Approximations

The IPC2251 specification document lists the approximation for a symmetrical stripline with the same thickness of dielectric, H1, above and below a trace with a width W1 and the trace thickness, T1, is:

$$Z_0 = \frac{60\Omega}{\sqrt{Dk}} \ln\left(\frac{4 \times (2 \times H1 + T1)}{0.67\pi(0.8 \times W1 + T1)}\right) \qquad (14.2)$$

This assumes both dielectric layers above and below the signal layer are the same thickness and the same dielectric constant, Dk.

The only approximation in the IPC specifications provided is for this simple symmetrical case. Other approximations have been published, and the Clemson University website, developed by Prof Todd Hubing and his students, offers a variety of online calculators that implement these approximations.

We can compare these approximation-based predictions with the results of a 2D field solver. **Figure 14.5** shows the closest matching structure for this geometry, the Offset Stripline 1B1A.

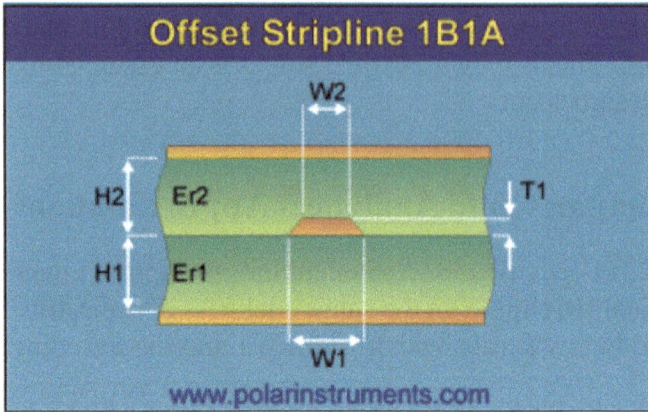

Figure 14.5 Structure in the Polar Instruments SI9000 tool that comes closest to the IPC approximation.

Using this model, we can sweep the linewidth and explore the impact on the characteristic impedance over a wide range, comparing the IPC approximation and the 2D field solver results.

We take as the nominal conditions:

- $H1$ = 10 mils (0.254 mm)
- $Er1$ = 4.0
- $H2$ = 10 mils (0254 mm)
- $Er2$ = 4.0
- $T1$ = 1-oz copper = 1.2 mils (0.031 mm)
- $W1 = W2$ = swept

This comparison is shown in **Figure 14.6**.

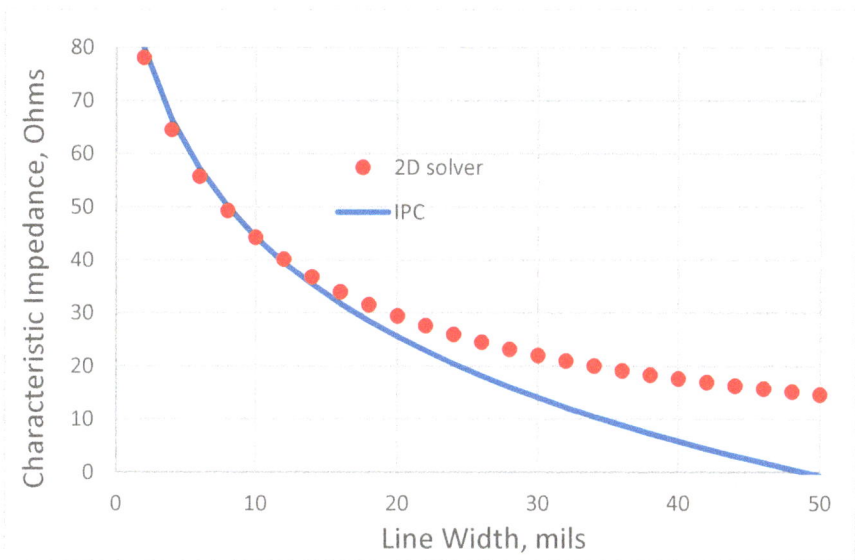

Figure 14.6 Comparing the IPC approximations with the results from a 2D field solver.

When the linewidth is narrow and the impedance is above 35 ohms, the analytical approximation is a pretty good match to the field solver. But as the line width increases and the impedance decreases, the agreement gets much worse.

This is another good reason to use a 2D field solver when accuracy is important. Of course, the IPC approximation is very limited in the structures that can be modeled.

14.3 A Design Example for Asymmetric Stripline

Usually there are two options in selecting the stack up for a multilayer circuit board;

- ✓ Using a standard stack up from the vendor
- ✓ Using a custom stack up

If you use a standard stackup, the dielectric thicknesses and Dk values of the laminates are already selected. The term to adjust to achieve the target impedance is the linewidth.

These designs are usually driven by two different requirements. If interconnect density is important, to keep the number of routing layers or the size of the board to a minimum, the narrowest linewidth is selected. In this case, you use whatever characteristic impedance for the traces that results. This is the approach for the most cost-sensitive designs.

If a specific target impedance is important, then the linewidth is selected to achieve the target impedance at the sacrifice of routing density.

Regardless of the driving force, the linewidth cannot be reduced below what the fab vendor can manufacture at either no cost adder, or an acceptable cost added.

The way to approach this design challenge is to use a 2D field solver with geometric and Dk values set by the stackup constrains from the fab vendor. The only design parameter to adjust is the line width. The linewidth is swept from narrow to wide to explore design space.

From this map of linewidth and characteristic impedance, the impedance-routing density trade off can be made.

For example, **Figure 14.7** is the standard stackup for a six-layer board from a popular internet fab shop, www.PCBway.com .

Thickness	Copper thick	Laminated chart
1.6mm±10%	1oz	

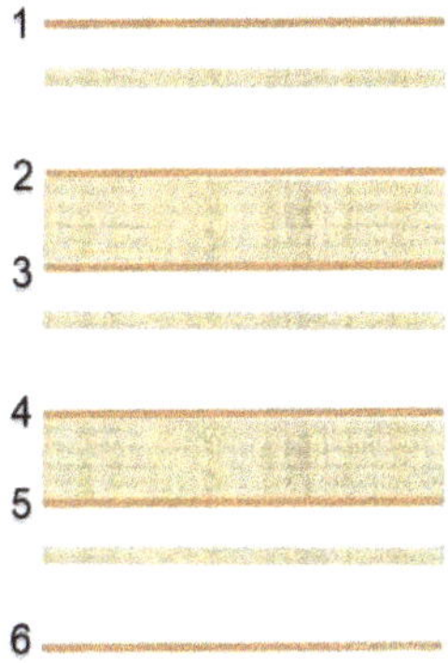

Copper   1   18 um–plating to 35um
Dielectric 1-2 0.11 mm dielectric constant 4.29
Copper   2   35 um
Dielectric 2-3 0.53 mm dielectric constant 3.96
Copper   3   35 um
Dielectric 3-4 0.11 mm dielectric constant 4.29
Copper   4   35 um
Dielectric 4-5 0.53 mm dielectric constant 3.96
Copper   5   35 um
Dielectric 5-6 0.11 mm dielectric constant 4.29
Copper   6   18 um–plating to 35um

Figure 14.7 A commonly used standard stackup for a six-layer board, used by PCBway with the layer numbers labeled. The thick dielectric layers are core and the other dielectric layers are prepreg.

In this stackup, the metal layers are counted from the top starting at 1. One approach to a stack up design is:

Layer 1: signal/power

Layer 2: ground

Layer 3: x-signal

Layer 4: y-signal

Layer 5: ground

Layer 6: signal/power

In this stackup, there are two planes both designated as ground. There are up to four signal layers with split power, but the power is not used as a return path. Two of the signal layers are microstrip on the outer surface. Two are in inner layers configured as asymmetrical stripline.

We can use a 2D field solver, such as Polar's SI9000, to analyze the impedance of the signal layer 4. Due to the symmetrical stackup, if the linewidths are the same on signal layer 3 and 4, they will have the same characteristic impedance. **Figure 14.8** is an example of the structure to use in the Polar SI9000 tools and the layer mapping.

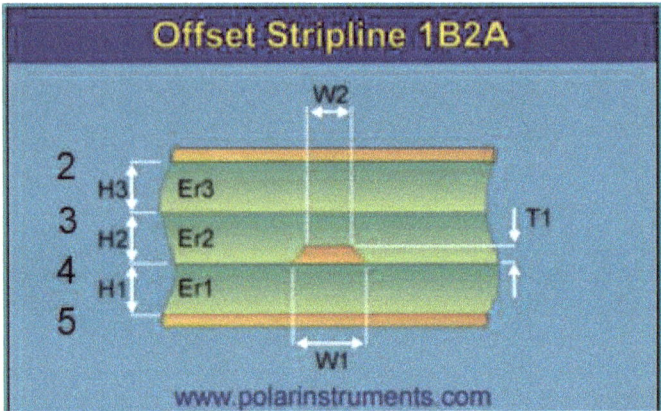

Figure 14.8 The example cross section to analyze the PCBway stackup. The top conductor is player 2, the signal layer is layer 4, and the bottom conductor is layer 5.

In this structure, the dimensions and material properties for each layer as per the data sheet are given here:

- H1 = 0.53 mm (21 mils)
- Er1 = 3.96
- H2 = 0.11 mm (4.3 mils)
- Er2 = 4.29
- H3 = 0.53 mm (21 mils)
- Er3 = 3.96
- W1 = W2, swept
- T1 = 1-oz copper = 0.035 mm (1.4 mils)

364

This problem setup is shown in **Figure 14.9**.

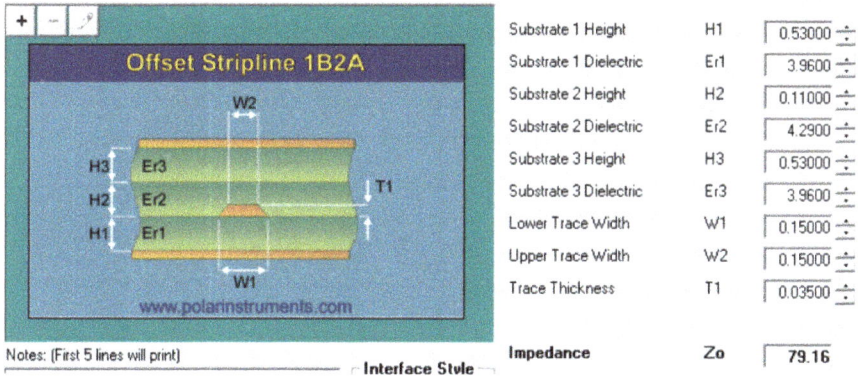

Substrate 1 Height	H1	0.53000
Substrate 1 Dielectric	Er1	3.9600
Substrate 2 Height	H2	0.11000
Substrate 2 Dielectric	Er2	4.2900
Substrate 3 Height	H3	0.53000
Substrate 3 Dielectric	Er3	3.9600
Lower Trace Width	W1	0.15000
Upper Trace Width	W2	0.15000
Trace Thickness	T1	0.03500
Impedance	Zo	79.16

Offset Stripline 1B2A

www.polarinstruments.com

Notes: (First 5 lines will print)

Interface Style

Figure 14.9 Setup in the Polar SI9000 tool to calculate the impedance on signa layers 3 and 4 for the nominal case of a 6-mil (0.152 mm)- wide trace in 1-oz copper.

For the nominal case of a 6-mil (0.152 mm) wide trace, the characteristic impedance is 79 ohms. This width is the narrowest this fab shop can do with no cost adder. It would result in the highest interconnect density at lowest cost, but would result in a relatively high characteristic impedance. If 79 ohms is acceptable, this would be a possible low-cost design.

To explore design space, the linewidth is swept, and the characteristic impedance calculated. **Figure 14.10** shows this mapping of design space.

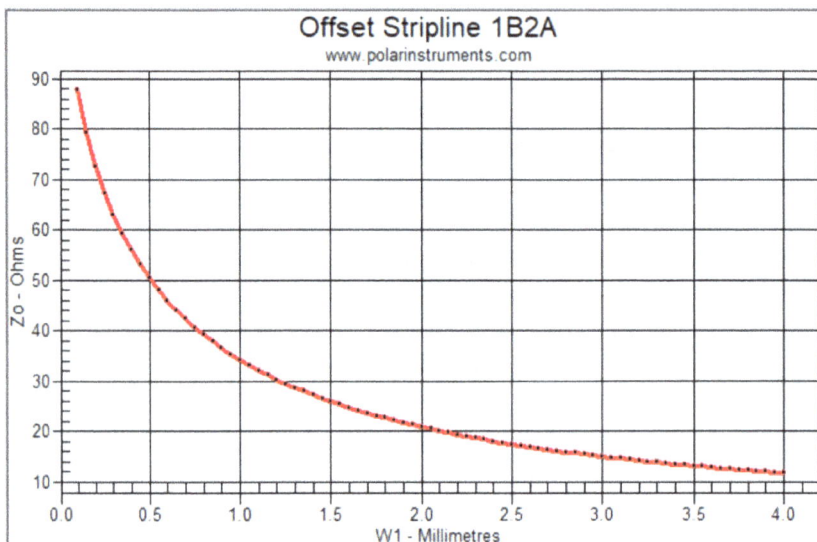

Figure 14.10 The calculated design space of the characteristic impedance as linewidth is increased. For this stackup, and 50 Ohms, a linewidth of 0.5 mm or 20 mils is required.

If achieving a target characteristic impedance of 50 ohms is important in an application, then a linewidth of 0.5 mm or 20 mils would be the required linewidth on either of the two inner routing layers. This would sacrifice interconnect density but if there are not many traces to route, it may not be an issue.

This design space is also a starting place to change the stackup.

If a target impedance of 50 ohms is required AND a 6-mil (0.152 mm)-wide trace, then some dielectric layer would have to change. Using the standard stackup, the characteristic impedance is 79 ohms for a 6-mil (0.152 mm)-wide trace.

To reduce the impedance, we clearly need to reduce a dielectric thickness.

In this stackup, layer H2 is a prepreg and is 0.11 mm or 4.4 mils thick, already rather thin. The best chance of reducing a dielectric

366

thickness is from the core layers H1 and H3. These are each nominally 0.53 mm or about 21 mils thick.

Keeping a symmetrical stackup to minimize board warpage, we can sweep design space and change the H1 and H3 layer thicknesses, keeping them the same. We sweep their thickness and explore design space to see what nominal core thickness would be required with a 6-mil (0.152 mm) wide trace to achieve 50 ohms.

The result of this analysis, using the sensitivity feature of the Polar SI9000 tool, is shown in **Figure 14.11** Characteristic impedance for the nominal case of a 6-mil (0.152 mm)-wide trace and changing both the H1 and H3 dielectric thicknesses.

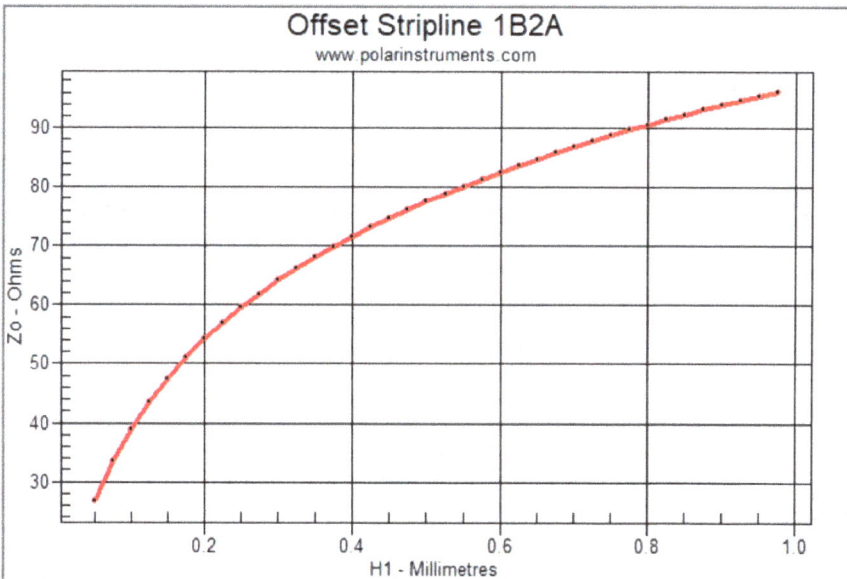

Figure 14.11 Characteristic impedance for the nominal case of a 6-mil (0.152 mm)-wide trace and changing both the H1 and H3 dielectric thicknesses.

In this example, to achieve a 50-ohm characteristic impedance and a 6-mil-wide trace would require two core layers of 0.175 mm or 7 mils each.

The standard stackup uses an H1 and H3 core thickness of 0.53 mm or 21 mils. To achieve 50 ohms at 6 mils wide, these layers would be reduced to 0.175 mm or 7 mils, a reduction of 0.35 mm or 14 mils.

If all else were kept the same, the total board thickness would be reduced by 0.7 mm (27.6 mils), from 1.6 mm (63 mils) to 0.9 mm (35 mils). If this is acceptable, then this would be a viable custom stackup.

14.4 When Does the Top Plane Not Matter?

In an asymmetrical stripline, return current flows in both the planes But, as one plane moves farther away from the signal line compared to the other plane, a smaller fraction of the return current flows in this plane. This means that at some point, the presence of the farther away plane does not affect the characteristic impedance of the signal line and it can be completely removed with no impact.

We can explore how far away the top plane needs to be so that it does not affect the impedance of the line by sweeping the top dielectric thickness H2 in an asymmetrical stripline and calculating the characteristic impedance. The setup for the nominal case is shown in **Figure 14.12**.

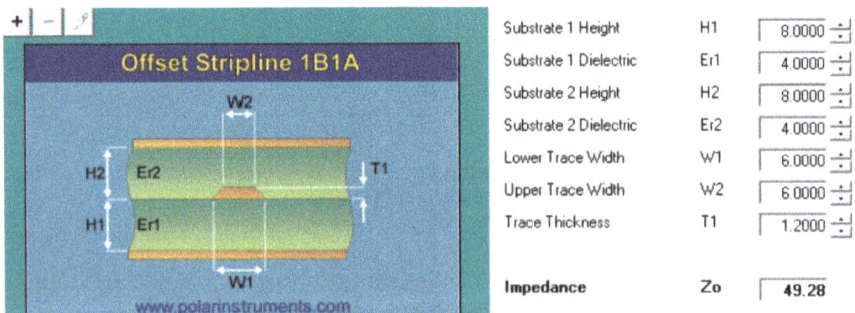

Substrate 1 Height	H1	8.0000
Substrate 1 Dielectric	Er1	4.0000
Substrate 2 Height	H2	8.0000
Substrate 2 Dielectric	Er2	4.0000
Lower Trace Width	W1	6.0000
Upper Trace Width	W2	6.0000
Trace Thickness	T1	1.2000
Impedance	Zo	49.28

Offset Stripline 1B1A
www.polarinstruments.com

Figure 14.12 Setting up the nominal case of an asymmetric stripline in a 2D field solver tool with a 6-mil-wide trace.

As the top layer thickness, H2 increases and the impedance increases, but reaches a saturated value. This behavior is shown in **Figure 14.13**.

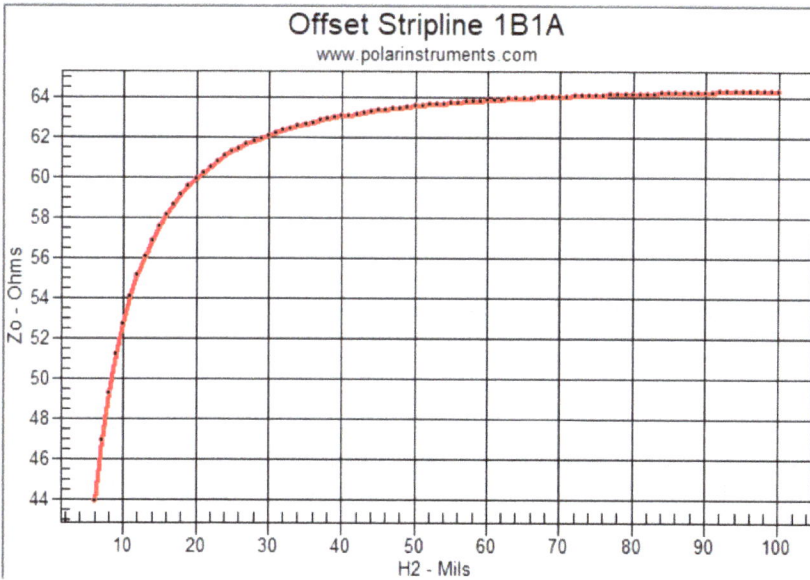

Figure 14.13 Calculated impedance of the asymmetric stripline as just the top dielectric layer increases using a nominal bottom thickness of 8 mils.

This says increase the top thickness about 40 mils and the presence of the top plane does not influence the characteristic impedance of the trace.

To see this experiment in a slightly different form, we can plot the relative percentage difference in impedance compared to the final value and the ratio of the top thickness to the bottom dielectric thickness. This replotting of the simulated result is shown in **Figure 14.14**.

369

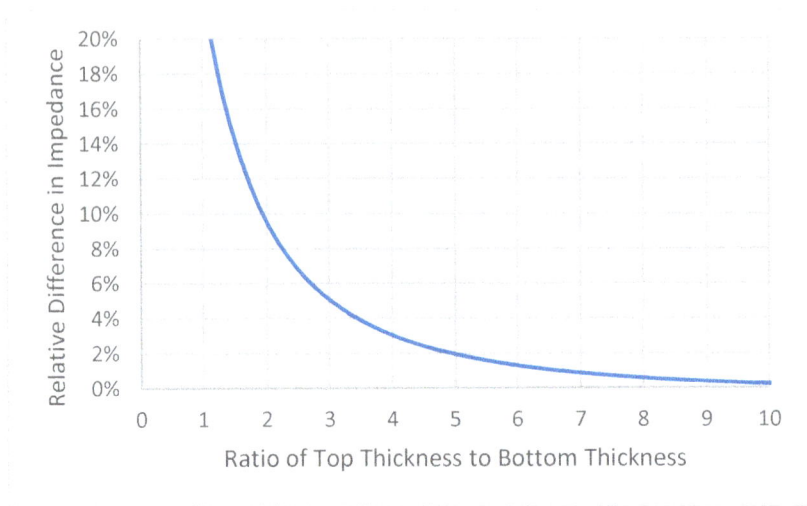

Figure 14.14 The percentage of the impedance compared to the final value as the top plane is moved away.

This says that when the top plane is about 4× farther away from the trace as the bottom plane is, the impact of the top layer is dramatically reduced, and it does not affect the characteristic impedance of the trace by more than 3%. This is a direct measure of the influence of the top plane.

This is also a rough indication of the amount of return current in the top plane. When the top plane is more than 4× farther away than the bottom plane from the trace, there is also a negligible amount of return current. At this distance away, the top plane could be a split plane or have other discontinuities and would not cause much impedance impact. This is an important observation.

14.5 First- and Second-Order Factors

For the case of a symmetrical stripline we can apply the same analysis method as in the case of a microstrip to explore the impact from first- and second-order factors influencing the characteristic impedance.

In a symmetrical stripline, there are only eight terms that influence the characteristic impedance, as identified in the setup variables for the 2D field solver. Of these, we would expect the impact on small variations in the etch back to have the smallest effect. Due to the slight asymmetry of the stackup, we would expect the upper half features H2 and Er2 to have a slightly larger impact than the lower half terms. This is because in this stackup, the trace is actually embedded in the upper half dielectric.

Using the same analysis technique as for microstrip, we can vary each term about the nominal by 10% and evaluate the change in the characteristic impedance.

The nominal condition for a 50-ohm line is shown in **Figure 14.15**.

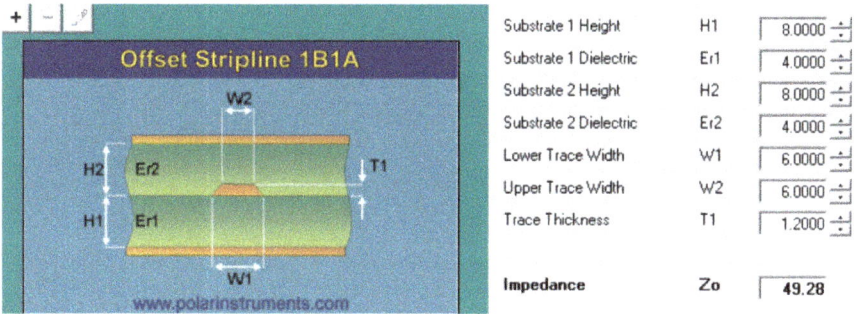

Figure 14.15 Nominal condition from which each term is varied by 10% to see the impact on the characteristic impedance.

The relative sensitivity, defined earlier as

$$\text{relative sensitivity} = \frac{\dfrac{\Delta Z_0}{Z_0}}{\dfrac{\Delta W}{W}} \tag{14.3}$$

for each term, is

371

H1:	0.2
Er1:	-0.2
H2:	0.31
Er2:	-0.27
W1 = W2	0.41
W1 – W2	-0.04 (if etch back is only 1 mil)
T1	-0.14

The term with the largest sensitivity is the linewidth. Each of the other terms have half the sensitivity or less except for the etch-back, as expected. For each design parameter, a 10% change in any term results in less than a 3% variation in the impedance, except for the linewidth.

14.6 Review Questions

1. If the dielectric thickness is 20 mils in each layer of a symmetric stripline, roughly what is the line width for a 50-ohm line?

2. What is one of the assumptions about the geometry in the IPC approximation for stripline?

3. What is the advantage of using the narrowest linewidth the vendor can fabricate?

4. How far away should the top plane be from the trace compared to the bottom plane in order to not have much return current in the top plane?

5. What are the most important terms influencing the characteristic impedance of a stripline?

Chapter 15 Differential Signaling and Differential Impedance

So far, all the interconnects considered have been single-ended. The signals propagating on the interconnects have been between the signal and the return path. In this chapter, we expand the type of signals to include differential and common signals.

There are two very different applications for differential pairs with very different design goals: transporting low voltage, noise sensitive analog signals and transporting very high-speed, higher-voltage digital signals.

We first consider the use of differential pairs for low bandwidth, noise-sensitive analog signals and then high-speed digital applications.

15.1 Differential Pairs for Low-Noise Analog Signals

All voltages are always a *voltage difference* between two specific points. But we often refer to the voltage at just one point in a circuit. Even then, there is always a second point from which we are measuring the voltage difference. Often, this reference point is not well identified, except by its name, ground.

We never measure the voltage at a point. We *only* measure the voltage difference between two points. This is why all amplifiers are really differential amplifiers in that they are measuring a voltage difference between two points. When we measure a single-ended signal, the second point is the local reference point that we call ground. In a single-ended amplifier, this ground reference point is one of the pins of the IC and is connected to the conductor on the circuit board labeled ground right at the IC pin.

There is nothing special about a ground point except that we all agree to use this point in the circuit as the reference conductor from which to measure the voltage to all other points.

Here are four important types of ground, each with a different preface and a different use.

Earth ground is a conductor that is tied with a low resistance path to a copper pipe stuck into the literal ground, buried at least 3 feet deep. This is an important safety feature. Most residential and commercial buildings have building codes that specify how large a pipe and how deep it is set in the ground.

In the electrical power wiring in your home with three-hole plugs in the wall, the bottom round hole is directly connected to earth ground. If the outside of any appliance is also connected to this earth ground, then all metal surfaces that could be touched will be at the same earth potential and there is less chance of a user being electrocuted.

Chassis ground is the connection to the metal enclosure of an electronic product. For safety, the chassis ground should be connected to earth ground. This is an Underwriters Laboratory (UL) requirement for many products. This reduces the chance for a user to be electrocuted using the product.

If a device is NOT connected to earth ground, we usually refer to this product as floating. When the AC power is transformer coupled inside the product and NOT connected to earth ground, it usually is floating. This can be a good or bad feature depending on the application. It can also be potentially dangerous.

Within a circuit, there is often identified a *digital ground* conductor. This is a little bit of a misnomer as it should be called the digital return current conductor. In addition, digital ground is also often used as the reference point for digital signals. Digital ground is usually a ground plane on the circuit board.

In some circuits, there is another ground labeled as *analog ground*. This is the reference point to which all analog voltages are measured. It also carries the return current of analog signals, but these currents are generally much smaller than digital signals.

There is also the ground associated with the power return, but it is not distinguished from digital or analog ground.

Except in a few very specialized situations, the analog ground and digital ground should be the same solid, continuous plane on one or more layers of the circuit board.

In a circuit board, even with a solid and continuous ground plane, there may be a voltage difference between two different locations on the ground plane. Even though we call every point on this plane the same name and use any point on the plane as an equivalent reference point, the voltage between one point and a farther away point on the same plane may vary.

This occurs for two reasons. First, when large DC currents flow in the return plane, as from power returns or even when signals carry a lot of current, these currents flowing through the small but finite resistance in the planes will cause an IR voltage drop.

For example, the sheet resistance of a 1-oz copper plane is 0.5 mohms/square. The resistance between two locations a few inches apart could be a milliohm. With 1 A of current flowing through these two points, the voltage between the two points on the same ground plane a few inches apart could easily exceed 1 mV.

If a sensor signal, such as from a microphone, is connected at one end of a board and its signal output is routed to the input of an amplifier at another location, the 1 mV of noise in the ground path would be in series with the signal voltage. **Figure 15.1** illustrates this example.

Single-ended Voltage Measurement

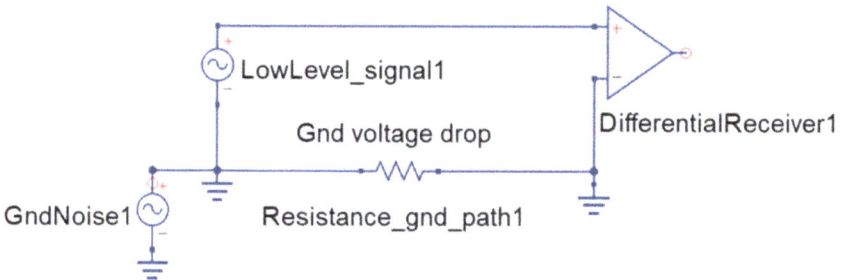

Figure 15.1 An example of a sensor signal generated remotely with IR drop noise in the return path back to the amplifier.

In a 12-bit ADC, there are 4095 different voltage levels. If the dynamic range is 0V to 3.3.V, then each level is 3.3V/4095 = 0.8 mV. Any ADC with a bit sensitivity of 12 bits or larger will be sensitive to this level of IR drop noise. When the signal levels are in the few uV level, the IR drop on the ground planes can completely swamp the signals.

In addition to the IR noise, which will fluctuate with the DC current from the power returns or digital returns, there is also the possibility for voltage differences from one region of the ground plane to another, due to the loop mutual-inductance of multiple signal-return paths and the sensor's signal-return path loop. We sometimes refer to this as LdI/dt noise since it is related to an inductance and the transient or changing currents. It is also a type of *ground bounce*.

The ground bounce noise in a plane can sometimes be more than 10× larger than the IR drop noise.

While one fix for this problem is to use a separate ground path for the sensor, a much cleaner solution is to carry the lower voltage side of the sensor in a separate conductor and NOT use the local ground plane as the reference.

This routing is illustrated in **Figure 15.2**. Here, the two connections from the sensor are routed in separate conductors back to the differential receiver. The differential receiver will measure the voltage difference between the two ends of the sensor and NOT include the ground plane noise. We call these two conductors from the sensor to the receiver a *differential pair*.

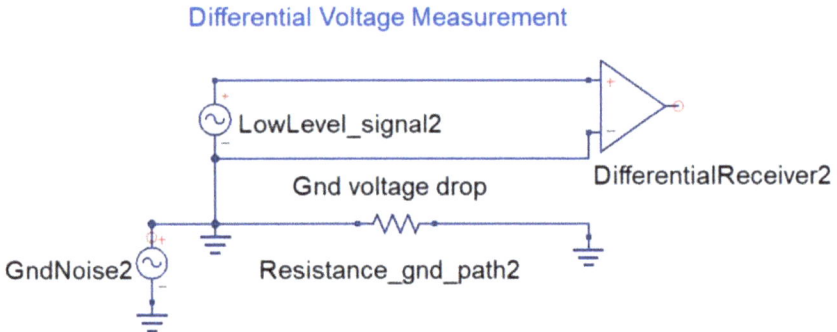

Figure 15.2 Routing a sensor signal with a differential pair back to the differential receive avoids the noise on the ground plane.

A differential pair is nothing more than two conductors routed as signal traces. In this application of carrying low-level signals the goal in routing these traces is to reduce the noise picked up between them. This usually means routing the traces close together so they see the same local electrical environment.

This application of transporting low-level signals is generally low current and low bandwidth and the characteristic impedance of either line is not important. The requirements on the differential pair design to carry low-level, low-bandwidth analog signals are not important, other than routing the traces close together.

Another application for differential pairs is when the signal is a higher-level, higher -bandwidth digital signal. In this case, the electrical properties of the interconnect are very important.

15.2 Differential and Common Signals in High-Speed Digital Signaling

All high-speed serial links, such as LVDS, USB, PCIe, SATA, Ethernet, and HDMI, use differential signaling. This means a single bit of information is carried as the difference in voltage between two traces from the transmitter to the receiver.

The bit level is determined by the difference in voltage between the signals on the two traces. This is illustrated for a low-voltage differential signaling (LVDS) interface in **Figure 15.3**.

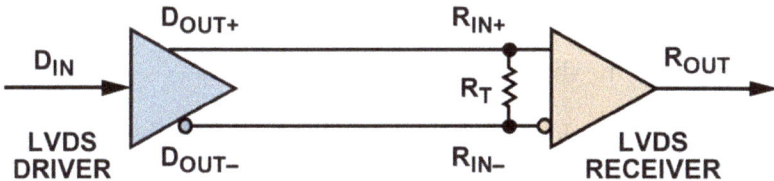

Figure 15.3 An example of the connections for LVDS between a transmitter and receiver.

There are generally four important benefits of differential signaling over single-ended signaling:

1. A differential receiver is more sensitive than a single-ended receiver.

2. A differential receiver can have a larger noise margin than a single-ended receiver.

3. The net power rail current from a differential transmitter is mostly constant and does not generate as much ground bounce as a single-ended transmitter.

4. The differential signal is less sensitive to noise on the ground plane and to ground plane discontinuities such as vias and connectors.

379

Each signal and return path that carries the differential signal between the transmitter and receiver is a *single-ended transmission line*. When two of them are used to carry a differential signal, we call the two, single-ended transmission lines *one differential pair*.

To distinguish the two lines, we often refer to one line as the p for positive line and the other as the n for negative line. In some documents these are p for plus and m for minus.

The p- and n-lines are literally just single-ended transmission lines each with a signal and return path. This means there is a voltage on each between their signal line and the return. The difference between the voltage on the p- and n-lines, at the transmitter and at the receiver, is the differential signal.

For example, in an LVDS signal, the output voltage between the p-line and its return path is roughly a swing between 1.05 V to 1.35 V. At the same time, the voltage between the signal and return connection on the n-line is 1.35 V to 1.05 V. The output voltage on each of the p and n-lines with respect to their return path is shown in **Figure 15.4**. These are the single-ended voltages that would be measured on the two lines on an LVDS driver with respect to their return paths.

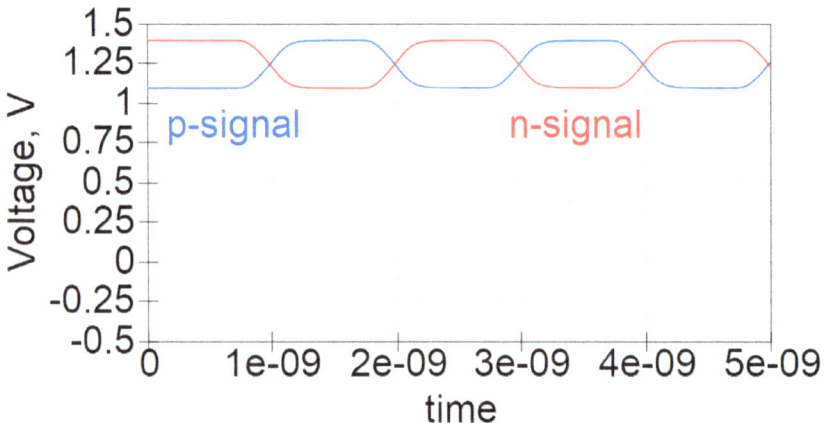

Figure 15.4 Simulated voltages on the p- and n-lines from an LVDS driver at 1 Gbps.

These are the voltages on each individual transmission line that make up the differential pair. While we usually pay attention to the *difference* voltage or the *differential* signal, there is also another signal, the *common* signal.

Any two arbitrary signals on two different transmission lines can be separated or decomposed into a differential signal component and a common signal component. Given two voltages on two different transmission lines, V1 and V2, the differential and common signal components are defined by:

$$V_{diff} = V1 - V2$$
$$V_{comm} = \frac{1}{2}(V1 + V2)$$

$$(15.1)$$

The differential signal component is the difference voltage. The common signal component is the average voltage. Both are present and independent of each other. From the differential voltage and common voltage, we can also back out the V1 and V2 signals from:

$$V1 = V_{comm} + \frac{1}{2}V_{diff}$$
$$V2 = V_{comm} - \frac{1}{2}V_{diff}$$

$$(15.2)$$

A differential receiver is sensitive to the difference voltage. It is much less sensitive to the common signal. All differential receivers have a figure of merit that describes how less sensitive they are to the common signal, referred to as the common mode rejection ratio (CMRR).

This term, usually in dB, describes how much common signal at the input would be recorded as differential signal component. For example, a CMRR of 40 dB means 1% of the common signal would be converted into differential signal. A common signal modulation of 0 V to 1 V would result in a recorded differential signal of 1% x 1 V or 10 mV. Generally, the CMRR decreases with higher frequency.

In the LVDS signal, the differential signal fluctuates between -0.3 V to +0.3 V. At the same time the common signal component is constant at 1.25 V. **Figure 15.5** shows the individual voltages of the two signals and the extracted differential and common signal components.

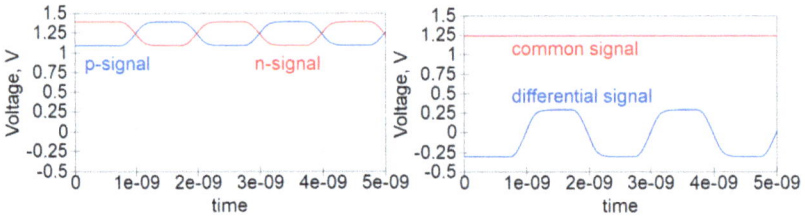

Figure 15.5 Two representations of the same signals: the single-ended signals and the decomposed differential and common signal components.

When the signals on two transmission lines have no common component, Vcomm = 0, we call this signal a pure differential signal. For the common signal to be 0 means the average is zero. A pure differential signal has an average of 0V.

Likewise, a pure common signal means no differential signal component, Vdiff = 0. This means the V1 and V2 voltage are identical. A pure common signal has exactly the same voltage on the two signal lines.

15.3 Differential and Common Impedance with Traces Far Apart

Each of these signals, the differential and the common signals, propagate on the differential pair transmission line independently and see a different electrical environment.

When the two lines that make up the pair are the same cross section, we call this a *symmetrical* or *balanced* differential pair transmission line. The analysis can be simplified in this case.

If the two transmission lines that make up the differential pair are not the same cross section, we refer to them as an *asymmetric* or *unbalanced* transmission line.

There can still be a differential or common signal on an unbalanced transmission line, but the analysis of the electrical environment they see is more complicated.

In the following analysis we will assume the two transmission lines that make up the differential pair are both identical transmission lines, or a balanced differential pair.

> *Generally, the electrical properties of the differential pair will always be optimized to transport a differential signal if we use a balanced differential pair. We always want to engineer a balanced differential pair for the best performance.*

The instantaneous impedance the differential signal sees as it propagates down the differential pair is the differential impedance. The instantaneous impedance the common signal sees as it propagates down the differential pair is the common impedance.

When the two transmission lines that make up the differential pair are far apart and there is no coupling between them, the differential signal sees the two independent single-ended transmission lines.

As an example, assume each transmission line is 50 ohms and is driven from the end by a pure differential signal. This means if one signal turns on from 0 to 1V, the other signal must turn on from 0 to -1V so the average is 0. As a pure differential signal, there is no common component and the average voltage is 0.

At the entrance to the transmission lines, the p signal sees a 50-ohm transmission line and the n signal sees a 50-ohm transmission line.

When the differential signal drives the front of the differential pair, the voltage between the two transmission lines creates a large electric field between the traces. Both the p- and n-signals, and the differential signal, propagate down the transmission line. This condition is illustrated in **Figure 15.6**.

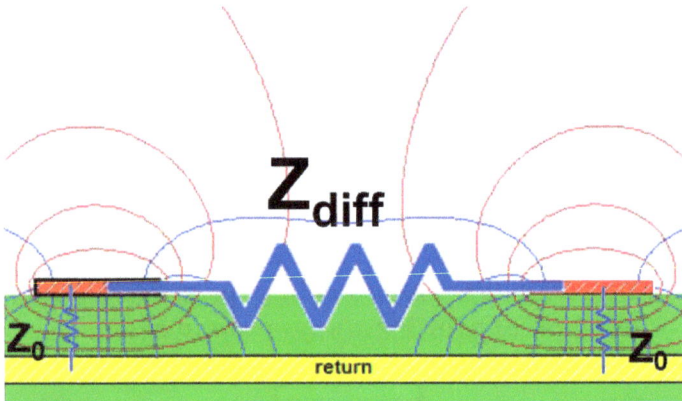

Figure 15.6 A differential pair with traces far apart, driven by a differential signal.

The impedance the differential signal sees is the impedance the signal sees between the p and the n lines. Since these two transmission lines are far apart and there is no coupling between them, the p and the n signals each see just the characteristic impedance between their signal line and the return, Z0.

The impedance between the two signal lines, what the differential signal sees, is the series combination of the impedance between the p-line and its return, connected through the return plane and going up through the impedance of the n-line to the top conductor.

This means the differential impedance the differential signal sees the impedance between the p and the n-line, is just $2 \times Z0$. If the single-ended impedance is 50 ohms, the differential impedance of the pair is 100 ohms.

With no coupling this is a simple analysis. The differential impedance is Zdiff = 2 × Z0.

If a pure common signal drives the pair, this means there is the same voltage on either p- or n-line with respect to the return path. The two single-ended signals on the two isolated transmission lines are in parallel.

The impedance the common signal sees is the parallel combination of the two single-ended impedances of the two p- and n-lines in parallel. This is ½ × Z0. This is illustrated in **Figure 15.7**.

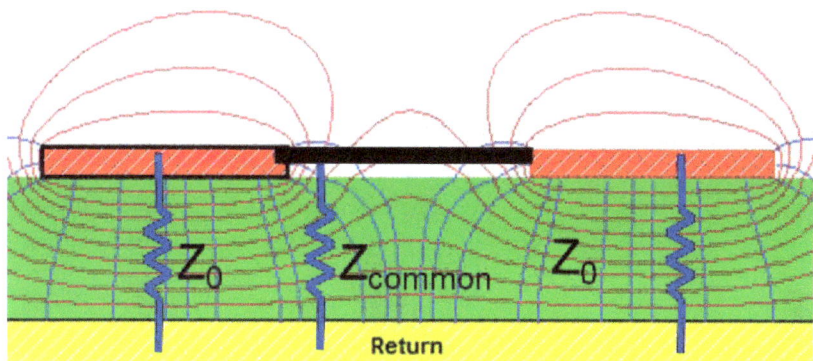

Figure 15.7 When the traces are far apart and there is no line-to-line coupling, the impedance the common signal sees is the parallel combination of the two single-ended impedance of either line.

In the absence of coupling the differential impedance and common impedance are:

$$Z_{diff} = 2 \times Z_0$$
$$Z_{comm} = \frac{1}{2} \times Z_0$$

(15.3)

385

We can confirm these simple results using a 2D field solver.

Watch this video and I will walk you through the fundamental principle of what differential impedance means.

15.4 Calculated Differential and Common Impedances with No Coupling

We construct a simple single-ended microstrip transmission line that is close to 50 ohms. **Figure 15.8** shows the nominal cross-section features. The single-ended characteristic impedance is 49.79 ohms.

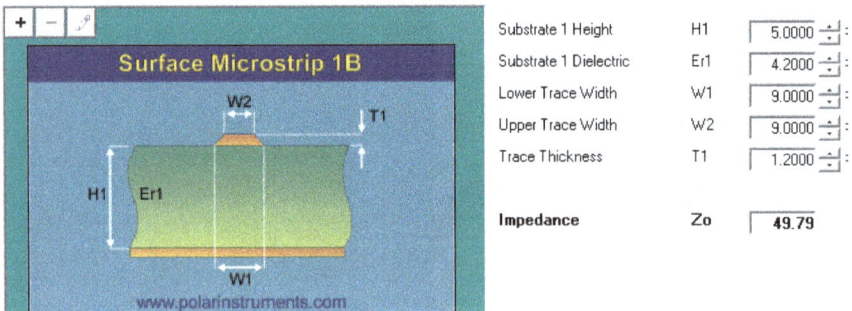

Substrate 1 Height	H1	5.0000 :
Substrate 1 Dielectric	Er1	4.2000 :
Lower Trace Width	W1	9.0000 :
Upper Trace Width	W2	9.0000 :
Trace Thickness	T1	1.2000 :
Impedance	**Zo**	**49.79**

Surface Microstrip 1B

www.polarinstruments.com

Figure 15.8 Construction of a single-ended microstrip transmission line of about 50 ohms.

We would expect if two of these transmission lines make up a differential pair and they were far enough apart so coupling between them does not play a role, the differential impedance will be $2 \times 49.79 = 99.58$ Ohms.

In **Figure 15.9**, these two, 9-mil-wide transmission lines are combined with a spacing of 100 mils. This is $20 \times$ the dielectric

386

thickness and results in very little line-to-line coupling. The calculated differential impedance is 99.50 ohms. This is precisely what was expected.

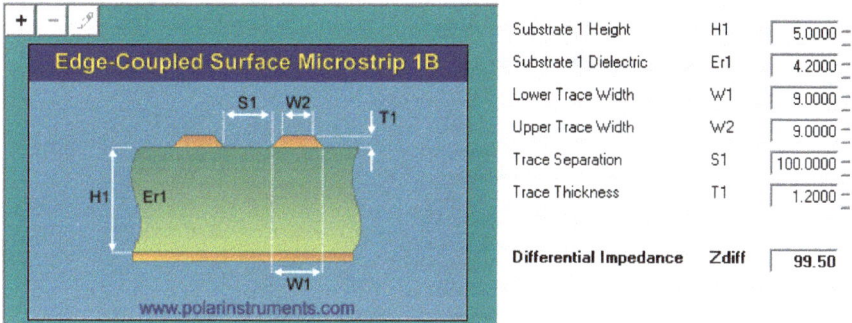

Figure 15.9 Setup for the differential impedance of two single-ended transmission lines showing 99.50 ohms.

From the single-ended impedance of 49.79 ohms, the expected common impedance is 24.9 ohms.

To also display the common impedance in the Polar SI 9000 field solver tool,, it is necessary to press the more button under the calculate button. The calculated common impedance for this differential pair is shown in **Figure 15.10** as 24.99 ohms, again, matching what is expected.

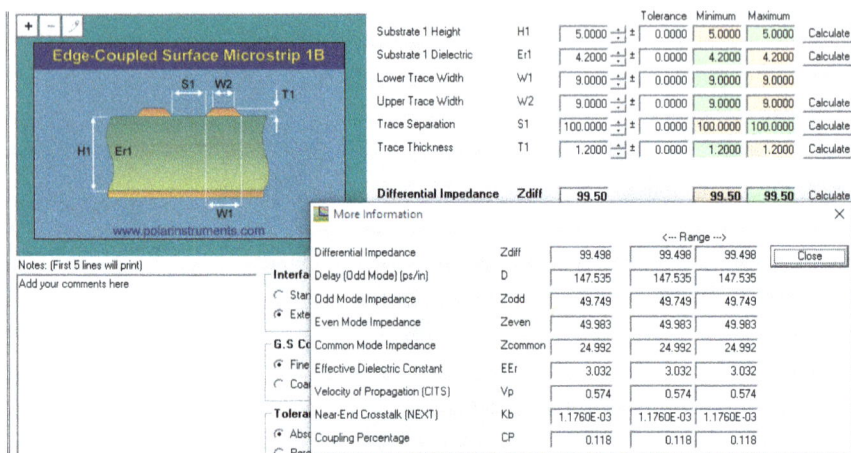

Figure 15.10 The output results from the SI9000 tool under the more button showing a common impedance of 24.99 ohms.

When there is no coupling between the two transmission lines, calculating the single-ended impedance, the differential impedance, and the common impedance is very easy. The challenge is when the traces come closer together and there is coupling.

15.5 Displacement Current and the Origin of Impedance

Fundamentally, impedance is always the ratio of voltage to current:

$$Z = \frac{V}{I} \qquad (15.4)$$

When a signal is launched into the beginning of the transmission line, as from a TDR step voltage, the V is the applied voltage, and the I is the current that flows from the signal to return path.

There is an insulating dielectric between the signal and return paths. How does current flow between the signal and return path to create the instantaneous impedance at the beginning of the transmission line?

The answer is through displacement current. This is the current introduced by Maxwell to provide continuity of current. Displacement current, in contrast to conduction current, is not the motion of free charges, but is fundamentally a different type of current, just as real and just as capable of creating magnetic fields and displaying all the other properties of conduction current.

Maxwell defined displacement current as:

$$I_{displacement} \sim \frac{dE}{dt} \tag{15.5}$$

It is directly proportional to how fast the electric fields change. The larger the rate of change of the electric field lines, the larger the displacement current.

It is through the electric field lines between two conductors, due to a voltage difference between them, that displacement current will flow when the fields change. When you see the electric field lines between the signal and return conductors in a microstrip, as in **Figure 15.11**, think of the displacement current flowing along these field lines, when the voltage changes, which is when the field line strength changes.

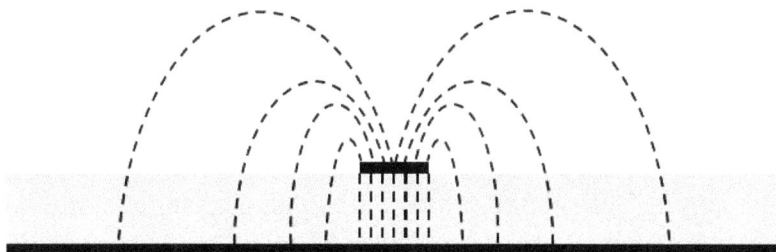

Figure 15.11 Exaggerated fringe electric field pattern between the signal and return conductors with a single-ended signal applied. It is through these fringe field lines displacement current will flow when the voltage changes.

The field lines define the path the displacement current takes, if the voltage were to change. The more field lines, the more the displacement current. The faster the dV/dt, the more the displacement current.

In this representation, it's possible to visualize where the current will flow between the signal and return conductors by looking at the electric field distribution.

Displacement current will flow along ANY changing electric field lines. If there is a voltage difference between the signal line and any other conductor, such as an adjacent signal line, there will be displacement current when these field lines change.

It is by displacement current that we can analyze the current flowing from one signal line when it is part of a pair and the other line in the pair has a different voltage pattern on it. All that is necessary is to visualize the electric field lines between the conductors. The displacement current, which defines the impedance, will flow along these electric field lines.

15.6 Impedance of One Line When Part of a Pair

The differential impedance of a differential pair is the series combination of the impedance of one line to the return path, when the differential pair is driven by a differential signal.

Likewise, the common impedance of a differential pair is the parallel combination of the impedance of each line when the pair is driven by a common signal.

To evaluate the impact on the differential or common impedance from coupling, we need to evaluate what happens to the impedance of one line as the pair is moved closer together.

To begin the analysis, we start with a differential pair very far apart and look at the impedance of one line as the other is brought closer and closer. We consider three cases based on how the pair is driven.

- ✓ Case 1: the second line is kept low (grounded)
- ✓ Case 2: the second line is driven opposite (a differential signal is applied to the pair)
- ✓ Case 3: *the second line is driven the same (a common signal is applied to the pair)*

15.7 Case 1: The Second Line is Kept Low

In the first case, we will look at the impedance of the first line when the second line is brought in from far away and it is tied low, connected to ground. We assume the line width is 5 mils and the other dimensions are adjusted to make a 50-ohm transmission line. This is illustrated in **Figure 15.12**.

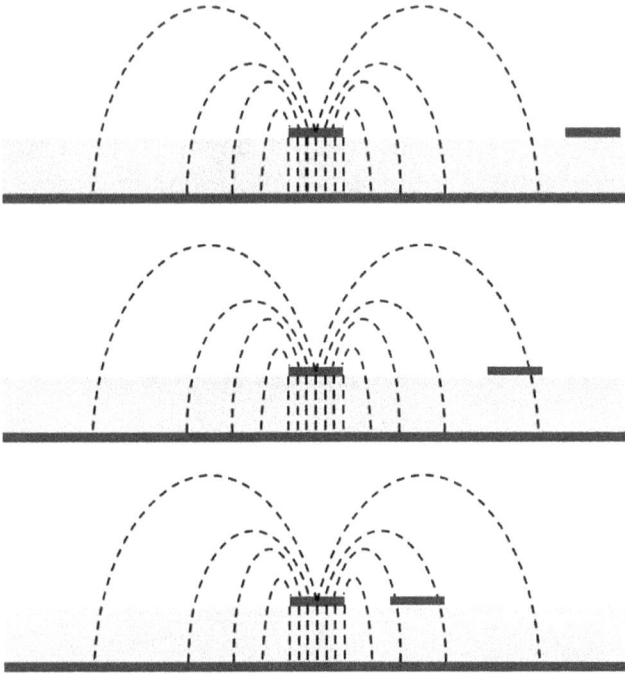

Figure 15.12 Looking at the impedance of one line when the other line is tied to ground and brought closer from far away. The fringe electric field lines are exaggerated. Top to bottom; the second line, on the right is brought closer and closer to the first line whose impedance we are measuring.

What would happen to the impedance of the first line when the second line is brought closer?

Here is a simple way of analyzing the impedance of the first line. To measure the input impedance of the first line, we can send a step voltage pulse in and look at the current that flows between the signal and return path. The ratio of the voltage to the current is literally the instantaneous impedance.

The current that flows due to the initial dV/dt is the displacement current that flows through the changing electric field lines from the signal line to the return plane. The more electric field lines that change, the more the current and the lower the impedance.

392

When the second trace is far away, the impedance of the first trace is due to the displacement current flowing through all the field lines under the signal trace and the fringe field lines that go from the signal line to the ground plane below.

As the second trace is brought closer, it begins to interact with the fringe field lines. The field lines that were going to the return plane are now going to the second trace. But the second trace is also connected to ground. This means the displacement current is mostly unchanged. It will still flow from the signal trace to the return plane, even as the second trace gets closer.

It is not until the second trace is very close that it begins to change the fringe field distribution and add more fringe field lines to the signal trace.

This analysis suggests that as the second trace, tied to ground, is brought in from far away, it will not change the impedance of the first line until it is very close and then will start to increase the field lines to ground, increase the displacement current, and decrease the impedance of the first line.

Since this impedance is due to fringe field coupling, the only way to calculate the impact on the impedance of the first line from the proximity of the second line is using a field solver. **Figure 15.13** shows the calculated impedance of the first line as the second line is brought in from far away.

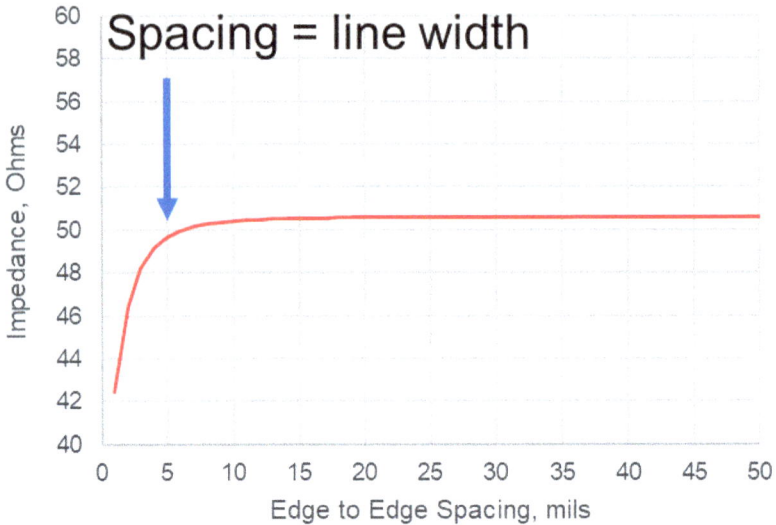

Figure 15.13 Calculated impedance of the first line, as the second line is brought closer and is tied to ground. Simulated with the Polar SI9000 field solver.

This calculation shows that the impedance of the first line is very constant and unaffected by the presence of the second line when it is tied low, even when it approaches a spacing equal to the linewidth. The impact on the characteristic impedance of the first line is less than 1.5% when the spacing is as close as a linewidth.

Of course, when the edge-to-edge spacing gets very small, less than the 5 mil of a trace width, the impedance does start to decrease, as expected.

This is mapping the behavior of the single-ended impedance of one line when a second line, grounded, is brought in from far away. Basically, there is no change in its impedance as long as the spacing to the other line is farther than a linewidth.

Watch this video and I will walk you through the impedance of one line when the second line is tied low.

15.8 Case 2: The Second Line is Driven with an Opposite Signal

In the second case, we apply an opposite signal to the second line. In other words, if we apply a 0 to +1V to the first line, we apply a 0 to -1V signal to the second line. This is a differential signal applied to the pair.

However, we are still only looking at *the impedance of the first line*, when it is part of a pair and the pair is driven with a differential signal.

As the second line is brought in from far away, it will begin to see the fringe field lines. The displacement current between the signal line and the return path travels on the fringe field lines. As the second trace, which is driven opposite, sees these fringe field lines, some of the fringe field lines will be intercepted by the trace and the displacement current will flow to the second trace.

But the displacement current is related to how fast the electric fields are changing. Those fringe field lines that go between the first trace and the second trace will see $2 \times dV/dt$ since the voltage difference between the two line is a +1V from the first line and a -1V on the second line. This means all the displacement current from fringe field lines going to the second trace will *double* the displacement current that would be there if the fringe fields just went to the ground plane.

As the second trace approaches the first trace, the displacement current from the first trace and the ground will increase. For the same voltage, there is more current, which means the impedance of the first line decreases.

As the second trace is brought closer from far away and the pair is driven with a differential signal, the impedance of the first trace will decrease. Exactly how much is hard to estimate since it is all about the fringe fields. But using a field solver we can calculate this decrease in impedance of the first line. **Figure 15.14** shows the results from the SI9000 field solver.

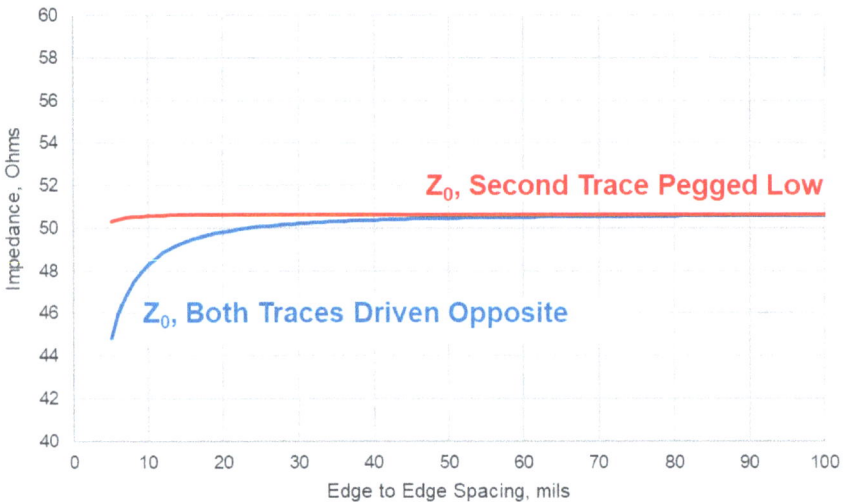

Figure 15.14 Calculated impedance of the first line as the second line is brought closer from far away, and driven opposite, in blue, and grounded in red.

When the spacing between the two lines is equal to the linewidth, 5 mils, the impedance of the first line has decreased to 45 ohms, when the second line is driven opposite, and the pair is driven with a differential signal.

Watch this video and I will walk you through what happens to the impedance of the first line when the pair is driven with a differential signal.

15.9 Case 3: The Second Line is Driven with the Same Signal

In the third case, the second line is driven with the same voltage as the first line. This means the differential pair is driven with a common signal.

As the second line is brought closer, it begins to interact with the fringe field lines from the first line. However, because the two lines have a pure common signal on them, their voltages are exactly the same. This means there cannot be an electric field between them.

The impedance when the second line is far away is due to the displacement current flowing through the fringe electric field lines. When the second line moves closer, its presence turns off the electric field lines from the signal line to where the second line is positioned. This means there are less electric field lines from the first line to ground and less displacement current.

As the second line approaches the first line, it blocks some of the fringe electric field lines, decreases the displacement current, and increases the impedance of the first line.

The closer the second line comes to the first line, the more fringe electric field lines are blocked and the higher the impedance it will have. The increased impedance of the first line as the second line approaches, when they are driven with a common signal, can be calculated using a 2D field solver. The result is shown in **Figure 15.15**.

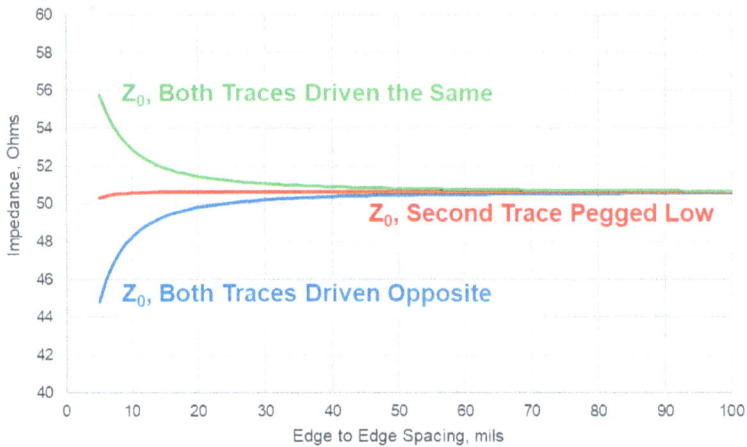

Figure 15.15 The three impedances the first line has depending on how the pair is driven.

In each case, we have looked *only* at the impedance of the first line as the second line is brought closer from far away. We found that the first line has three different impedances depending on how the pair is driven.

It is the same line in all three cases, yet its impedance depends on the proximity of the other line and how the pair is driven.

- ✓ If the second line is connected to ground, so that we are applying a single-ended signal to the first line when we measure its impedance, such as with a simple single-ended TDR measurement, we see the impedance of the first line is relatively constant.

- ✓ If the pair is driven by a differential signal, the impedance of the first line decreases as the second line is brought closer.

- ✓ If the pair is driven with a common signal, the impedance of the first line increases as the pair is brought closer together.

398

We need a clear and concise set of labels to call each of these three different impedances the first line has. We leverage the terms already used in RF applications based on *modes*.

Watch this video and I will walk you through what happens to the impedance of the first line when the pair is driven with a common signal.

15.10 Odd and Even Modes and Impedance

In RF applications, we describe the state a symmetrical, balanced differential pair based on the pattern of the electric fields on it.

Earlier in this chapter, we saw that any two arbitrary voltages, V1 and V2, on the two transmission lines that make up a differential pair can be described in terms of a differential signal component and a common signal component.

When the balanced differential pair is driven by a pure differential signal component, there are larger electric fields between the two lines. In RF applications, this state of the differential pair, when driven by a differential signal component, is called the *odd mode* of the differential pair.

We say the differential pair is driven in the odd mode when a differential signal is applied.

When the balanced differential pair is driven by a common signal component, there are fewer electric field lines between the two transmission lines. In RF applications, the state of the differential pair is called the *even-mode* of the differential pair.

We say the differential pair is driven in the even mode when a common signal is applied. The electric fields for the same differential pair, driven in the odd and even modes, is shown in **Figure 15.16**.

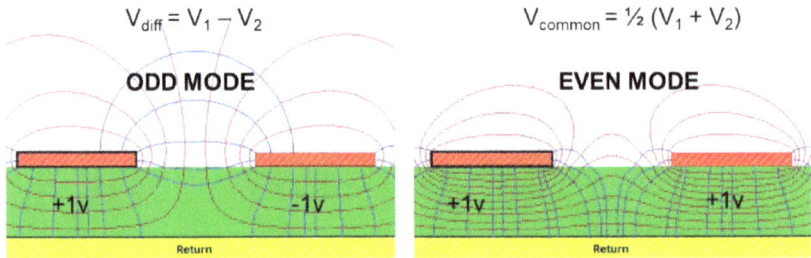

$$V_{diff} = V_1 - V_2 \qquad V_{common} = \tfrac{1}{2}(V_1 + V_2)$$

ODD MODE   **EVEN MODE**

+1v   -1v   +1v   +1v

Return   Return

Figure 15.16 The state of the same differential pair when driven in the odd-mode by a differential signal (left) and in the even mode by a common signal (right).

In RF applications, we refer to the impedance of each line in the balanced differential pair depending on the mode of the transmission line.

When driven by a differential signal in the odd mode, the impedance of either line to the ground plane in the differential pair is labeled as the *odd-mode impedance*.

When driven by a common signa in the even mode, the impedance of either line of the differential pair to the ground plane is labeled as the *even mode impedance*.

In signal integrity applications, we adopt these labels to describe the three impedances of one line in a differential pair depending on how the pair is driven.

In the example of calculating the impedance of one line when the second line was brought closer and changing the way the pair was driven, we were really calculating the single-ended impedance, the odd-mode impedance. and the even-mode impedance of the first line.

✓ When the second line is driven by ground, the impedance of the first line is called the single-ended impedance.

✓ When the pair is driven by a differential signal, the impedance of the first line is the odd-mode impedance.

✓ When the pair is driven by a common signal, the impedance of one line is the even-mode impedance.

The behaviors we identified for one line in the pair when driven with different signals can be labeled using the correct terms of single-ended impedance, odd-mode impedance, and even-mode impedance. Their behaviors, as the second line in the pair is moved closer, are relabeled in **Figure 15.17**.

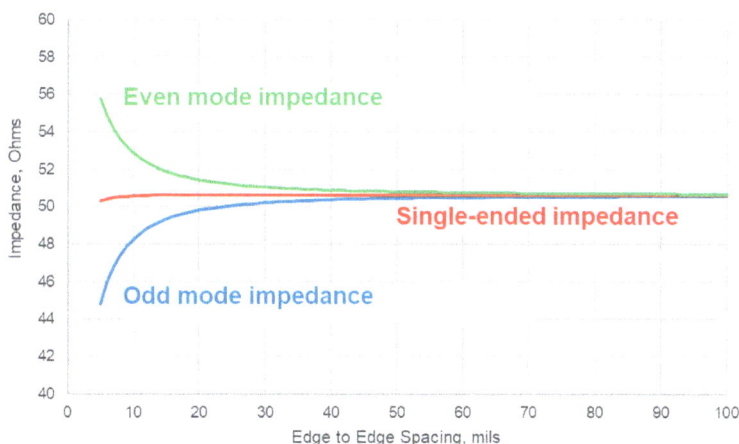

Figure 15.17 The behavior of the single-ended, odd-mode, and even-mode impedances as the two lines in a balance transmission line are brought closer together.

15.11 Important Properties of Odd- and Even-Mode Impedances

When the two transmission lines that make up a differential pair are far apart and there is no coupling, the shape of the field lines for the odd mode and even mode are the same. The signals we apply to the pair to drive them into the odd mode or even mode are still a differential and common signal, but the field patterns happen to be identical. We have two independent transmission lines with no interactions and they each have the same signal imposed on them.

This means the odd-mode impedance and even-mode impedance are the same. And, since either term refers to the impedance of one line, when the traces are far apart, each line also has a single-ended impedance.

When the traces are far apart and not coupled, the odd-mode impedance = even-mode impedance = single-ended impedance.

This is an important observation.

If the two lines that make up the differential pair are uncoupled, and each have a single-ended impedance of 50 ohms, the odd-mode impedance is 50 ohms, and the even-mode impedance is 50 ohms.

If the two lines are brought closer together, the odd-mode impedance will decrease and the even-mode impedance will increase.

15.12 Relating Odd-Mode, Even-Mode Impedance, and Differential and Common Impedance

We saw earlier that differential impedance is the series combination of the impedance of each line. In the first example, we assumed the traces were far apart and not coupled.

When we applied a differential signal to the pair, we drove the differential pair in the odd mode to see the differential impedance. While we assumed the impedance of each line was the single-ended impedance, we should have called it the odd-mode impedance, since we were driving the pair with a differential signal in the odd mode.

When the traces were far apart, the single-ended impedance was equal to the odd-mode impedance, so the analysis was correct for this special case.

In general, the differential impedance is the series combination of the impedances of the two lines. By definition, we call the impedances of either line, when part of a differential pair and driven by a differential signal into the odd mode, the odd-mode impedance.

This means the general definition of the differential impedance is that it is the sum of the odd-mode impedances of either line. In a balanced differential pair, the odd-mode impedances of either line are the same, so,

$$Z_{\text{diff}} = 2 \times Z_{\text{odd}} \qquad (15.6)$$

Likewise, the impedance the common signal sees is the parallel combination of the impedance of either line. When driven with a common signal into the even mode, the impedance of either line is the even-mode impedance. For a balanced differential pair, this makes the common impedance, by definition,

$$Z_{\text{comm}} = \frac{1}{2} Z_{\text{even}} \qquad (15.7)$$

What is actually measured or simulated in most situations is really the odd- and even-mode impedances. From these, the common and differential impedances are calculated in a field solver, or in a TDR instrument. This usually happens under the hood, so is transparent to the end user.

As the two traces in a differential pair are brought closer together, the differential impedance of the pair will decrease, and the common impedance of the pair will increase. **Figure 15.18** shows the calculation for this example for 5-mil-wide traces as the second trace is moved closer showing the four different impedances.

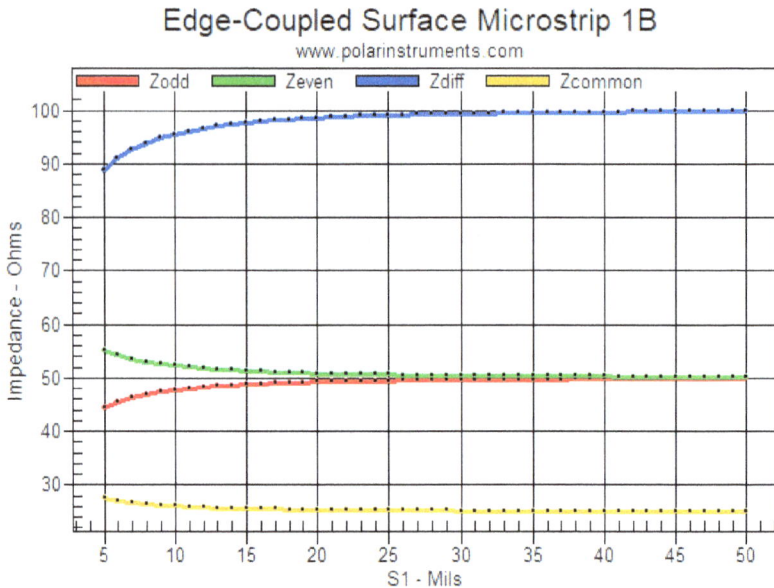

Figure 15.18 Calculated impedances of one line in the pair and the resulting differential and common impedances as the second line is brought closer, changing the edge-to-edge spacing. Simulated using the SI9000 2D field solver.

15.13 How Not to be Confused about Differential Impedance

One of the most confusing aspects of differential pairs is the language that is used. Unfortunately, this is a combination of the terms that are used in RF applications, in EMC applications, and in signal integrity applications.

The words are sometimes the same, but the definitions are sometimes different because the applications are different. Each application space has its own jargon.

This is not to say that how the terms are used in one application space is wrong and in another it is correct, it's just that it can be very confusing shifting back and forth between the applications.

For example, in EMC applications, it is common to refer to the impedances the differential and common signals see as the differential-mode impedance and the common-mode impedance. The signals themselves are referred to as differential-mode signals and common-mode signals.

But in the RF world, there is only odd-mode impedance and even-mode impedance and odd-mode signals and even-mode signals.

And in the signal integrity world, we care about the impedance the differential signal sees, the differential impedance, and the impedance the common signal sees, the common impedance.

These different terms refer to similar behaviors but are not exactly the same. If the differential pair is a balanced differential transmission line, the odd-mode is driven by a differential signal and the even-mode is driven by a common signal. However, if the differential pair is asymmetric, or unbalanced, a differential signal will drive the pair into a combination of its odd modes and even modes.

Can it get any more confusing?

If we assume the differential pair is balanced, then we should restrict ourselves to a set of vocabulary that reduces the confusion yet still provides an unambiguous way of referring to all the important behaviors.

To reduce the confusion, I recommend never using the combinations *differential-mode signal* and *common-mode signal*.

Instead, if we are referring to signals on a differential pair, or the modes of a differential pair, we should refer to the *differential signal* and the *common signal*.

Adding the term *mode* to these phrases that refer to signals does not add any new information and mixes up the concepts of the modes of the differential pair, the odd or even mode, or the voltages on the two lines that make up the differential pair.

When we refer to the modes of the transmission lines, we should refer to the odd mode and the even mode, not the differential mode and common mode.

This recommendation is emphasized in **Figure 15.19**.

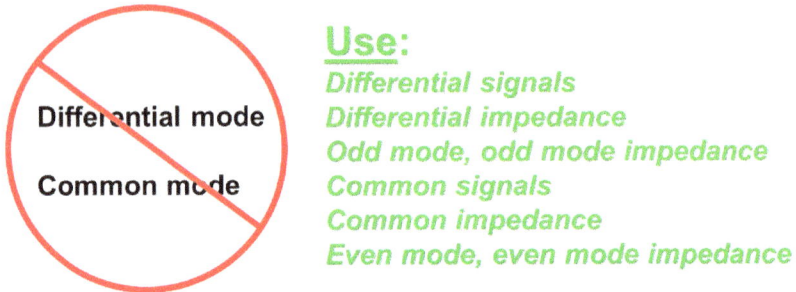

Use:
Differential signals
Differential impedance
Odd mode, odd mode impedance
Common signals
Common impedance
Even mode, even mode impedance

Figure 15.19 To reduce confusion, try to avoid using the terms differential mode and common mode. Instead use the terms differential signals, differential impedance, odd mode, common signals, common impedance and even mode.

Here is the correct way to describe the behaviors of a differential pair to reduce the confusion. To reduce the ambiguity and confusion, get in the habit of using these terms in this way:

✓ A *differential pair* is any two signal-ended transmission lines, that may have some coupling between them.

✓ A *balanced or symmetric differential pair* is composed of two identical cross-section single-ended transmission lines.

✓ A *uniform, balanced differential pair* has the same cross section of its two single-ended transmission lines and this cross section is constant down the length of the transmission line.

✓ The *differential impedance* is the instantaneous impedance a differential signal would see propagating down the differential pair.

✓ A *differential signal* drives the odd mode of a differential pair.

406

✓ The *odd-mode impedance* is the impedance of one line when the pair is driven in the odd mode by a differential signal.

✓ The *common impedance* is the instantaneous impedance a common signal would see when propagating down the differential pair.

✓ A common signal drives the *even mode* of a differential pair.

✓ The *even-mode impedance* is the impedance of one line when the pair is driven in the even mode by a common signal.

15.14 Review Questions

1. What are the two applications for differential pairs?

2. Why are the requirements different for these two applications?

3. What is the difference between earth ground and chassis ground?

4. What is a better name for digital or analog ground?

5. What is the difference in the signals on differential pairs in analog signaling vs digital signaling?

6. What are two important benefits of differential signaling?

7. What is a pure differential signal? A pure common signal?

8. What is the difference between a balanced and unbalanced differential pair?

9. What is differential impedance or common impedance?

10. When the p and n lines are far apart and each has a single-ended impedance of 50 ohms, what is their differential impedance and their common impedance?

11. What happens to the input impedance of a single-ended line when an adjacent single-ended line tied to ground is brought in from far away, but never closer than $2 \times w$?

12. What happens to the input impedance of a signal line when an adjacent signal line driven opposite is brought in from far away?

13. What happens to the input impedance of a signal line when an adjacent signal line driven the same is brought in from far away?

14. What is the odd mode of a differential pair?

15. What is the even mode of a differential pair?

16. What is the difference between the odd-mode impedance and differential impedance of a differential pair?

17. What is the difference between the common impedance and the differential impedance?

18. When the traces in a differential pair are far apart, what is the single-ended impedance, the odd-mode impedance and the even-mode impedance of a 50-ohm impedance transmission line?

19. Why is it confusing to refer to the impedance of a line as just 45 ohms?

Chapter 16   Differential TDR

In previous chapters we looked at how a TDR instrument can measure the instantaneous impedance profile of a real transmission line. While it was not stated explicitly each time, what we were really measuring was the single-ended impedance profile.

When we have a differential pair, what the TDR actually measures is the odd-mode impedance of either line in the differential pair. The sum of the two odd-mode impedances of both lines is the differential impedance of the pair.

Not all TDRs display all the information the same way. But by using a specific TDR as an example, the principles can be applied to all TDRs that measure a differential impedance.

Any differential pair has five impedances associated with it:

1.   The single-ended impedance of either line

2.   The odd-mode impedance of either line

3.   The differential impedance of the pair

4.   The even-mode impedance of either line

5.   The common impedance of the differential pair

The T3SP15D DTDR is only able to measure the first three impedances.

16.1 Measuring the Odd-Mode Impedance

The way most TDR instruments measure a differential impedance of a differential pair is by measuring the odd-mode impedance of both lines in the pair. The differential impedance is the sum of the odd-mode impedances.

409

The way to measure the odd-mode impedance of one line in the pair is by measuring the TDR response of the line while the pair is driven with a differential signal. This means that the other line needs to be driven with an opposite signal.

In a differential TDR, channel 1 outputs a 1V to 0V signal while channel 2 outputs a -1V to 0V signal. The average is 0, which means there is no common signal and it is a pure differential signal. These signals are shown in **Figure 16.1**.

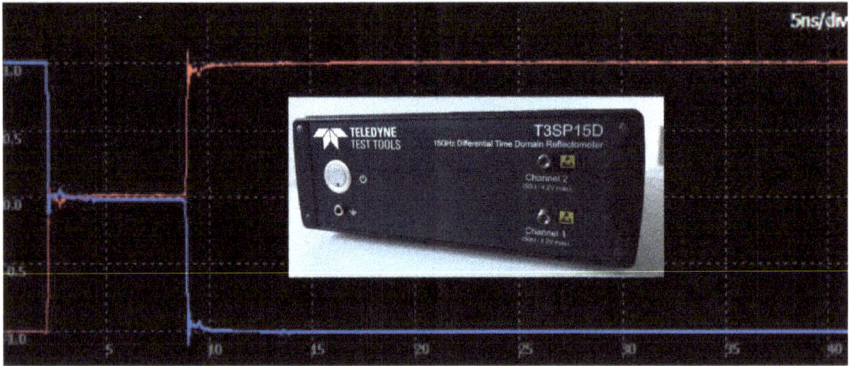

Figure 16.1 An example of the output signals on the two channels of this specific TDR, the Teledyne Test Tools T3SP15D TDR. One channel transitions from -1V to 0V and the other 1V to 0V. The average value is always 0.

For linear, passive, time-invariant interconnects, in principle, it is not necessary for both channels to output the signal on channel 1 and channel 2 at the same time. It is also possible to sequentially generate the signals on channel 1 and 2, but measure the response on channel 1 alternately when channel 1 drives and then channel 2 drives. Then the signals on channel 1 are combined to calculate the odd-mode impedance.

However, in practice, there is a little better signal-to-noise ratio in some interconnects if both signals are generated on channels 1 and 2 at the same time and measured on channel 1 and channel 2 at the same time.

In the T3SP15D DTDR, there is always a differential signal being generated on each of the two channels.

When only one channel is connected to the p-line of the differential pair, the TDR response of channel 1 is the measured single-ended impedance. This is the case for the differential pair shown in **Figure 16.2**.

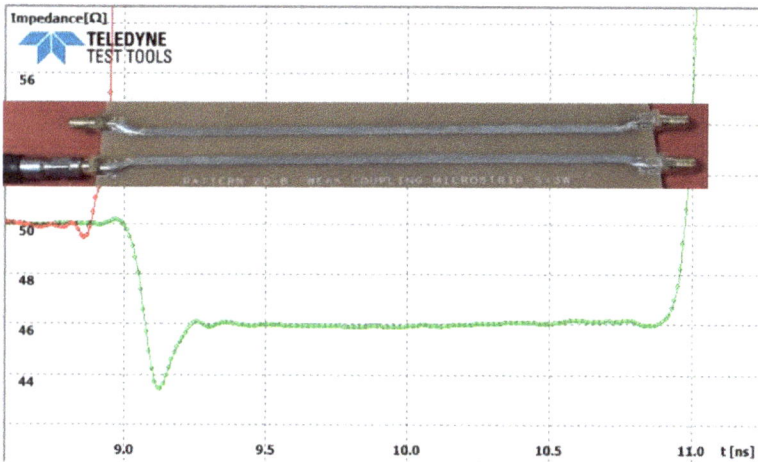

Figure 16.2 Measured single-ended impedance of one line in a loosely coupled differential pair.

The measured single-ended impedance of the p-line is about 46 ohms.

We expect that with no coupling between the two lines that make up the differential pair, the single-ended and odd-mode impedances should be the same, and the differential impedance will be 2 x the odd-mode impedance.

When the second channel is connected to the differential pair, and a differential signal is applied to the pair *but only the reflected signal on channel 1 is measured*, what is measured on channel 1 is the odd-mode impedance of that line. It is the impedance measured on one line when the pair is driven with a differential signal.

411

When the two lines in the differential pair are far apart and uncoupled, the odd-mode impedance is nearly the same as the single-ended impedance. This is the case for the example shown in **Figure 16.3**.

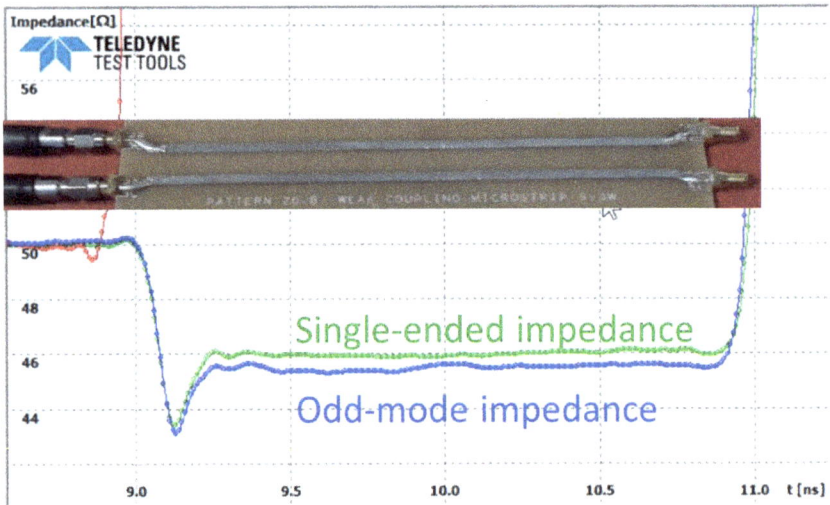

Figure 16.3 When both channels of the TDR are connected to the differential pair, it is driven with a differential signal. When only one channel is measured, it is the odd-mode impedance that is measured.

In this differential pair with a spacing equal to 3 linewidths, the single ended impedance is 46 ohms and the odd-mode impedance is 45.4 ohms, very close.

In this specific interconnect, with traces about 120 mils wide, the impedance profile is very constant. The linewidth does not vary much and the electrical environment of the trace is very constant down the line.

The differential impedance is measured when the odd-mode impedance from both channels is measured and added together. We expect, for this differential pair, the differential impedance should be 2 x 45.4 ohms = 90.8 ohms. This is very close to the

measured differential impedance of 90.74 ohms, as shown in
Figure 16.4.

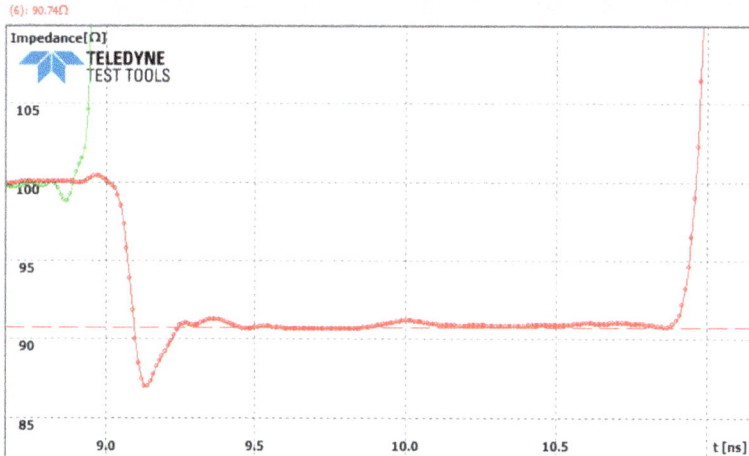

Figure 16.4 The measured differential impedance from the sum of the measured odd-mode impedances of both lines. The value of 90.74 ohms is close to the expected value of 90.8 ohms.

The differential impedance is also very constant. Generally, for narrower lines, the variation in geometry and local materials' Dk will cause more variation in the differential impedance.

Watch this video and I will walk you
through measuring the odd-mode
impedance of a line in a differential pair.

16.2 Differential Impedance for a Tightly Coupled Differential Pair

When the otherwise uniform traces are moved closer together and their line widths are kept constant, we expect the single-ended impedance to not change very much, the odd-mode impedance to decrease, and the differential impedance to decrease.

When the traces are brought close together, the higher coupling will decrease the odd-mode impedance compared to the single-ended impedance.

In the following example, the traces are 20 mils wide and spaced 20 mils as well. This is tightly coupled. The narrower line will show more variation down the length of the trace in the impedance profiles of the single-ended impedance, odd-mode impedance, and differential impedance. **Figure 16.5** shows the structure and the single-ended and odd-mode impedance profile.

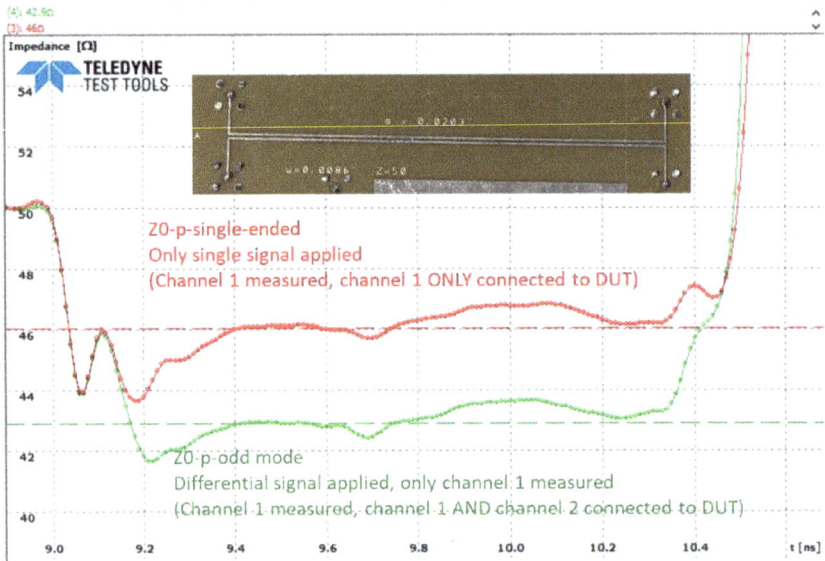

Figure 16.5 A tightly coupled differential pair showing the single-ended impedance and odd-mode impedance of one line.

The single-ended impedance is about 46 ohms and the coupling reduces the odd-mode impedance to about 43 ohms.

This impedance variation down the length of the trace is typical of many circuit board traces. It is due to the linewidth variation and the glass weave variation causing a variation in the local dielectric constant.

The differential impedance should be about 2 x 43 ohms = 86 ohms.

When the odd-mode impedance of each line is measured and added, the differential impedance is measured as 85.7 ohms, close to the expected value of 86 ohms. This is shown in **Figure 16.6**.

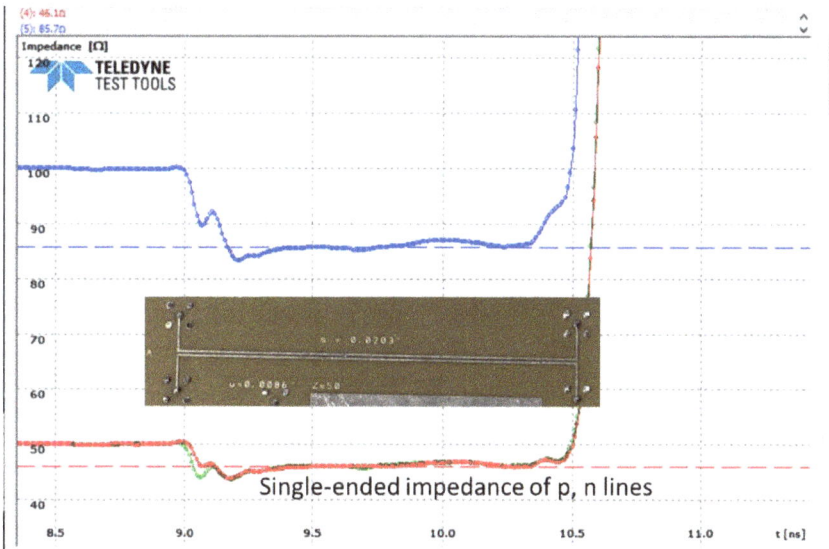

Figure 16.6 The measured differential impedance of the pair. Inset is the picture of the actual differential pair with the impedance profiles mapped to the spatial location of the transmission line.

16.3 Changing Coupling with Constant Linewidth

In the example shown in **Figure 16.7**, the spacing between two traces was decreased in the middle, but the linewidth was kept the same down the length of the trace. We would expect the single-ended impedance to be constant, since the linewidth is constant.

415

The odd-mode impedance will decrease where the coupling increases.

In this measurement, the single-ended impedance is constant down the length of the line, at about 45 ohms, while the odd-mode impedance decreases where the two lines are more tightly coupled. The coupling decreased the odd-mode impedance, exactly as expected.

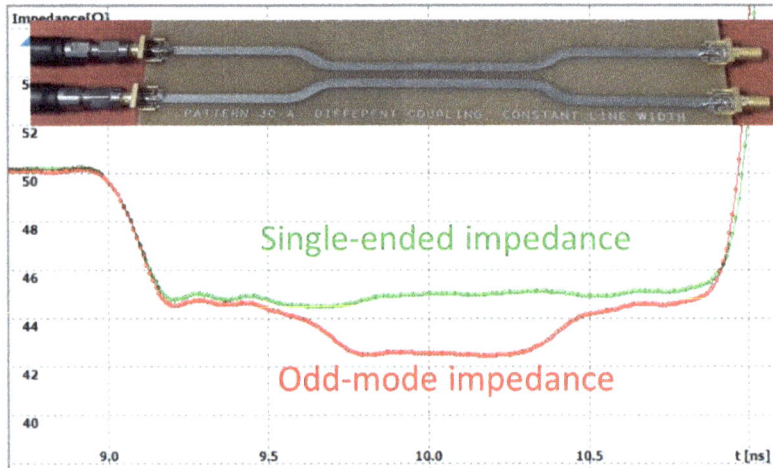

Figure 16.7 The single-ended and odd-mode impedance of this differential pair which has a tightly coupled region, but the linewidths are kept that constant down the trace.

In this differential pair, the differential impedance will decrease when the coupling increases. **Figure 16.8** shows the differential impedance profile of these two traces with a coupling that increases in the middle, compared to a uniformly spaced, loosely coupled differential pair with the same linewidth.

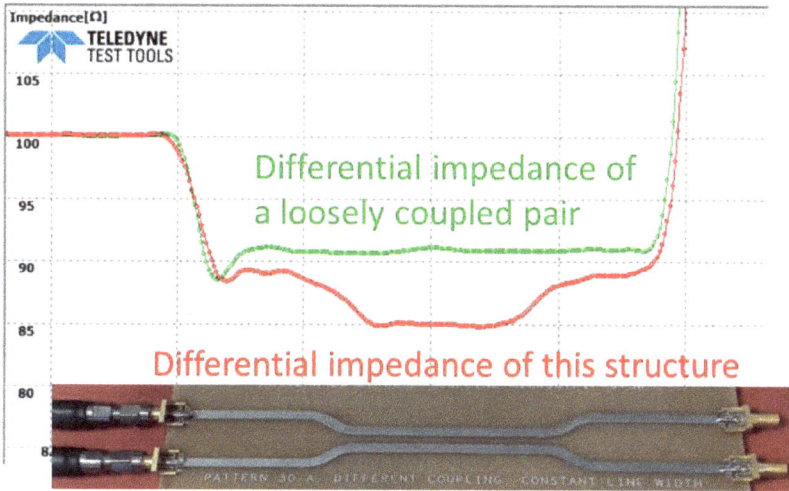

Figure 16.8 The differential impedance profile of a uniform, uncoupled differential pair and with a varying coupling. The differential impedance profile identifies features of the trace width and coupling.

16.4 Changing Coupling with Constant Differential Impedance

If all we do is bring two traces closer together, the differential impedance decreases. If we want to bring the two traces closer together, and at the same time keep the differential impedance constant, we can decrease the linewidth to compensate for the higher coupling. This will increase the odd-mode impedance back to the target value.

If the linewidth is adjusted to keep the differential impedance constant, the linewidth will decrease when the coupling increases. This means the single-ended impedance will increase in the coupled region where the linewidth is narrower. If done correctly, the odd-mode impedance will be constant, and the differential impedance will be constant.

An example of a differential pair structure in which the linewidth is adjusted to keep the differential impedance constant is shown in **Figure 16.9**.

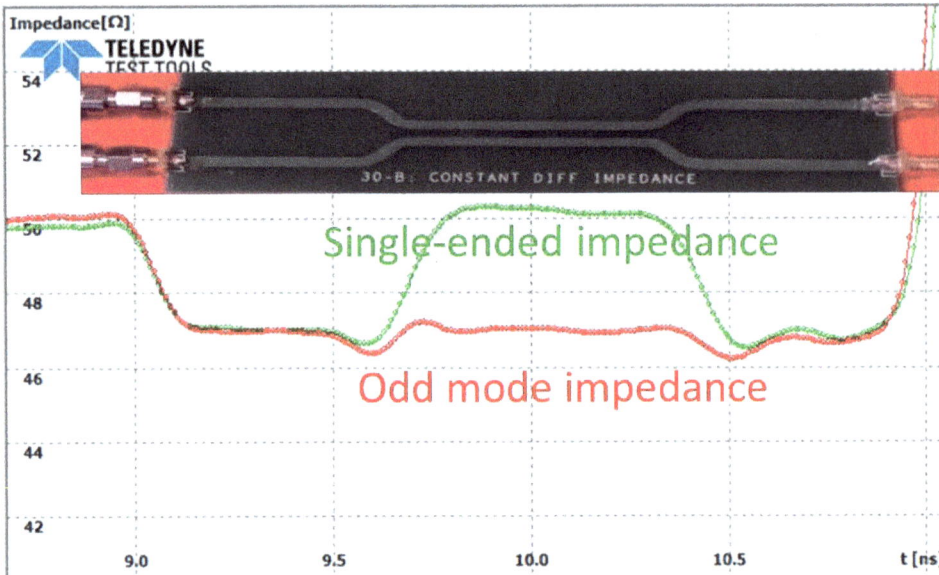

Figure 16.9 A differential pair with increased coupling but adjusted linewidth to keep the differential impedance constant. The odd-mode impedance is constant while the single-ended impedance increases.

This is an example of measuring the impedance of one line, which varies depending on how the pair is driven. The odd-mode in the tightly coupled region is 47 ohms. We expect to see a differential impedance of 2 × 47 ohms = 94 ohms. This is exactly what is measured by the differential TDR, as shown in **Figure 16.10**.

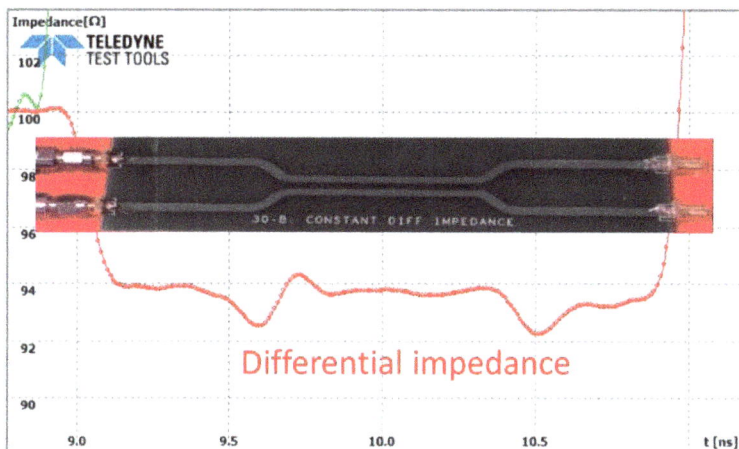

Figure 16.10 The differential impedance profile of a differential pair designed for constant differential impedance.

16.5 Impedance Profile of a Very Tightly Coupled Differential Pair

It is possible to engineer a differential pair with extremely tight coupling. The same principles of single-ended impedance, odd-mode impedance, and differential impedance still applies.

When the traces are brought very close together, the odd-mode impedance can drop considerably. To keep the differential impedance constant, the linewidth can be engineered narrower.

In this case, the single-ended impedance would increase due to the narrower line, but the odd-mode impedance would stay the same, as would the differential impedance. In **Figure 16.11** is an example of a very tightly coupled differential pair engineered for constant differential impedance.

419

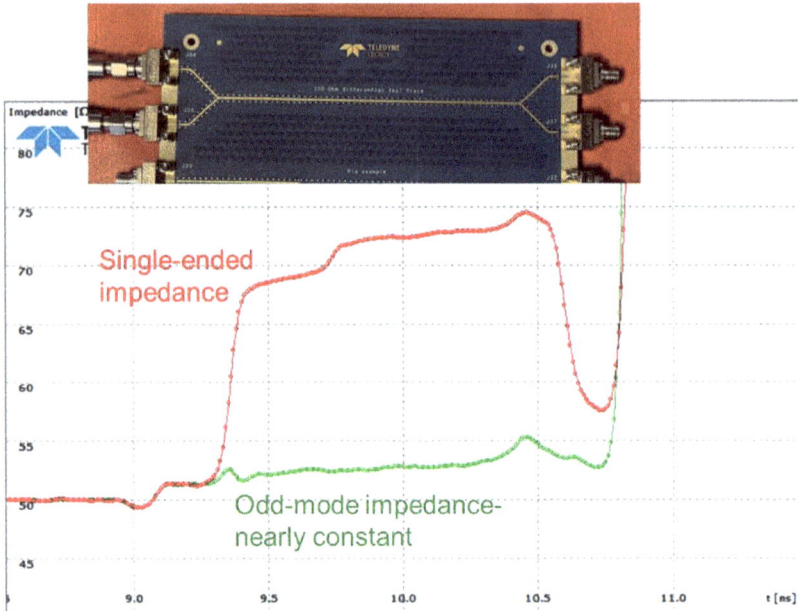

Figure 16.11 The single-ended impedance and odd-mode impedance in an extremely tightly coupled differential pair. The slight increase in impedance down the trace is due to the series resistance of the narrow traces.

In this example, the single-ended impedance and the odd-mode impedance in the region where the traces are uncoupled is about 52 ohms. When the traces are tightly coupled, the single-ended impedance would have stayed the same, but in this structure, the linewidth was dramatically decreased, increasing the single-ended impedance to about 70 ohms.

The combination of the tighter coupling and the narrower line, if done correctly, can result in a constant odd-mode impedance, as shown in the measurement.

In the measured odd-mode impedance profile, there is a slightly higher impedance right where the coupling increases. This suggests there is some underlying mechanism that causes a higher impedance where the coupling increases.

420

A close-up of the coupled region clearly shows the root cause of this peak in the odd-mode impedance or peak in the differential impedance, as shown in **Figure 16.12**.

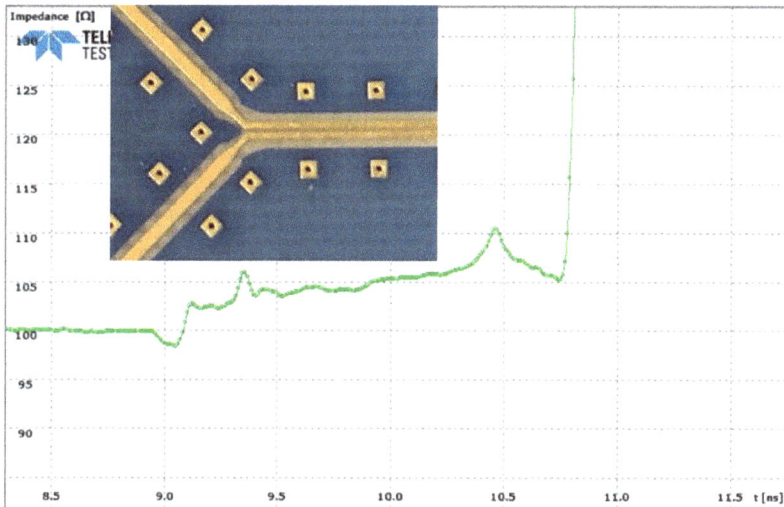

Figure 16.12 The close-up of the coupling region shows that the linewidth decreases *before* the coupled region. This is where the odd-mode impedance would increase before the coupling brings it down. Note the higher impedance in the differential impedance profile as measured by the TDR.

This is an example of how very small features in the differential pair design can have a noticeable impact on the differential impedance profile.

By understanding how the geometry influences the differential impedance profile, it is possible to engineer a constant impedance given the design constraints of the stackup and manufacturing conditions.

16.6 Review Questions

1. How does a TDR measure the single-ended impedance of a transmission line?

421

2.   How does the TDR measure the odd-mode impedance of a differential pair?

3.   How does the TDR measure the differential impedance of a differential pair?

4.   As the coupling in a differential pair increases, what happens to the difference between the single-ended impedance and the odd-mode impedance?

5.   If you want to maintain a constant differential impedance and the coupling increases, what can you do to the linewidth?

6.   If the odd-mode impedance of a line in a differential pair is 45 ohms, what is the differential impedance?

Chapter 17 Exploring the Properties of Differential Pairs

We can explore how the physical design of a differential pair affects the electrical properties of the differential pair using a 2D field solver. Especially when dealing with complex structures, it is always important to apply rule #9 and apply our new intuition to anticipate, before we run the simulation, how the electrical behavior will be affected by the geometry.

17.1 Engineering a Microstrip with a Constant Differential Impedance

A commonly used differential pair structure is a differential microstrip. If we start out with a geometry when the traces are far apart, they will act like two single-ended transmission lines.

As we bring the traces closer together and the traces interact with the fringe field lines, we expect the following to happen:

✓ The odd-mode impedance will decrease

✓ The differential impedance will decrease

✓ The even-mode impedance will increase

✓ The common impedance will increase

We can use a 2D field solver as a virtual prototype environment to explore design space. **Figure 17.1** shows the setup for a microstrip differential pair with 50-ohm single-ended impedance using a nominal 10-mil-wide trace, starting from far apart.

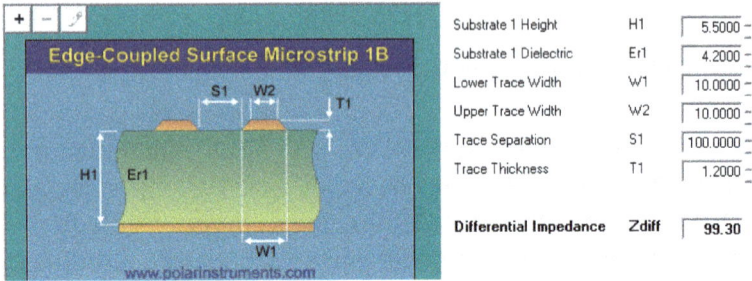

Figure 17.1 Nominal conditions for a 99-ohm differential microstrip with traces far apart.

Using the SI9000 tool, we can explore the impact on the four different impedances of this differential pair as the traces are brought closer together.

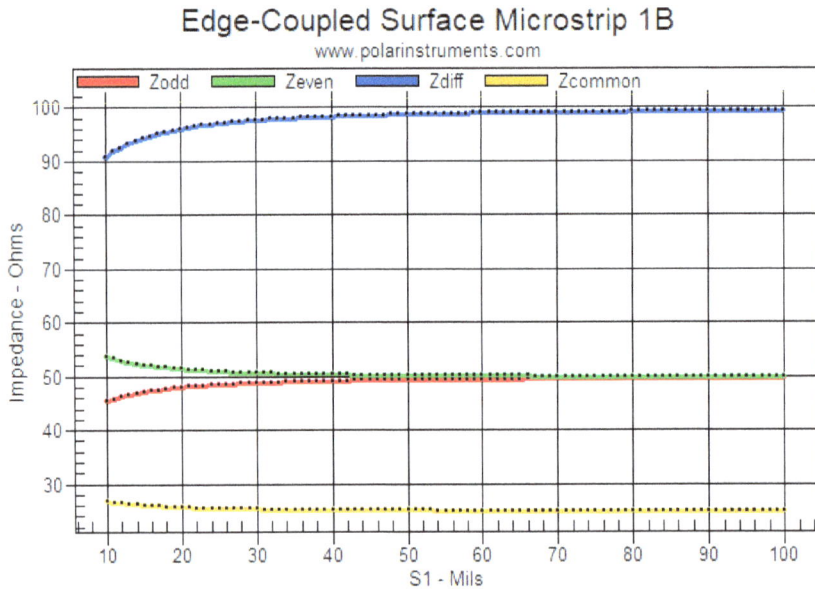

Figure 17.2 Nominal conditions for a differential microstrip, with 10-mil-wide traces and initially far apart and brought closer together.

In this example, we see that when the traces are far apart, their spacing has no impact on the impedances. It is only when they are

424

close together where the coupling between the signal line and the fringe fields plays a role.

When they are uncoupled, the odd-mode impedance = the even-mode impedance = 50 ohms = the single-ended impedance.

We can identify three levels of coupling from this behavior:

Tightly coupled is when the traces are as close together as they can be easily fabricated, usually when the spacing = the linewidth. In this example, this is a spacing = linewidth = 10 mils.

Uncoupled is when the traces are far enough apart so there is very little influence on the impedance from the spacing. In this case, when the spacing is 30 mils or 3 × w, there is less than 1.5% impact on the differential impedance compared to when they are infinitely far apart.

Loosely coupled is somewhere in between. Halfway between is a spacing = 20 mils or s = 2 × w.

These represent the three terms that characterize the level of coupling.

When designing a differential pair, which is better? Tightly coupled, loosely coupled, or uncoupled?

17.2 Which is Better: Tightly Coupled or Loosely Coupled Differential Pairs?

We saw in measurement examples that it is possible to adjust the linewidth to compensate for the decrease in impedance from coupling to keep the differential impedance constant.

This means that the differential impedance can be kept constant as coupling changes to maintain a constant differential impedance. To first order, the differential signal doesn't care what the coupling is. All it cares about is the differential impedance.

Starting with this nominal case, we can bring the traces closer together, but adjust the linewidth to keep the differential impedance constant. Using the SI9000 field solver, we can calculate the specific linewidth to keep the differential impedance constant. The result for this first test case is shown in **Figure 17.3**.

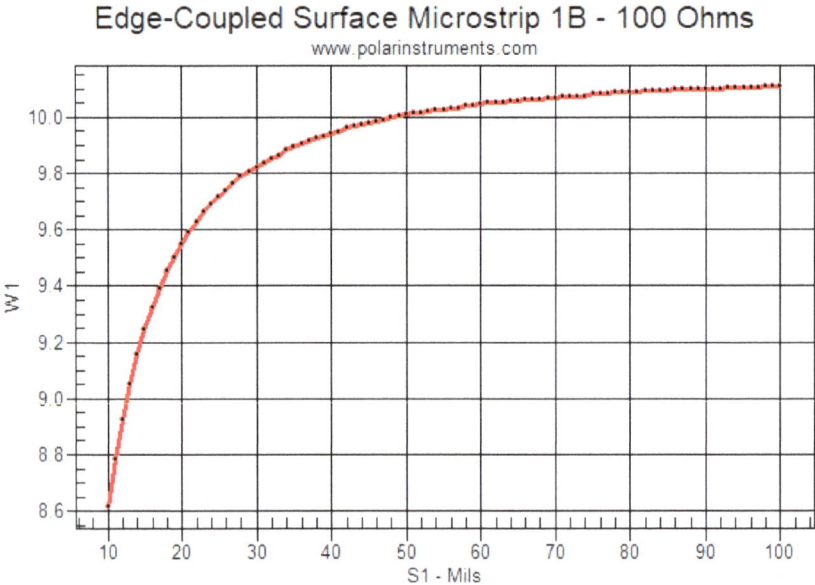

Figure 17.3 The linewidth needed for a differential microstrip to maintain 100 ohms as the spacing changes.

This curve defines *design space* for the specific values of dielectric thickness and dielectric constant. As long as a design walks this line and changes width as mapped out with the spacing, the differential impedance of the differential pair will be kept constant at 100 ohms.

A differential signal propagating down this differential pair, which may have a spacing that changes but a linewidth compensating, would see a constant differential impedance.

For example, if it were uncoupled at 30 mils apart, the linewidth could be about 9.8 mils to achieve 100-ohm differential impedance. In a region where the traces needed to be brought closer together, with a spacing of 10 mils, when tightly coupled, the linewidth could be reduced to 8.6 mils and still achieve 100 ohms.

This principle of walking the line, is also referred to as the Johnny Cash principle. The first album Johnny Cash produced that went gold was based on his title song, "I Walk the Line," shown in Figure 17.4.

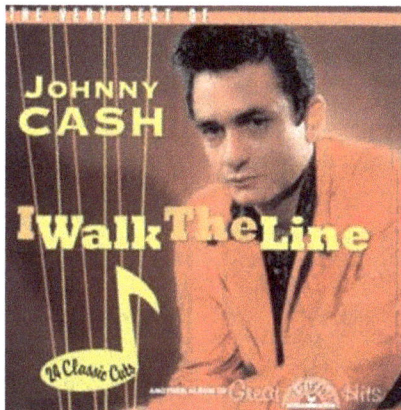

Figure 17.4 Johnny Cash's first album that went gold.

If a differential pair can be fabricated with the same differential impedance either uncoupled or tightly coupled, which is better?

In principle, it doesn't matter; an uncoupled differential pair can be just as good a differential pair as a tightly coupled differential pair.

In practice, there are a few trade-offs to consider.

The advantage of a tightly coupled differential pair is that the interconnect density is highest. This could result in a fewer number of routing layers or a smaller board size. In both cases, using

427

tightly coupled differential pairs means potentially a lower-cost board.

The downside of a tightly coupled differential pair is that the line width is narrower that it might be. As long as the narrower trace can be fabricated with no cost adder, the narrower trace can be acceptable.

However, the other consequence of a narrower line is higher series resistance. This will contribute to more frequency-dependent loss that contributes to the collapse of the eye at the receiver. An example of an eye at the receiver through a low-loss channel and a lossy channel is shown in **Figure 17.5**.

Through a low loss channel   Through a lossy channel

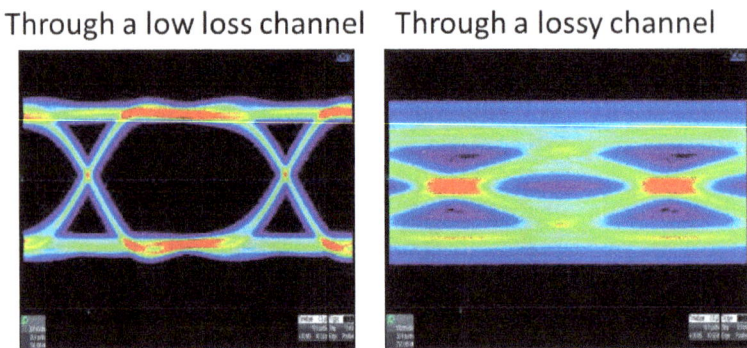

Figure 17.5 Examples of the collapse of the eye at the receiver due to frequency-dependent losses in the channel.

Usually the losses in a differential pair on circuit boards are not an issue except at data rates over 1 Gbps. Above this rate, reducing the losses in the interconnect can be as important a design goal as achieving a target differential impedance.

At data rate applications above 10 Gbps, the losses in the differential pair are often the most important driving force in a design. This is the situation where it is important to do everything possible to reduce the frequency-dependent losses.

428

To reduce the conductor losses, it is important to use as wide a line as practical. This is achieved by:

- ✓ Use as low a differential impedance as practical
- ✓ Use as low a dielectric constant as practical
- ✓ Use as thick a dielectric layer as practical
- ✓ Use uncoupled traces, which results in a wider line

To achieve the lowest conductor loss, and use as wide a line as practical, the optimum coupling is uncoupled. Beyond the range of spacing > 3 × w, there is no advantage to using a larger spacing. In fact, the difference in-line width from using a spacing = 2 × w and 3 × w for the same differential impedance is only about 3%.

It will often be the case that the best trade-off between wider line for lower loss, but tighter coupling for higher interconnect density, is a loosely coupled pair, with a spacing = 2 × w. This is why many high-speed serial link designs suggest loosely coupled differential pairs when loss is important.

This analysis suggests a simple decision tree to decide between tightly coupled and uncoupled differential pairs:

- ✓ If the data rate is < 1 Gbps and loss is not important, always use tight coupling. This will result in the lowest cost board.

- ✓ If the data rate is > 10 Gbps, consider using loosely coupled differential pairs. This will result in a good balance between interconnect density and lower loss.

In general, though, at data rates above 1 Gbps, where loss may be a concern, it is always important for every engineer to do their own analysis.

Watch this video and I will walk you through the question of which is better tightly coupled or loosely coupled differential microstrip.

17.3 Speed of a Differential and Common Signal

In a single-ended microstrip, with air above the trace, the signal that propagates sees the bulk dielectric constant below the trace and also some amount of air due to the fringe field lines that extend above the traces. This means the signal sees an effective dielectric constant that is less than the bulk dielectric constant of the laminate layer.

This also means the speed of the signal is faster in a microstrip trace than if the trace were in a stripline structure and the only dielectric constant the signal would see is the bulk dielectric constant.

In a differential microstrip, the fringe field lines of the differential signal also extend more in the air than for a common signal. This means the differential signal will see a lower dielectric constant than the bulk dielectric constant and will propagate faster than in a stripline of the same bulk laminate.

When a common signal propagates on the differential pair, the voltage on the p- and n-lines are the same; they have no field lines between them and most of the electric field lines are in the bulk material. The effective dielectric constant the common signal sees is close to the bulk dielectric constant and higher than for the differential signal.

The difference in the field distribution between a differential signal and a common signal is shown in **Figure 17.6**.

430

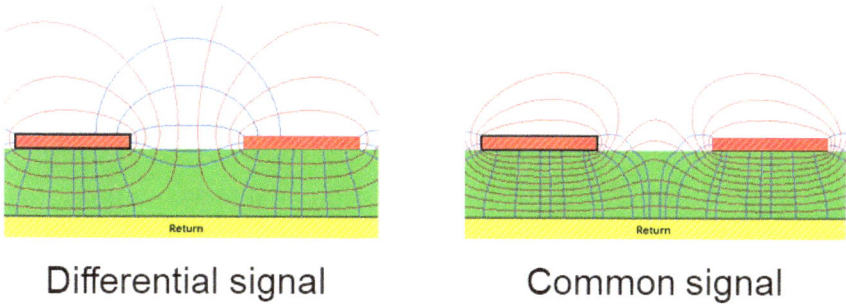

Differential signal   Common signal

Figure 17.6 The field distribution for a tightly coupled microstrip showing a differential signal and a common signal. The differential signal has more field lines in air and a lower effective Dk.

The consequence of the difference in effective dielectric constant the differential signal sees compared to the common signal is that the speed of a differential signal will be faster than a common signal.

An example of the measured transmitted differential and common signals through a 10-inch-long tightly coupled microstrip differential pair is shown in **Figure 17.7**.

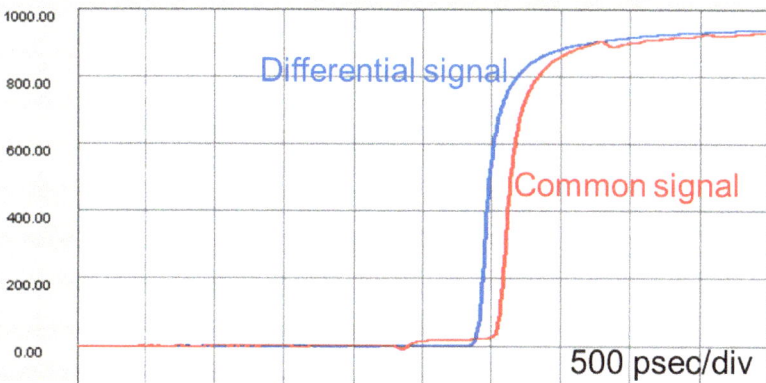

Figure 17.7 Measured differential and common signal transmitted through a 10-inch-long tightly coupled differential microstrip. The differential signal is faster and comes out before the common signal.

431

When the traces are uncoupled, there is little field distribution difference between a differential signal and a common signal and the differential and common signals travel at the same speed.

In a stripline, where the fields of the differential and common signals see exactly the same material, the speed of a differential and common signal are exactly the same.

17.4 Stripline First-Order Factors

In a symmetric differential stripline, there are only a few terms that define the geometry. **Figure 17.8** shows the setup for a typical geometry. The line widths are 10 mils, dielectric thickness for both layers are 13 mils, and spacing is far apart.

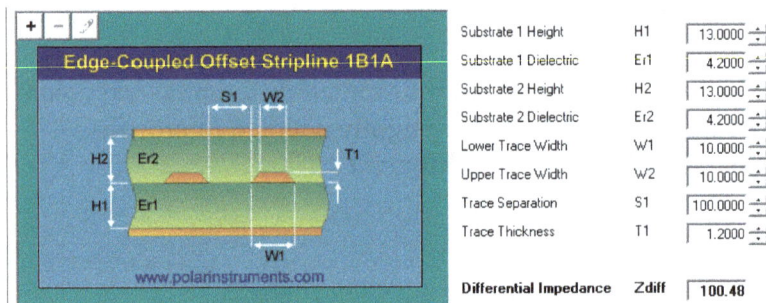

Figure 17.8 A nominal design for a 100-ohm differential stripline.

The impact on the differential impedance from the spacing between the signal lines is similar to the behavior in microstrip. The farther apart the signal lines, the less the coupling and the less impact on the differential impedance.

As an example, **Figure 17.9** shows how the differential impedance varies for this nominal case when the spacing is changed. The condition for uncoupled, when the spacing does not affect the differential impedance anymore, is also 30 mils, which is a spacing

432

$= 3 \times w$. This is the same rule of thumb as with microstrip that defined the range for uncoupled.

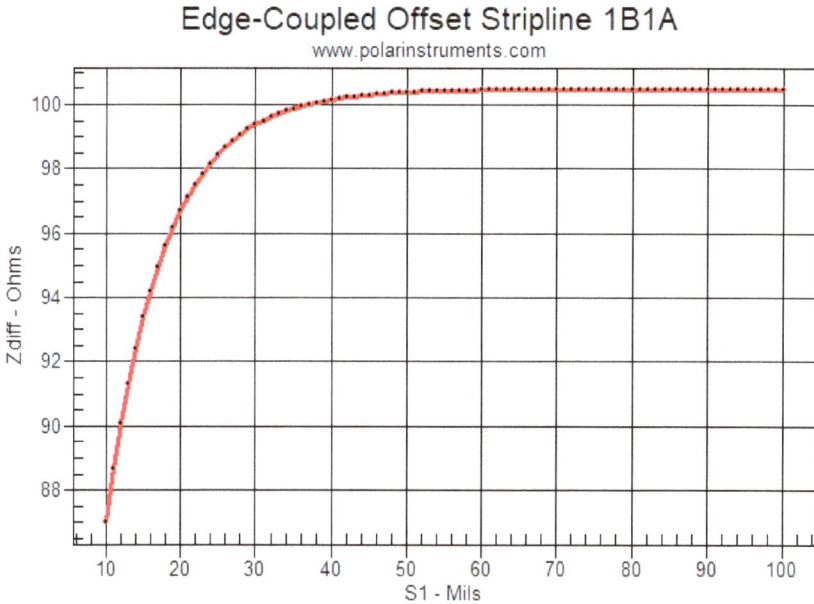

Figure 17.9 Calculated differential impedance as the spacing between the two traces changes.

As expected, the differential impedance decreases as the coupling increases. To compensate for the increased coupling, there are two design knobs that could be adjusted:

✓ The dielectric spacing can be increased

✓ The line width can be decreased

In this example, the linewidth is fixed at 10 mils. As the spacing decreases, the dielectric thickness of both layers can be increased to keep the differential impedance constant. **Figure 17.10** shows the design space of constant differential impedance.

433

Edge-Coupled Offset Stripline 1B1A - 100 Ohms

www.polarinstruments.com

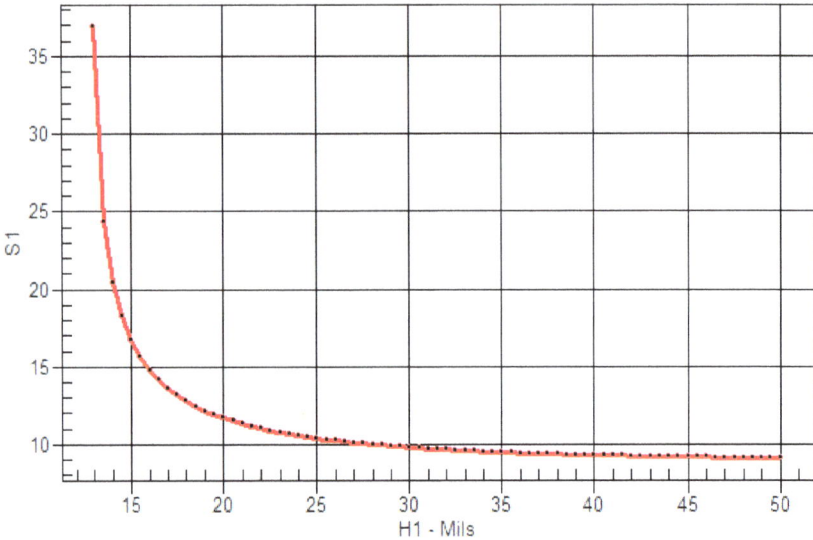

Figure 17.10 Design space for constant differential impedance stripline and constant linewidth.

As the distance between the traces decreases, the coupling increases and the dielectric spacing to the planes must increase to maintain 100 ohms. When the spacing is farther apart than 30 mils, the uncoupled regime, the dielectric thickness stabilizes at about 13 mils.

This analysis also indicates that for a tightly coupled pair, with a spacing close to the linewidth, the dielectric thickness to each plane grows very fast. This means that a board with a tightly coupled differential stripline would have to be very thick.

This is a significant limitation with stripline differential pairs. To keep board thickness to an acceptable thickness requires using loosely coupled stripline differential pairs.

For example, in the stripline above, with a loosely coupled pair, and spacing = 20 mils = 2 × w, the dielectric thickness should be about 14 mils above and below the traces. This is a total thickness between the planes of 28 mils.

But when the coupling increases to tightly coupled and a spacing of 10 mils, the dielectric thickness must increase to 25 mils per layer. This is a total thickness of 50 mils between the planes. This would contribute to a very thick board. This is sometimes just not practical.

Keeping the dielectric thickness layers to a manageable thickness is a strong reason to use uncoupled or loosely coupled stripline differential pairs.

Alternatively, the dielectric thickness can be held constant and the spacing between the traces increased. If nothing else changes the differential impedance will increase. To keep it constant, the line width will increase. This is the Johnny Cash principle.

The calculated design space that maps out the linewidth required for different coupling to maintain 100-ohm differential impedance is shown in **Figure 17.11**.

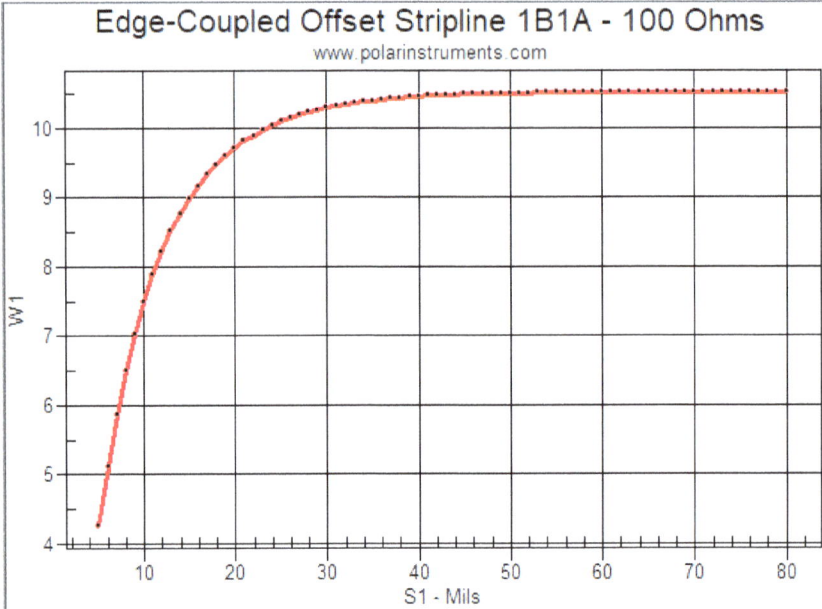

Figure 17.11 Design space for a changing line width as the spacing increases for a constant 100-ohm differential pair.

17.5 Stripline Second-Order Factors

There are two important second-order features that might affect the differential impedance of a loosely coupled stripline: the trace thickness and the slight resin-rich region between the two traces.

A starting place configuration is shown in **Figure 17.12**. The line width is 10 mils and spacing is 20 mils with a 15-mil dielectric thickness above and below the traces.

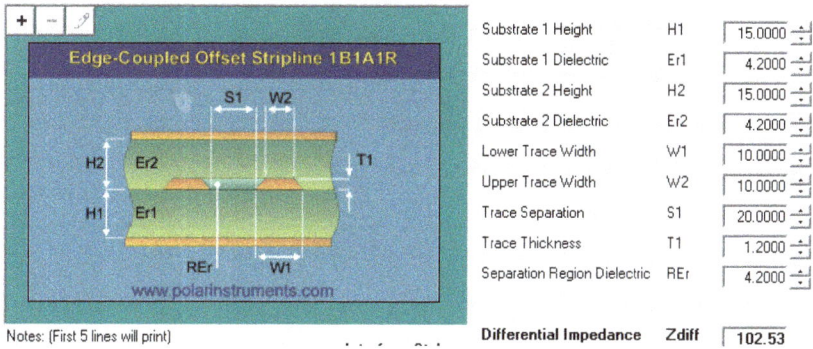

Figure 17.12 The starting condition of a typical loosely coupled differential stripline.

From this nominal condition, the trace thickness is swept. As the trace thickness increases, the top surface of the trace gets closer to the top plane and the differential impedance will decrease. Exactly how much is related to the specific fringe fields between the traces and the planes. It can only be analyzed by a 2D field solver.

Using the Polar SI9000 field solver, the impact on the differential impedance for different trace thicknesses is calculated and shown in **Figure 17.13**.

Edge-Coupled Offset Stripline 1B1A1R
www.polarinstruments.com

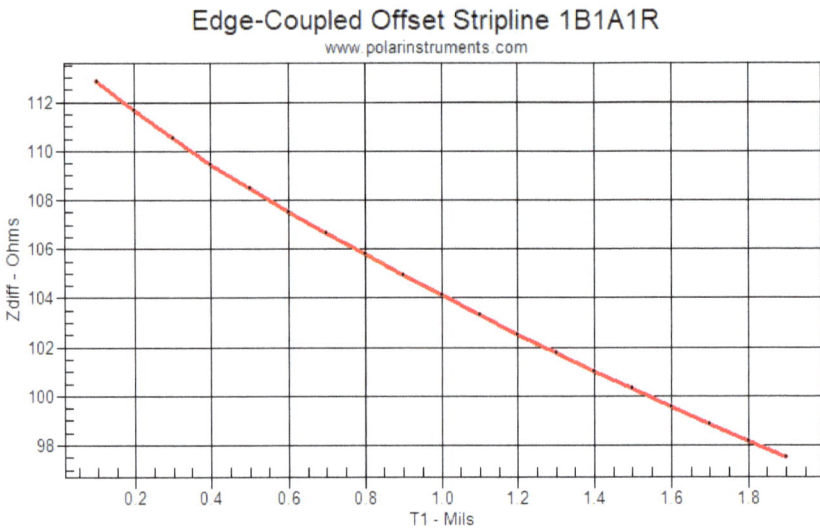

Figure 17.13 Differential impedance of a loosely coupled different stripline as the trace thickness is increased.

From a nominal value of 1 mil to 1.1 mil increase, a 10% change, results in a 1% decrease in the differential impedance. This is a second-order factor influencing the differential impedance.

The second term that has a small impact on the differential impedance is the dielectric material between the two traces. The laminate materials are usually composed of composites of a glass yarn with a Dk ~ 6.7, embedded with a cured polymer resin, with a Dk ~ 3.2.

When the laminate layers are pressed together, sometimes the glass yarn is prevented from filling in the gap between the traces and it becomes filled with just resin. The resin has a lower dielectric constant than the glass-resin combination.

As the dielectric constant of the small region between the traces decreases, the differential impedance will increase, but not by much. Using the SI9000 field solver, the impact on the differential impedance as the dielectric constant of the region between the two traces increases is shown in **Figure 17.14**.

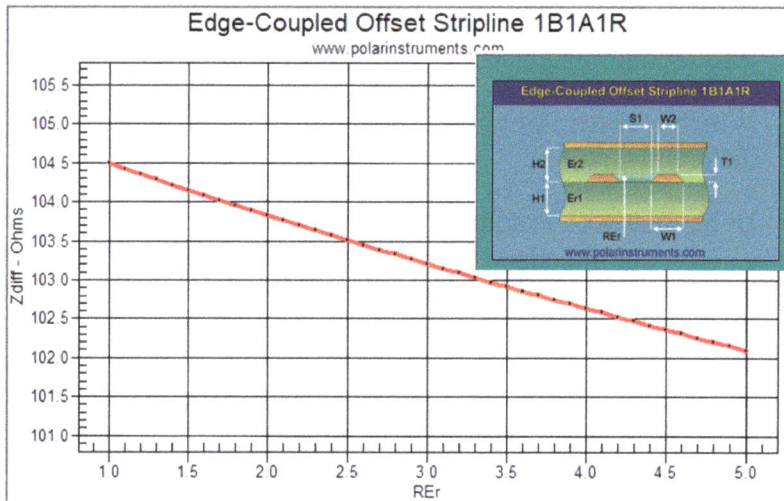

Figure 17.14 The differential impedance as the dielectric constant of the region between the two traces increases. Note the impedance scale is 0.5 ohms/div.

In this example, from the nominal case of a Dk of 4.2, and a differential impedance of 102.5 ohms, the differential impedance would increase to only 103.5 ohms, about a 1 ohm change, if the resin-rich region were to decrease from 4.3 to a typical value of resin of 3.0. This is a very small effect.

While it is often pointed out as an area of concern, thinking the large electric fields between the differential pair must be more sensitive to the dielectric constant of the material in the region between them, we see it has a very small impact when analyzed with a field solver.

17.6 Review Questions

1. What is the difference between a tightly coupled and uncoupled differential pair?

439

2. If all you know is the differential impedance or the odd-mode impedance of a differential pair, how do you estimate the coupling between the two lines that make up the differential impedance?

3. Which is better, a tightly coupled or an uncoupled differential pair?

4. If the linewidth is adjusted in a differential pair to keep the differential impedance constant as coupling changes, what will a differential signal see propagating down such a differential pair?

5. What is the Johnny Cash principle and why is this important when designing a constant differential impedance pair?

6. What is an advantage of using a tightly coupled differential pair?

7. What is an advantage of using an uncoupled differential pair?

8. As coupling increases and the differential impedance decreases, what happens to the common impedance?

9. If the coupling in a differential pair increases and the linewidth is decreased to keep the differential impedance constant, what happens to the common impedance?

10. When are losses in a differential pair important?

11. What are the two loss mechanisms in a transmission line?

12. For lowest conductor loss at a target impedance, what linewidth should be used?

13. What are three ways of engineering the widest possible linewidth at a fixed impedance?

14. What is the advantage of a loosely coupled differential pair?

15. What determines the speed of a differential signal or a common signal?

16. In a differential microstrip, which travels faster, a differential signal or a common signal?

17. In a differential stripline, which travels faster, a differential signal or a common signal?

18. For linewidths of about 10 mils and 50-ohm single-ended impedance transmission lines, how close can the lines get before the differential impedance drops by more than about 1%?

19. What is the disadvantage of a tightly coupled differential stripline pair?

20. Why is it more effective to decrease the linewidth for tightly coupled differential stripline pairs than increase the dielectric thickness?

21. How strong an impact is there on the differential impedance in a stripline geometry from the resin-rich region between the traces? Is this a design issue to worry about?

Chapter 18 Differential Pairs with No Return Paths

So far, all the differential pairs we've looked at have a return plane in close proximity. One of the important features of a differential pair is that in some cases, a differential signal can propagate just as well with or without a continuous return path in close proximity.

For single-ended signals, this is not the case. The return plane plays a critical role. In this chapter, we explore when we can forego the return plane and what the properties of the differential pair are without a return plane.

18.1 The Single-Ended Impedance and Location of the Return Plane

The characteristic impedance of a microstrip transmission line depends on the distance between the signal line and return plane as a first-order term.

The impedance of the line is directly related to the displacement current through the electric field lines between the signal and return conductors. There are no other conductors in a single-ended transmission line.

As the dielectric thickness increases, the electric field lines will continue to decrease in magnitude due to the increase in distance over which the voltage difference extends, decreasing the displacement current and increasing the characteristic impedance in a nonlinear way. We can explore this behavior using a 2D field solver. The fundamental assumptions are that the dielectric laminate thickness increases and the return plane is infinite in extent. **Figure 18.1** shows this impedance increase with dielectric thickness for a 5-mil-wide trace.

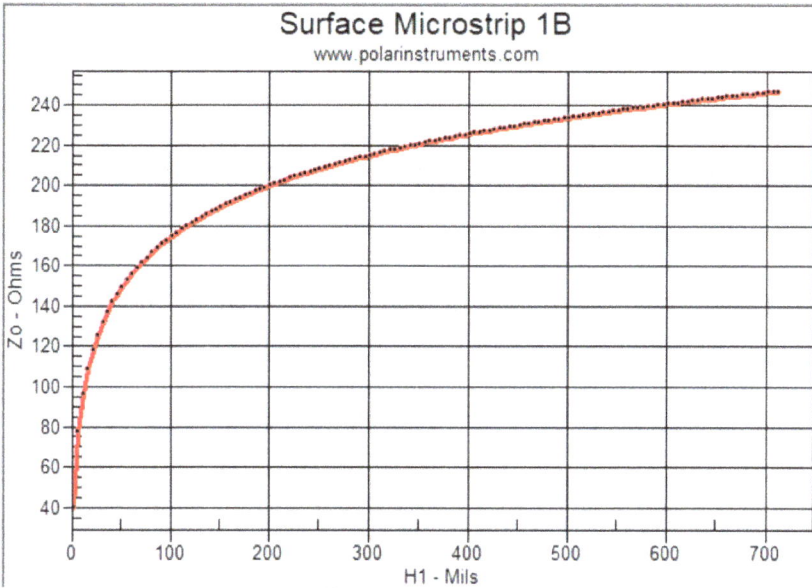

Figure 18.1 The impedance of a 5-mil-wide microstrip as the dielectric thickness increases.

The single-ended impedance will continue to increase as the thickness increases. The return plane is a critical element to a single-ended transmission line.

18.2 The Common Impedance and Location of the Return Path

In a differential microstrip transmission line pair, the single-ended impedance of each line is sensitive to the distance to the return path.

The common signal, which sees the parallel combination of the two transmission lines driven in the even mode, behaves very similar to the single-ended impedance.

The common signal drives the same signal into the p- and n-lines of the pair. There is no electric field between the p and n signal

443

lines, since they are at the same voltage. The only displacement current is between the two signal lines and the return plane. The currents have the same polarity and are in parallel to the return plane. While these two signals on the p- and n-lines interact, they are similar, but not identical, to the current distribution of the single-ended signal.

When the return plane is far away, the common signal looks like a single-ended signal. **Figure 18.2** shows the common impedance of a differential microstrip transmission line as the dielectric thickness increases.

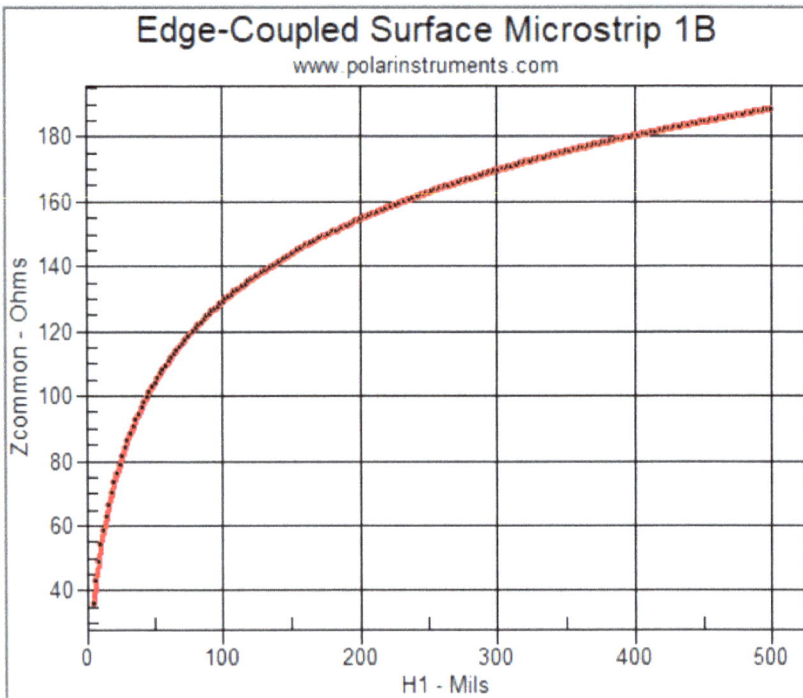

Edge-Coupled Surface Microstrip 1B

www.polarinstruments.com

Figure 18.2 Common impedance of a microstrip transmission line as the dielectric thickness to the return plane increases.

The common impedance will continue to increase as the return path moves farther away. The presence and location of the return plane is of first-order importance influencing the impedance the common signal sees.

However, this is not the case for a differential signal.

18.3 Coupling Between the p- and n-Lines and the Return Plane

A differential signal drives a differential pair in the odd mode. The odd-mode impedance is the impedance of one line when the pair is driven with a differential signal. Each line has an odd-mode impedance. The differential impedance of the pair is the sum of the odd-mode impedances of the two lines.

When we look at the odd-mode impedance of the p-line, its impedance is related to the applied voltage and the electric field lines between the p-line *and all other conductors.*

This is why the presence of the n-line can influence the odd-mode impedance of the p-line. When a differential signal is applied to the pair, the n-line will have the opposite voltage change, which means potentially large electric field lines between the p-line and the n-line.

The precise shape and magnitude of the electric field lines between the p- and n-lines are related to the presence of the plane below. **Figure 18.3** shows the fringe electric field lines between the p-line and the return plane and the adjacent n-line.

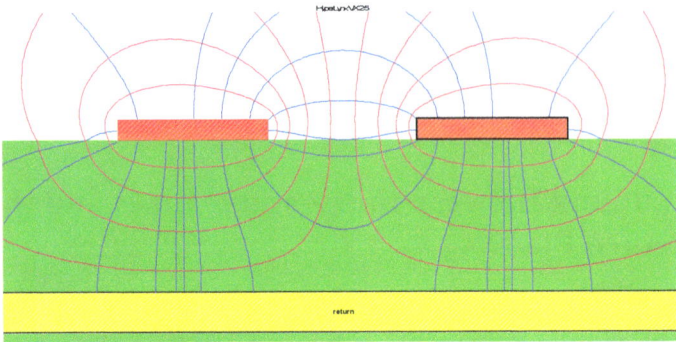

Figure 18.3 The electric field lines between the p-line and the plane and adjacent n-line, on an exaggerated scale with a differential signal applied. The electric field lines are the blue ones.

The number of electric field lines between two conductors is a direct measure of the coupling between them. A lot of field lines connecting them means more coupling between them. More current would flow through the higher coupling path with more field lines.

When the electric fields between the p-line and the plane are larger than the electric fields between the p-line and the n-line, the p-line is more tightly coupled to the plane than to the n-line. The presence of the plane dominates the odd-mode impedance and the proximity of the n-line is less a factor.

When the electric fields between the p-line and the n-line are larger than the electric fields between the p-line and the plane, the coupling between the p- and n-lines is larger than the coupling between the p-line and the plane. The presence of the n-line dominates the odd-mode impedance and the plane is less important. **Figure 18.4** shows this case.

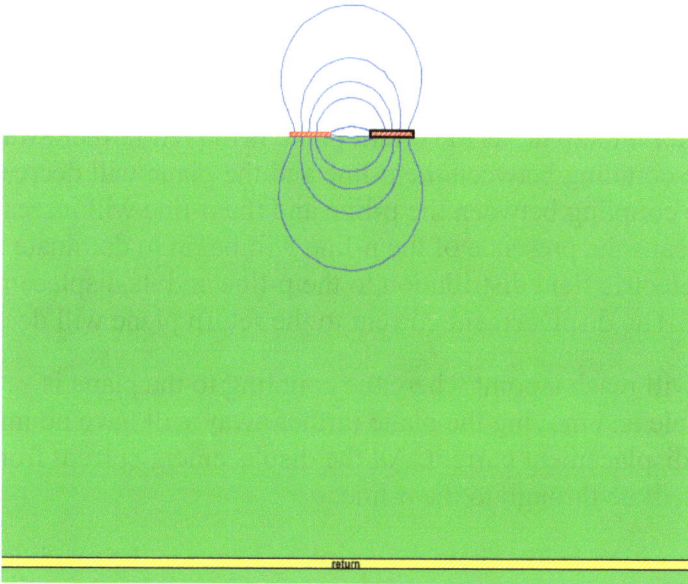

Figure 18.4 When the plane is far away all the electric field lines are between the p- and n-lines. The displacement current is dominated by the proximity between the traces, not the plane location.

When the plane is so far away that all the field lines are between the p- and n-lines and none to the plane, the current between the p-line and the return, in the odd mode, is dominated by the other line in the pair. It is not related to the plane.

In this case, the current loop of the signal into the p-line flows between the signal and return at the source, enters the p-line, flows to the n-line, and returns back to the source, where it returns to the return path at the source.

Effectively, the current pattern for the differential signal is the signal current goes into the p-line and returns back out the n-line. This is as though the p- and n-lines carry a single-ended signal.

In this case, when the return plane is far away, the differential impedance is independent of the presence or absence of the distant plane.

447

18.4 Differential Impedance of a Differential Pair and Location of the Return Plane

In a differential pair, as the return plane is moved farther away, the relative coupling between the p-line and the plane will decrease and the coupling between the p-line and the n-line will increase. This means the presence of the n-line will begin to dominate the fringe electric field distribution to the p-line and its displacement current. The displacement current to the return plane will decrease.

There will reach a point where the coupling to the plane is negligible and moving the plane farther away will have no impact on the displacement current. All the displacement current from the p-line will be through to the n-line.

This transition point where the coupling between the two lines dominates over the coupling to the plane is roughly when the plane is farther away than the span between the two signal lines.

In the case of a 5-mil-wide line and 5-mil-space, the differential impedance should be mostly constant when the dielectric thickness is greater than the span of the two lines, or 5 mils + 5 mils + 5 mils = 15 mils. This is exactly the behavior shown in the simulation with the Polar SI9000 field solver tool in **Figure 18.5**.

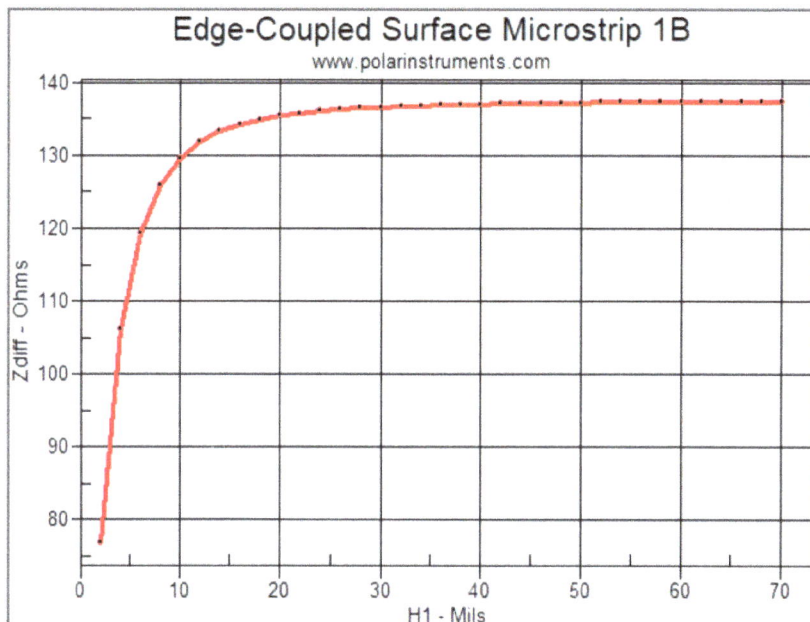

Figure 18.5 Differential impedance as the distance to the return path increases.

We expect that when the plane is 15 mils away from the trace layer, the coupling between the p-line and the plane will be less significant than the coupling between the two signal lines. Beyond 15 mils away, the presence of the return plane should not affect the differential impedance.

This is precisely what we see in the simulation. As the distance increases between the signal traces and return plane, the differential impedance does not change. It stabilizes at a value of 138 ohms, in this example. Of course, the common impedance and single-ended impedances continue to increase, but not the differential impedance.

When the return plane is far away, the differential impedance is not about the return plane, it is about the coupling between the p- and n-lines, which is directly related to the spacing between the two traces. When the plane is very far away, the differential impedance

449

between the two signal lines increases as the coupling between the two traces decreases and their distance increases. **Figure 18.6** is the simulation of the differential impedance as the spacing between the two traces increases while the return plane is so far away that it plays no role.

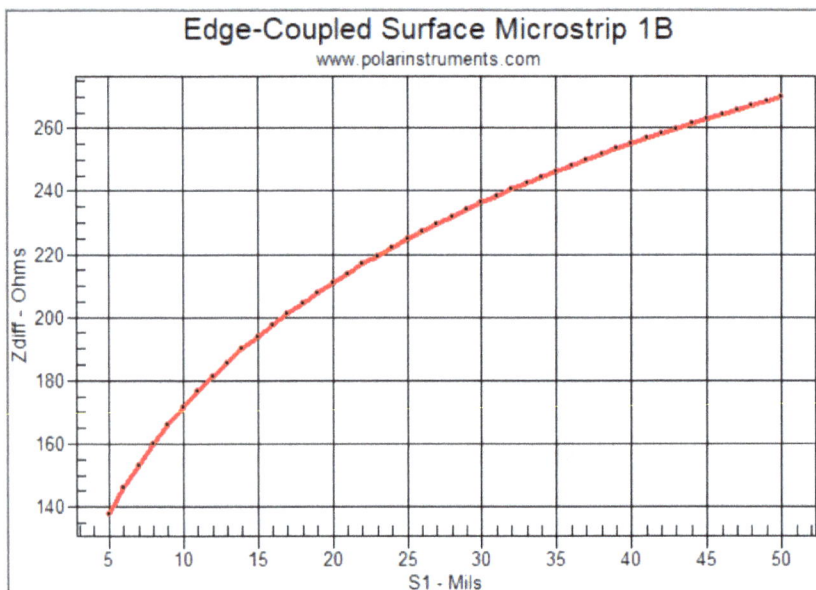

Figure 18.6 Differential impedance of the pair of traces as they are moved farther apart and the plane is very far away. Simulated with Polar's SI9000.

There are two extreme geometries in a differential pair. In the examples explored for differential pairs on circuit boards, the coupling between the two signal lines is smaller than the coupling between either line and the plane. The presence of the plane dominates the differential impedance.

The second extreme case is when the return plane is moved so far away that the coupling between the two lines that make up the differential pair dominates the displacement current and the differential impedance is about the geometry of the two lines and the presence or absence of a return plane plays no role.

450

This is the case when there is a gap in the return plane, or when there is no return plane at all, as in unshielded twisted pairs (UTP).

18.5 Return Currents in a Differential Pair with the Plane Far Away

The concept of relative coupling between the two signal lines or between the signal line and the plane is a useful description to understand the role of the plane in the differential impedance of a differential pair.

But what does this mean in terms of the presence of the return current distribution in a differential pair? If the return plane is far away, or even nonexistent, where is the return current for a differential pair?

When a differential transmission line is driven by a differential signal, the voltage signal into the p-line and the n-line have opposite polarity. A 0V to 1V signal travels into the p-line and a 0V to -1V signal travels into the n-line.

Both voltage wavefronts in the p- and n-lines propagate in the same direction down the line. And both current wavefronts in the p- and n-lines propagate in the same direction down the transmission line. These directions of propagation are independent of the polarity of the signals.

What is different is the direction of *circulation* of the currents in the p- and n-lines.

In the p-line, with the 0V to 1V signal, the current wavefront circulates from the signal line to the return path. This means, at the beginning of the transmission line, current is flowing into the p signal trace and flowing out of the p return path.

In the n-line, with the 0 V to -1 V signal, the current wavefront circulates from the return path up to the signal line. This means that at the beginning of the transmission line, current is flowing into the return plane and flowing up to and back out of the signal trace.

This is a very confusing aspect of signal propagation on differential pairs, yet it is one of the most important principles:

The direction of propagation of the current wavefront ALWAYS travels in the same direction as the voltage wavefront. The direction of propagation of the voltage wav front and current wavefront is exactly the same into the p and n transmission lines. The direction of circulation of the current wavefront will be different in the p- and n-lines as it depends on the polarity of the voltage signal.

At the beginning of the differential pair, with a differential signal applied, the directions of the current into and out of the signal traces and the return plane are shown in **Figure 18.7**. This is simulated at 100 MHz using Ansys HFSS. Red means going into the page and blue means coming out of the page.

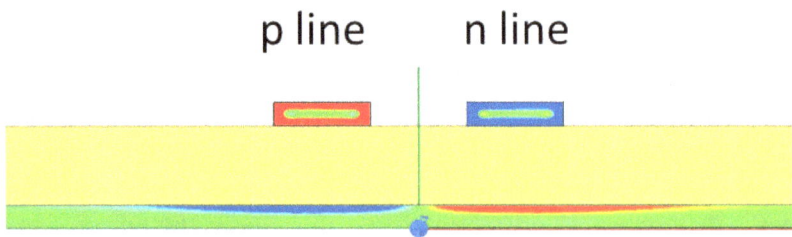

Figure 18.7 The current distribution in a differential microstrip driven with a differential signal. Red means current traveling into the page. Blue means current traveling out of the page.

At the beginning of the differential pair, the return current in the p transmission line flows out of the return plane and the return current in the n transmission line flows into the return plane.

The current in the signal trace is constrained to flow in the signal trace conductor. The return current in the return plane, at very low frequency, would flow uniformly throughout the return plane. However, as frequency increases, the return current redistributes, driven by the current taking the path of lowest impedance, dominated by the path of lowest loop inductance. The principle that describes the current redistribution is referred to as the *skin depth* effect.

At higher frequencies, the current distribution in the signal and return paths is a balance of the two actions. Within each conductor, the current will move as far away from itself as it can to reduce the self-inductance of the current path. This causes the current to spread out to the outside in the constrained signal path. In the return path, this action causes the current to spread out in the return conductor.

However, there is a second action. Between the signal conductor and the return conductor in one transmission line, the two opposite direction currents will move closer to each other to increase their mutual inductance.

These two actions sculpt the current distribution in the return conductor to be mostly in the surface close to the signal conductor, but spread out.

The farther away the signal conductor from the return conductor, the more the current in the return conductor will spread out. The extent to which the return current spreads out in the return path is roughly proportional to the spacing between the signal and return conductors.

When the signal trace is close to the return plane, the return current distribution is spatially localized. As the return plane is pulled farther and farther away, the return current spreads out in the return plane.

The direction of the return current in the return plane under the p-line and the n-line are in opposite directions. In the region in which they overlap, these currents, *propagating* in the same direction but *circulating* in opposite directions, will cancel out.

The farther away the return plane, the more the return currents overlap and the more they cancel out. This behavior is shown in **Figure 18.8**. The return distribution in the plane for a differential signal, simulated in HFSS, overlaps and eventually cancels out as the return plane is pulled farther away.

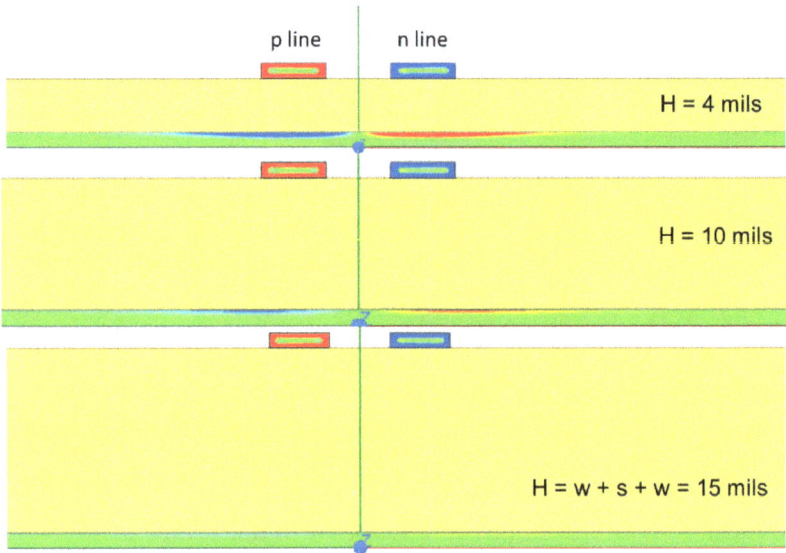

Figure 18.8 Current distribution in the same differential pair for a differential signal as the return plane is moved farther away.

As a rough rule of thumb, when the return plane is as far away as the distance between the outer edges of the two signal lines, return currents are so spread out and overlapping in the return plane there is effectively no return current in the return plane.

This means that when the return plane has been pulled far enough away, pulling it even farther away does not influence the return current in the return plane, and the return plane can be eliminated with no impact on the differential impedance of the differential pair.

So where is the return current in a differential pair, with a return plane far away, when driven by a differential signal?

In the special case of a balanced differential pair with the return plane farther away than the span of the traces, the return current of one trace is carried by the other. This is evident looking at the current distribution of a differential signal in this geometry, as simulated in **Figure 18.9**.

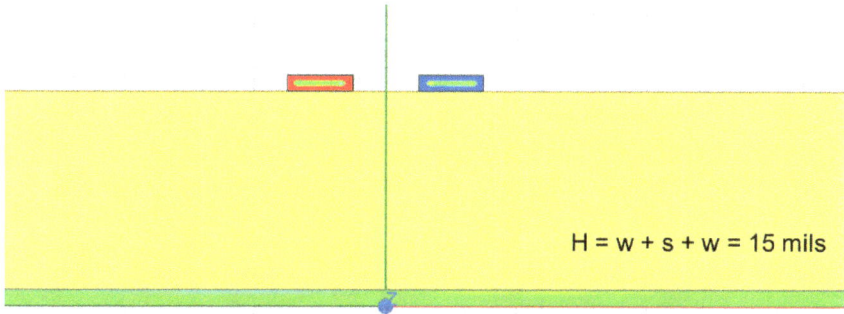

$H = w + s + w = 15$ mils

Figure 18.9 Simulated current distribution in the differential pair. At this distance away, the return currents from the p- and n-lines overlap and cancel out in the return plane.

This current distribution is identical to the case of a coplanar transmission line consisting of a two identical conductors on the same plane.

If the two traces are driven by a differential signal or a single-ended signal, the current distributions and the impedances will be the same. The single-ended impedance and the differential impedance will be the same.

18.6 TDR Response of a Differential Pair with a Short Gap in the Return Plane

The measured TDR response of a differential pair transmission line really has five different impedances, of which three can be measured with the T3SP15D TDR:

- ✓ The single-ended impedance of one line
- ✓ The odd-mode impedance of one line
- ✓ The differential impedance of the differential pair.

The other two impedances not measured are the even-mode impedance of one line and the common impedance.

When the differential pair crosses a short region where the return path is removed, each of these impedances will be affected differently. An example of a differential pair with an electrically short region with no return path is shown in **Figure 18.10**.

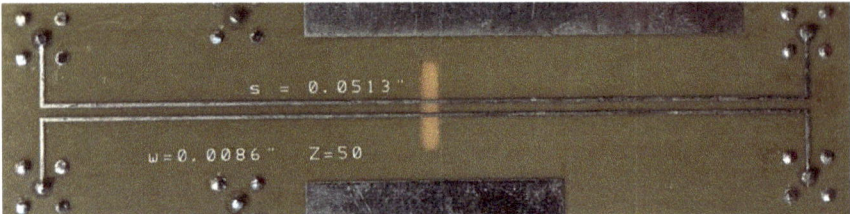

Figure 18.10 Example of a very short gap in the return path for a microstrip differential pair crossing the gap. The light-colored region is where the return plane has been removed.

The single-ended impedance is very sensitive to the lack of return path. In the region where the return path is far away, the single-ended impedance increases. Because this is electrically short, the higher impedance looks like an inductive discontinuity. An example of the single-ended impedance is shown in the measurement in **Figure 18.11**.

456

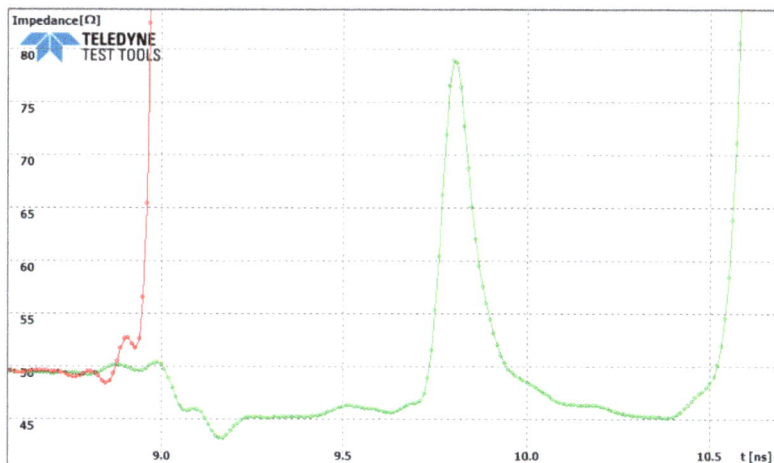

Figure 18.11 The single-ended impedance profile of the microstrip transmission line showing the large inductive discontinuity.

In this example, at a rise time of 65 psec, the short region with no return path shows a very large impedance discontinuity. The single-ended line is coupling to the return plane, and since the voltage on the other line is floating, it couples only a little to the adjacent line.

When the differential pair is driven with a differential signal, the impedance of one line is the odd-mode impedance. Even though we are only looking at the p-line, the pair is driven with a differential signal. This means there is twice the dV/dt between the p- and n-line than between the p-line and the return plane some distance away.

The odd-mode impedance is reduced from the single-ended impedance due to the extra coupling to the adjacent n-line. This comparison is shown in **Figure 18.12**.

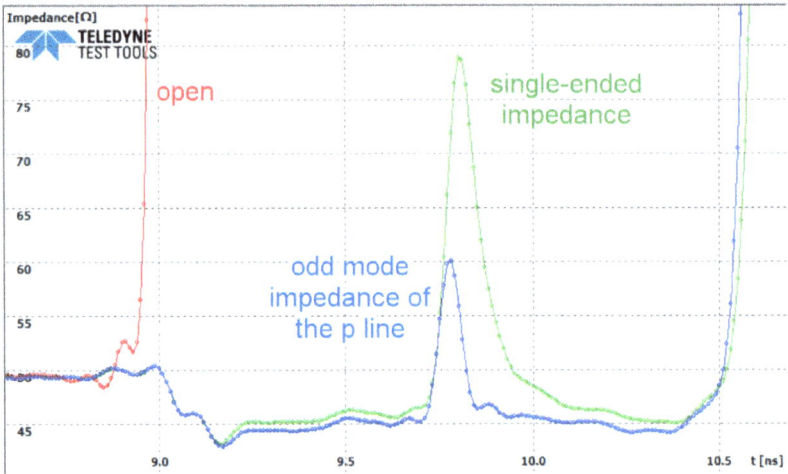

Figure 18.12 Comparing the single-ended and odd-mode impedance of the same line crossing a gap in the return plane.

When the gap size increases, the single-ended impedance will increase but the odd-mode impedance will begin to be dominated by the coupling to the adjacent n-line and not the distant plane. **Figure 18.13** shows the structure of a differential pair and measured single-ended and odd-mode impedance with the return plane removed over a long distance under the two signal lines.

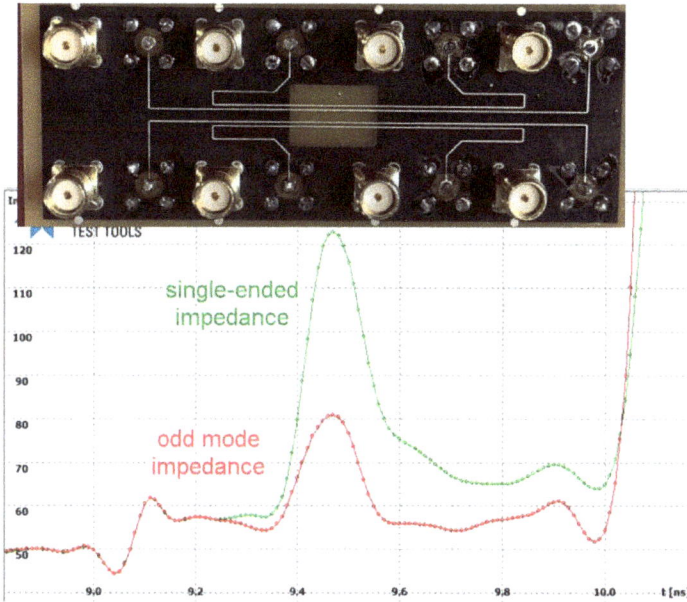

Figure 18.13 A microstrip differential pair with longer gap and the resulting measured single-ended and odd-mode impedance.

In the region of the gap in the return plane, the coupling between the p- and n-lines is larger than the coupling between the p-line and the plane. This means the spacing between the p- and n-lines will dominate the odd-mode impedance the p-line sees. While the odd-mode impedance will be higher if due only to the coupling to the n-line than when the plane is nearby, this impedance will still be much lower than due to the coupling to the far-away plane.

Watch this video and I will show you a TDR measurement of differential pair with a large gap in the return path.

18.7 TDR Response of a Differential Pair with a Short Gap in the Return Plane

When the region with no return plane is electrically long, the single-ended impedance will be very high, with variations due to reflections and resonances from the fringe fields extending far from the signal line.

If the n-line is floating, there will be little coupling to this line. The coupling of the p-line will be to any nearby conductors that might be connected to ground.

If the n-line is grounded, then there will be some coupling between the p-line and the n-line, but with only 1 dV/dt of voltage change.

In addition, any noise voltage coupled into the n-line will see the ground at the far end and reflect, contributing additional voltage variation back to the TDR source.

But when a differential signal is applied to the pair, the coupling to the adjacent n-line with its $2 \times$ dV/dt to the p-line, will dominate the coupling and the odd-mode impedance will be well controlled and much lower than the single-ended impedance.

There are three cases of the impedance of the p-line:

- ✓ Single-ended impedance with the n-line floating
- ✓ Single-ended impedance with the n-line grounded
- ✓ Odd-mode impedance of the p-line with the pair driven with a differential pair

460

These three cases are shown in **Figure 18.14** for the case of a differential pair with no return plane over a very long path.

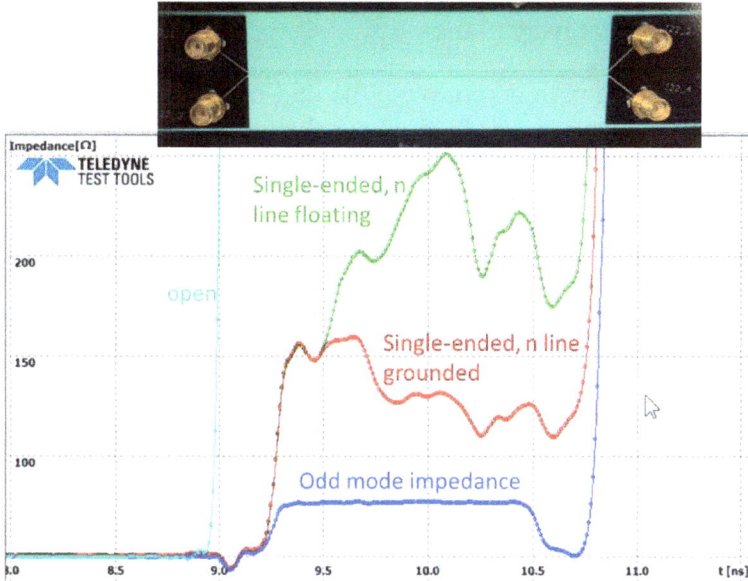

Figure 18.14 A differential pair with the return path removed over a long length, shown above, and the single-ended impedance, floating or grounded and the odd-mode impedance.

In the region with the return plane, at the beginning of the differential pair, the single-ended and odd-mode impedances are the same and close to 50 ohms.

In the region where the return plane is removed the light blue colored region in the photograph, the single-ended impedance goes very high and uncontrolled. There is little coupling between the p-line and the floating n-line. The coupling to the n-line when it is grounded reduces the single-ended impedance. The additional structure in the TDR impedance profile is due to multiple reflections from the grounded end.

But the odd-mode impedance is very constant and well behaved. The odd-mode impedance is 75 ohms, due to the tight coupling

461

between the p-line and the n-line, with its $2 \times dV/dt$ between them. In this region, the coupling to the adjacent line is larger than the coupling to the faraway plane.

Where the return path is removed, the odd-mode impedance, and the differential impedance is all about the coupling between the p- and n-lines.

18.8 TDR Response

The ultimate situation of having no return plane is the case of unshielded twisted pair. For example, CAT5 cable is a collection of four twisted pairs in one bundle. These are typically terminated in an RJ45 connector used for carrying ethernet communications.

Each of the twisted pairs are independent and separate from the other pairs. There is no shield in the CAT5 cable. These twisted pairs are designed to carry a differential signal.

By terminating one pair to two SMA connectors, it is possible to drive one line in the twisted pair by a single-ended signal or the pair with a differential signal. The UTP cable with the wires terminated to SMA connectors is shown in **Figure 18.15**.

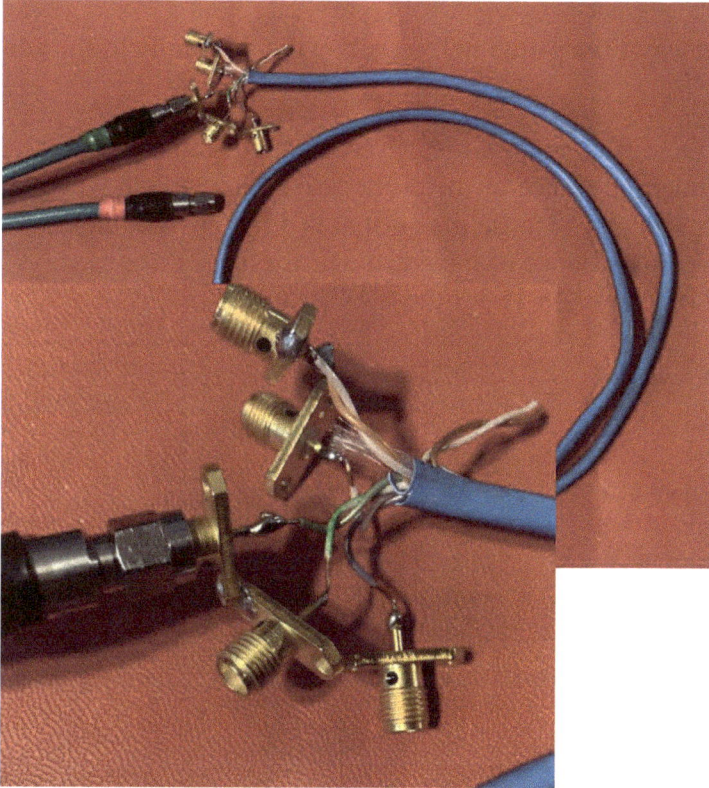

Figure 18.15 Example of the CAT5 cable with one pair connected to SMA connectors, shown in close up. The returns of the two SMA connectors are soldered together.

When the p-line is driven as a single-ended line and the n-line in the pair is floating, the other wires in the cable are literally just floating adjacent to the driven p-line. The current into the p-line, and its single-ended impedance, is due to the random fringe field coupling to any stray conductors near the cable, or from the TDR cable's shield coupling to the floating wire. This single-ended impedance will be very high and variable.

When the second line in the pair, the n-line, is grounded, the current into the p-line will be through the 1 dV/dt to the n-line. This will give the appearance of a constant impedance for the p-line related to the proximity to the n-line. The single-ended

463

impedance of the p-line will be set by the proximity to the adjacent n-line.

When the differential pair is driven with a differential signal, the odd-mode impedance of the p-line will be lower than when the n-line is grounded because the current is driven by $2 \times dV/dt$. The measured impedance of a twisted pair for these three cases is shown in **Figure 18.16** for the CAT5 cable.

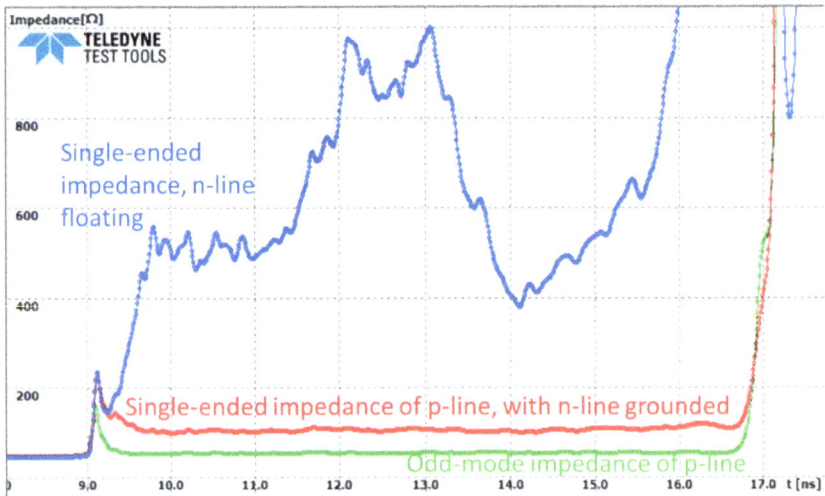

Figure 18.16 Measured impedance of the p-line in a UTP cable, with the second line under three different conditions.

To achieve a controlled impedance of one line when the plane is far away, the current will flow from the p-line, through the displacement current to the n-line, and back to the return of the signal source. The amount of displacement current will depend on the amount of coupling and the dV/dt between them.

When the return plane is far away, the return current from the p-line is carried by the adjacent line, either when it is grounded, or when it is driven with the other half of the differential signal.

464

Of course, the common impedance of the pair will be large and uncontrolled depending on the where any adjacent grounded metal may be positioned.

18.9 Review Questions

1. What is displacement current and when does it flow?

2. As the return plane is moved farther away, what happens to the single-ended and common impedance in a microstrip or stripline differential pair?

3. What is the difference in the electric field lines between the p-line and the n-line in a differential pair when driven by a differential signal and a common signal?

4. What is coupling between the p-line and other conductors a measure of?

5. What are two different explanations as to why there is no return current distribution in the return plane of a differential pair when the plane is very far away from the signal lines and driven with a differential signal?

6. In a differential microstrip, the signal lines are 10 mils wide and the spaced 20 mils apart. Roughly how far away would the plane have to be so that its presence did not influence the differential impedance?

7. In a differential pair driven by a differential signal, what is the direction of propagation of the voltage wavefront in each line of the pair? What is the direction of propagation of the current wavefront into each line of the pair? What is different in each line of the pair?

8. What is the fundamental principle that drives the current redistribution in a transmission line as frequency increases?

9. When the plane is very far away in a differential pair, what other impedance is the differential impedance the same as?

10. When a single-ended signal crosses a gap in the return path, what happens to the instantaneous impedance in the gap region and why?

11. When a differential signal crosses a gap in the return path, what happens to the instantaneous differential impedance in the gap region and why?

12. In an unshielded twisted pair, what do you expect to see as the single-ended impedance of one line with the other line grounded, the odd-mode impedance of one line, and the differential impedance of one line?

13. Where is the return current in a twisted pair in CAT5 cable?

14. CAT6 cable has a shield surrounding all four twisted wire pairs. The spacing between wires in each twisted pair is about 50 mils. The spacing from the twisted pair to the shield is about 200 mils. Where do you expect the return current to flow in a CAT6 cable when a pair is driven by a differential signal?

Chapter 19 Analyzing Discontinuities and Hacking Interconnects

All interconnects are transmission lines. Their cross section can be uniform or nonuniform. When a signal travels down a uniform transmission line, it sees a constant instantaneous impedance. If the rise time of the step is short enough, the TDR response will show a flat top or a flat bottom.

If the TDR shows a flat top or flat bottom, we describe the transmission line as electrically long. We can literally read the figures of merit of the interconnect, the Z0 and TD, from the front screen.

If the TDR response shows a peak or a dip, we describe this section as electrically short. It could be a uniform transmission line, or it could be a nonuniform transmission line.

In the TDR response of an interconnect, we refer to all electrically short interconnects that appear as peaks or dips as discontinuities. Analyzing discontinuities is a little more complicated than analyzing uniform transmission lines.

19.1 Electrically Short Interconnects as Discontinuities

Every TDR response of any interconnect shows two different types of structures, uniform transmission lines and discontinuous. An example of the TDR response of a backplane interconnect trace is shown in **Figure 19.1**. The interconnect is a collection of uniform transmission lines, with a constant instantaneous impedance and a flat region, and discontinuities as either peaks or dips.

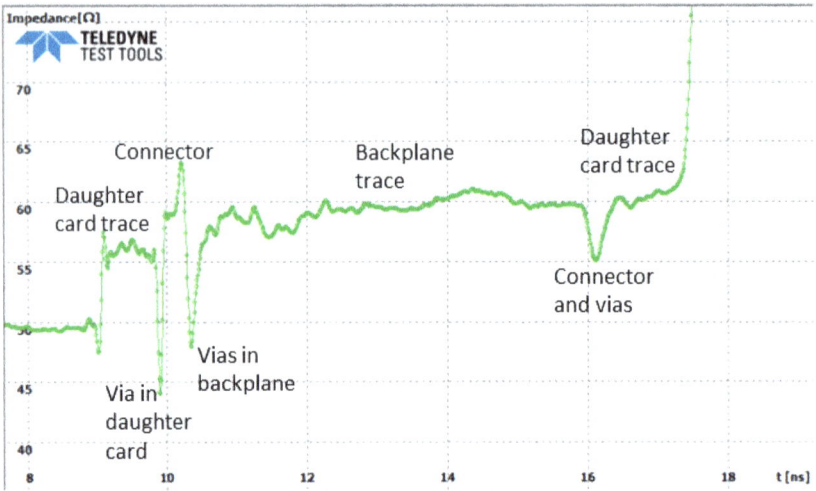

Figure 19.1 The measured impedance profile of a backplane interconnect showing a series of uniform transmission lines and discontinuities. Each feature of the interconnect is identified.

In this backplane interconnect, we can identify the physical features of the interconnect from the flat regions and discontinuities.

When the TDR response shows a flat region, we interpret this structure as a uniform transmission line with just two terms that characterize it: the characteristic impedance and time delay.

When the interconnect is electrically long, we can literally read these two figures of merit off the front screen of the TDR plot.

When the interconnect is electrically short it does not have a flat top or flat bottom. We cannot extract its characteristic impedance and time delay. We need to use a more involved process to extract these important figures of merit.

The condition for an interconnect being electrically long or short depends on the rise time of the TDR signal and the TD of the interconnection. In principle, electrically long means that the difference in time for the reflection from the front and the reflection from the back of the interconnect is longer than the rise time. This would allow a period of time for the edge to not be reflecting outside of the rise time. In principle, this condition for electrically long is:

$$2 \times TD > RT \tag{19.1}$$

Likewise, for an interconnect to be electrically short, the condition is

$$TD < \tfrac{1}{2} \times RT \tag{19.2}$$

In practice though, to have a flat top or bottom really requires the time delay of the interconnect to be a little bit longer. In practice, for an electrically short interconnect we require,

$$TD < RT \tag{19.3}$$

This is illustrated in **Figure 19.2**. In this example, a uniform transmission line was simulated with a region in the middle having a 5-ohm higher impedance. This discontinuity started out with a TD twice as long as the rise time of the signal. With its flat top, it clearly looks like a uniform transmission line. Then its TD was decreased to be TD = RT and TD = $\frac{1}{2}$ × RT.

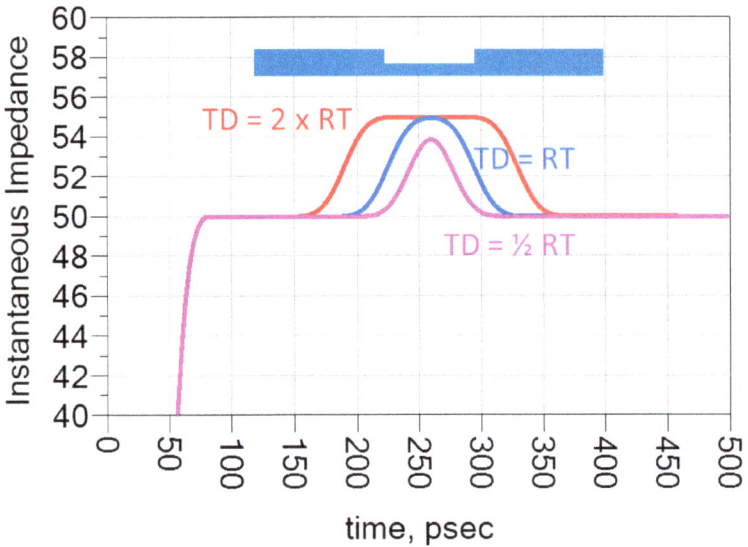

Figure 19.2 The simulated TDR response of a discontinuity with a decreasing TD compared to the rise time.

When the TD = RT, the top is not quite flat and it's hard to distinguish as a flat top. It would be identified as a discontinuity. That's why the practical condition is that the TD < RT for an electrically short discontinuity structure.

In an FR4-like laminate, this condition for an electrically short interconnect is:

$$TD < RT \qquad \frac{Len}{6 \, ^{in}/_{n\,sec}} < RT \qquad Len[inches] < 6 \times RT[n\,sec]$$

$$(19.4)$$

If the rise time is 0.05 nsec = 50 psec, then an electrically short interconnect is 0.3 inch or shorter.

Figure 19.3 is an example of the measured TDR response of a uniform transmission line with increasingly longer regions of a short length of uniform low impedance transmission line. Even

470

though the characteristic impedance of each section is exactly the same, the TDR response shows a lower impedance the longer the section. In this example, the rise time is 50 psec. The longest section is 300 mils long, at the threshold of being electrically short.

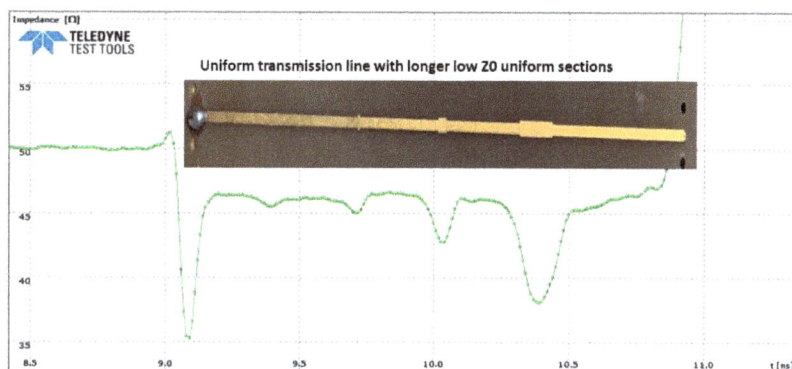

Figure 19.3 An increasingly longer section of a uniform transmission line, each still electrically short.

Ultimately, we would like to describe a discontinuity in terms of an ideal electrical circuit model which, if brought into a model of the rest of the system, would accurately predict the entire system's electrical performance.

A discontinuity could be described in terms of a uniform transmission line, an ideal lumped capacitor element or an ideal lumped inductor element or combinations. In addition, a transmission line element could be connected in series or in parallel to model the case of a stub interconnect structure.

Describing a discontinuity and extracting a few figures of merit that describe it electrical cannot be done from the front screen of the TDR directly. Instead, we have to go through a slightly more complicated process, referred to as *hacking*.

471

19.2 Hacking Interconnects

Hacking interconnects is the process of taking a measured response of the interconnect, building a scalable circuit topology-based model to describe it, and fitting the parameter values of the circuit so its simulated response matches the measured response. This process is also referred to as *measurement-based modeling.*

An example of the final result is shown in Figure 19.4. In this example, the measured TDR response is brought into a simulation environment and a circuit model is created. Its parameters are optimized so that the simulated TDR response matches the measured TDR response. Based on this agreement, we use the fitted parameters as the figures of merit of the interconnect and the discontinuity region.

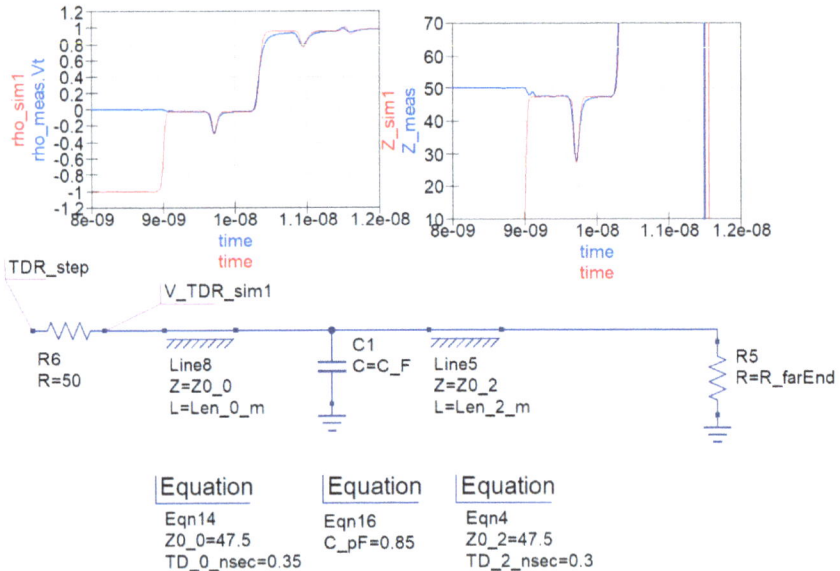

Figure 19.4 An example of the elements in hacking an interconnect.

In this example, the agreement between the measured reflection coefficient and impedance profile and the simulated reflection coefficient and impedance profile gives confidence that the circuit model created and the parameter values chosen match the electrical behavior of the interconnect.

In this specific structure measured, the first section is modeled as a uniform transmission line with a characteristic impedance of Z0 = 47.5 ohms and TD = 0.35 nsec.

The discontinuity in the middle is modeled as lumped capacitor with a C = 0.86 pF.

The last section is modeled as a uniform transmission with Z0 = 47.5 ohms and TD = 0.3 nsec.

Using this simple circuit, all of the multiple bounces within the real interconnect showing up after the reflection from the open end are clearly taken into account by the circuit model.

The process steps in hacking an interconnect are as follows:

1. Bring the measured TDR (step response) into a simulation environment

2. Display it as rho, Z(t)

3. Synthesize a transmission line/L/C circuit topology to match the measured structure with parameters: Z0, TD, C, L

4. Simulate the TDR response of the synthesized ideal circuit with the rise time of the measured TDR response

5. Adjust the parameter values to match the simulated Z(t) to the measured Z(t)

6. Refine the topology as needed

7. When the agreement is "close enough use the fitted parameters as the extracted figures of merit for the measured interconnect

19.3 The Simulation Environment: QUCS

The most important tool required to hack interconnects is a simulation tool that allows bringing in measured data and simulating TDR responses of ideal circuit elements with transmission line models.

Any SPICE-compatible tool can do this in principle. Some versions make the task easier than others. The most suitable, free, and open-source tool for hacking interconnects is Quite Universal Circuit Simulator (QUCS). It can be downloaded from a source forge using this link.

In QUCS, simulations are organized by projects. These are collections of schematic files, data sets, and graphs. **Figure 19.5** shows an example of the typical screen for a QUCS circuit with the most important elements.

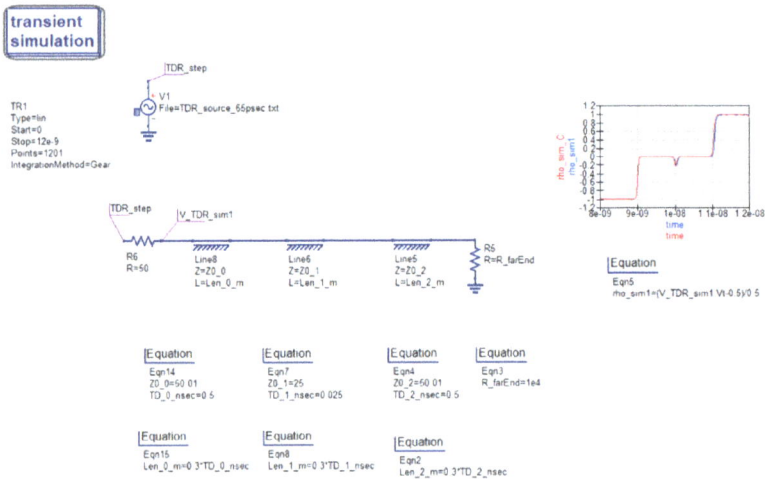

Figure 19.5 An example of a simulation page from QUCS with the typical elements used in hacking.

In QUCS, the following elements are commonly used:

Transient engine: used to set up the simulation conditions

Piecewise linear (PWL) voltage source: used to bring in measured TDR data by turning each voltage vs time data point into a voltage source that can then be plotted, used to stimulate circuits, or used in equations to be converted into something else.

Transmission line elements: an ideal circuit element describing a transmission line with a Z0 and TD.

Lumped elements: such as an ideal capacitor and inductor elements.

Equations and parameters: allow converting simulated terms or parameters into other terms.

Graphical response: displaying data in publication quality graphs automatically.

A circuit is really a graphical description of a series of differential equations. Each element in the circuit model describes how a voltage or current behaves at each node. When solving the differential equations, it is possible under some conditions that the solution becomes unstable and it either takes forever to converge, or errors crop in.

There are a few settings that will reduce the risk of a convergence error. The three most important settings to reduce convergence errors are

1. The integration method in the transient simulator engine

2. The interpolation method for the PWL voltage sources

3. The precise time delay lengths of the transmission line elements

In the transient simulation engine, select the Gear method for integration mode. When the simulation engine is double-clicked and opened, the option under properties for integration method should be selected for Gear. The order of integration should be selected as the highest, 6. Finally, the minimum time step should be selected as 0.1 psec. These selections are shown in **Figure 19.6**.

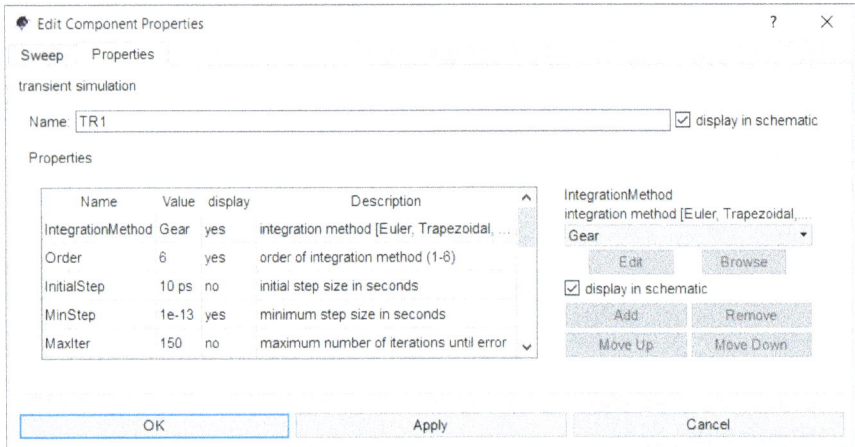

Figure 19.6 To increase the stability of a simulation, select the Gear method under integration with the other parameters as shown.

The Teledyne T3SP15D TDR measures a voltage at 10-psec intervals. When the measurements are exported, the data is in the form of time intervals and voltages. When setting up the transient engine, a time step of 10 psec should also be used. This is the optimal time step to reduce computation time and minimize any interpolation required.

The PWL voltage source will generate a voltage at each time step used in the simulation engine. Sometimes, the time step in the simulation is decreased when the change in voltage exceeds a threshold. In this case, the PWL source has to interpolate a voltage level to supply to the simulation engine.

For best stability, the interpolation method selected in the PWL circuit element should be cubic. To set this interpolation method, double-click on the PWL element and select the cubic method as shown in **Figure 19.7**.

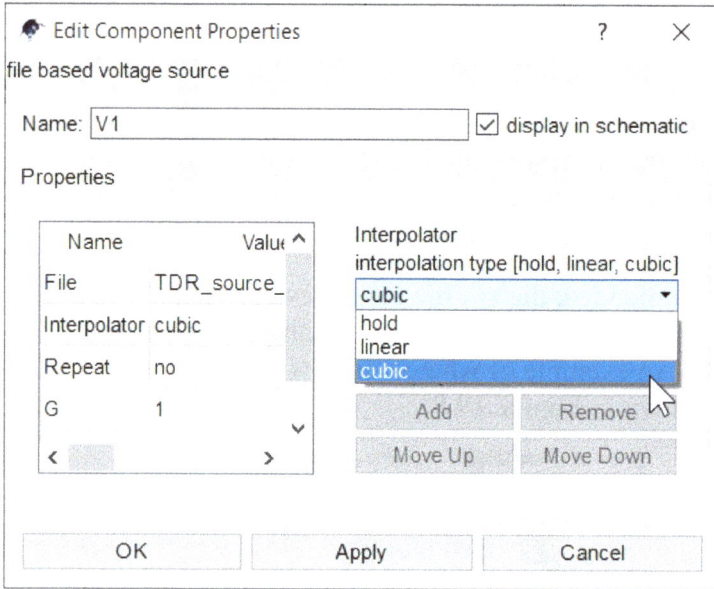

Figure 19.7 Select the cubic method of interpolation in the PWL circuit element.

Finally, there are occasions when even with these settings, the simulation takes more than 10 sec to complete and the results have way too much noise. This is an indication of an instability in the convergence of the solution to the differential equations. The root cause is usually that some reflection occurred superimposed on another edge and an error term got too large and propagated through the rest of the simulation and grew.

If this happens, the solution is to change the time delay of a transmission line a small amount, even as small as 1 psec or less. This will shift the instant of the reflected voltage a few iteration steps enough to remove the overlap and achieve stability. If the simulation takes longer than 10 sec, something is not converging. In this case, add a tiny delay to a transmission line element.

Using these three settings will dramatically improve the stability of the simulations.

477

19.4 QUCS Schematics Available for Download

To practice hacking interconnects and to provide a quick start template for a variety of models if you want to hack your own TDR measurements, all the examples in this chapter can be downloaded in a zip file from this location.

Once downloaded, the zip file can be extracted into a single project folder and placed into the directory with your other QUCS projects. An example of what your QUCS content list should look like is shown in **Figure 19.8**.

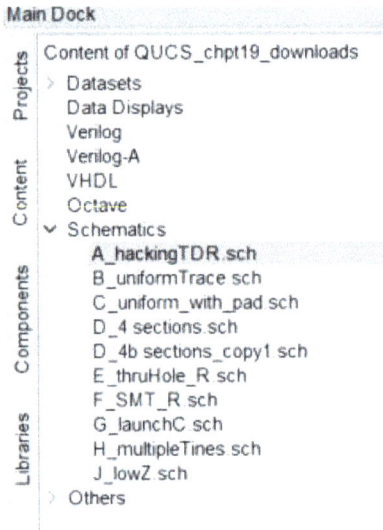

Figure 19.8 The list of nine schematic files provided to hack along and how they appear in the QUCS project directory after installation.

When opened, there are nine different schematics, each corresponding to a different measured TDR response and model interconnect.

In the schematics provided, the optimum model values have been selected so that the simulated response and measured response are well matched.

To practice hacking, you can set all the parameter values to a constant neutral value and then try to find the best values on your own.

In addition, a few measurement files are provided that are not solved. Try your hand at importing these measurements with a PWL voltage source, determine the best circuit topology, and then find the optimum parameter values.

The following circuits are provided:

A_hacking: examples of the three basic circuit elements

B_uniformTrace: a low-impedance uniform transmission line

C_unifrom_with_pad: a uniform transmission line and a small pad in the middle

D_4 sections: an interconnect with four different width transmission lines

E_thruHole_R: a 50-ohm transmission line with a 50-ohm axial lead resistor at the far end terminated to ground

F_SMT_R: a 50-ohm transmission line with a 50-ohm surface mount 1206 resistor at the far end terminated to ground

G_launchC: a uniform transmission line with a small capacitive discontinuity at the launch

H_multipleTlines: an interconnect with a few different line width traces

J_lowZ: a very low impedance uniform transmission line with two small pads in the middle

*Watch this video and I will walk you
through these circuit files provided.*

19.5 Simulating a TDR

Any SPICE simulator can simulate the reflections from
transmission lines and display the results as in a TDR. Just like in a
real TDR, a voltage source drives a step edge waveform with some
rise time into the circuit. There is a 50-ohm source resistance and
circuit elements that define the model for the DUT.

We simulate and record the *total* voltage after the 50-ohm resistor.
This is the *total* voltage. It is composed of an *incident* voltage due
to the voltage divider and any *reflected* voltage. By subtracting off
the *incident* voltage, we find the *reflected* voltage component.

The incident voltage continues to propagate down the interconnect
and interact with the rest of the circuit.

The reflection coefficient, rho, is calculated as the ratio of the
reflected voltage to the incident voltage. This is the starting place
of simulating the TDR response. From the reflection coefficient,
the first-order impedance profile, Z(t), is calculated. This is exactly
what is displayed on the real TDR as the impedance profile.

An example of an ideal TDR circuit and the equations to turn the
simulated voltage into a reflection coefficient and impedance
profile are shown in **Figure 19.9**.

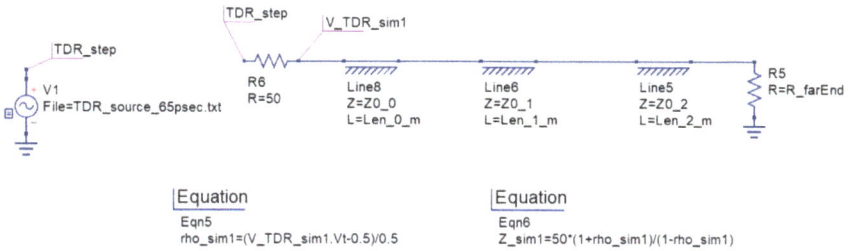

Figure 19.9 A typical TDR circuit with equations to convert the simulated voltage into rho and Z(t).

The TDR response from a discontinuity depends on the rise time of the incident step edge. This means that if the goal is to match a simulation to a specific measurement, the simulation must use the same rise time and step edge shape as the measurement source. This is accomplished by recording the reflected step edge from the TDR and using this waveform as the stimulation source in the simulated TDR.

For best correlation, the same rise time used to generate the TDR response of the DUT should be used in the open measurement.

A PWL voltage source based on the open response of the TDR is used to stimulate the circuit. What is recorded by the real TDR and exported as a text file is the reflection coefficient, which scales from 0 to 1. This is turned into a voltage source as the step response. Through the initial voltage divider, the incident voltage becomes 0.5 V.

The measured reflection coefficient from the TDR is brought into the simulation environment as a PWL voltage source. This is displayed as the reflection coefficient and with an equation, turned into an impedance profile.

It is artificially delayed by a about 100 psec in order for the opportunity of adding a short transmission line feed in the model to simulate the TDR cable launch into the DUT.

An example of the imported step response used as the simulation source, and the measured TDR response from a simple, low-impedance uniform transmission line, is shown in **Figure 19.10**.

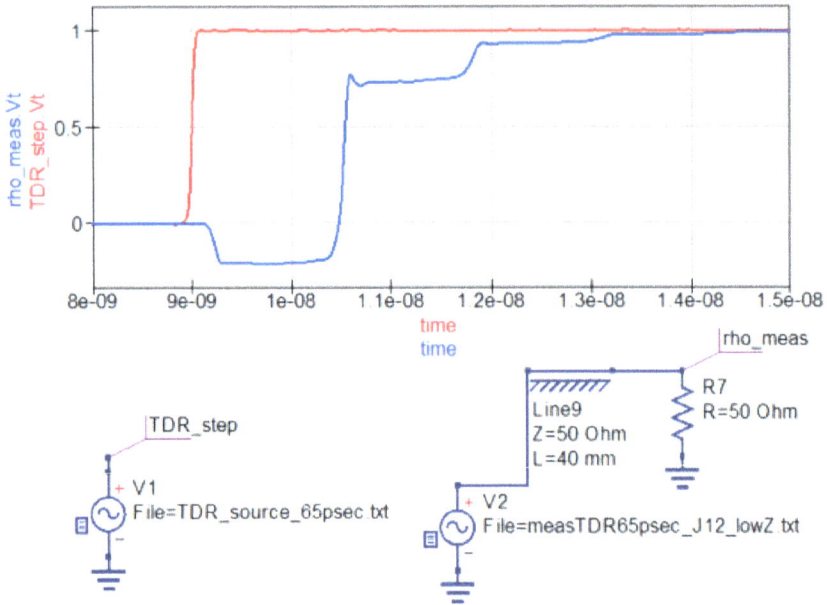

Figure 19.10 The imported step voltage and measured TDR response brought into the simulation environment. The slight delay of the onset of the TDR response from the DUT after the step edge signal is evident. This is the example in the schematic B_uniformTrace.

19.6 Models of Discontinuities

The first step in hacking an interconnect is to identify the circuit topology required to match a measured response. The uniform regions with a flat top or flat bottom to the impedance profile are easily measured by uniform transmission line elements with the two parameters, Z0 and TD.

Discontinuities that appear as peaks or dips can be modeled by three simple ideal circuit element models. Combinations of these three different elements can be used to describe very complex interconnect structures.

An important limitation in using QUCS is that it only has ideal lossless transmission line elements as building block elements. It cannot take into account the frequency-dependent losses as from dielectric loss or series resistance.

But given these limitations, it is still remarkable how well real interconnect structures can be modeled by simple ideal circuit elements.

A discontinuity can be a peak or a dip. Either of these can be described by an electrically short transmission line element with a Z0 and TD that describe it. The TD would be chosen as shorter than the rise time, so it is electrically short. A peak is a higher impedance than the rest of the line and a dip is a lower impedance than the rest of the environment.

If a dip is described by an electrically short, low-impedance transmission line, its TD must be shorter than the TDR source rise time. But the combination of values for the Z0 and TD are not unique. As the TD is made shorter, a lower value of Z0 can be found to give just as good agreement to the measurement as a longer TD with a different Z0.

Which one of the infinite possible combinations should be used?

If the model required is based on an assumed TD or an assumed Z0, then fixing one term will make the other value unique. If there is no compelling reason to use a transmission line for the discontinuity, using a C or L element might be a better choice. They each have just one figure of merit and would be a unique solution.

A dip, a low impedance, can also be described by an ideal capacitor element. It has just one parameter that defines it, the capacitance.

A peak, a high impedance, can also be described by an ideal inductive element. It has just one parameter that defines it, the inductance.

These three circuit elements are the building blocks from which more complicated circuits can be built. **Figure 19.11** shows how these elements are combined together in a circuit with uniform transmission lines. Also shown are examples of the simulated TDR response of a lower-impedance ideal transmission line element that matches the response of an ideal capacitor and the response of a higher-impedance ideal transmission line that matches the response of an ideal inductor.

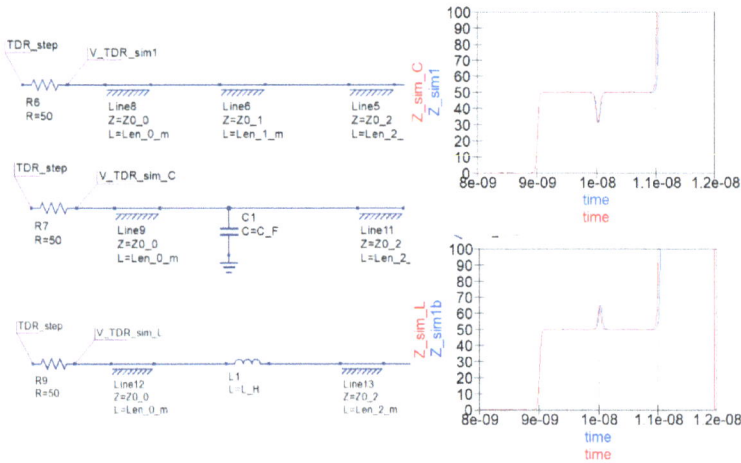

Figure 19.11 Examples of the three different ideal circuit elements used to model measured TDR responses. These examples are provided in the schematic file A_hackingTDR.

Watch this video and I will walk you
through circuit A.

Watch this video and I will walk you
through an example of hacking a simple
transmission line device under test.

19.7 An Example of Hacking an Interconnect with a Launch Discontinuity (Schematic File G_LaunchC)

The first step in hacking an interconnect is to measure a real interconnect and bring it into the simulation environment.

The second step is to look at the interconnect itself or the TDR response and decide on the circuit topology. This is literally as simple as using a uniform transmission line element for regions of the TDR response that are flat and a C element when it dips and an L element when it peaks.

Once the circuit topology is set up, it's a matter of adjusting the parameter values to match the simulated response to the measured response.

The simplest process is to start at the beginning, adjusting the parameters for each element. This way the impact on the TDR response in the following section from the previous sections is already included in the simulated TDR response.

When low-cost connectors are used to transform the coax geometry of the TDR cables into the PCB trace geometry, and the via pad stack into the board is not optimized, the launch often looks like a capacitive discontinuity. **Figure 19.12** shows an example of a simple stripline trace in a circuit board with low-cost SMA connectors on the ends. These are not optimized and cause a large launch discontinuity.

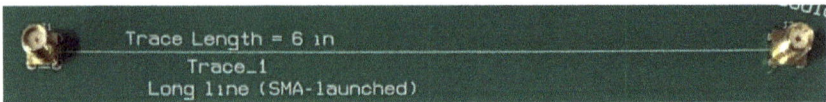

Figure 19.12 Simple example of a uniform stripline trace with SMA connectors on the ends. The measurement from this trace is in schematic file G_launchC.

The measured TDR response, as brought into QUCS, is shown in **Figure 19.13**.

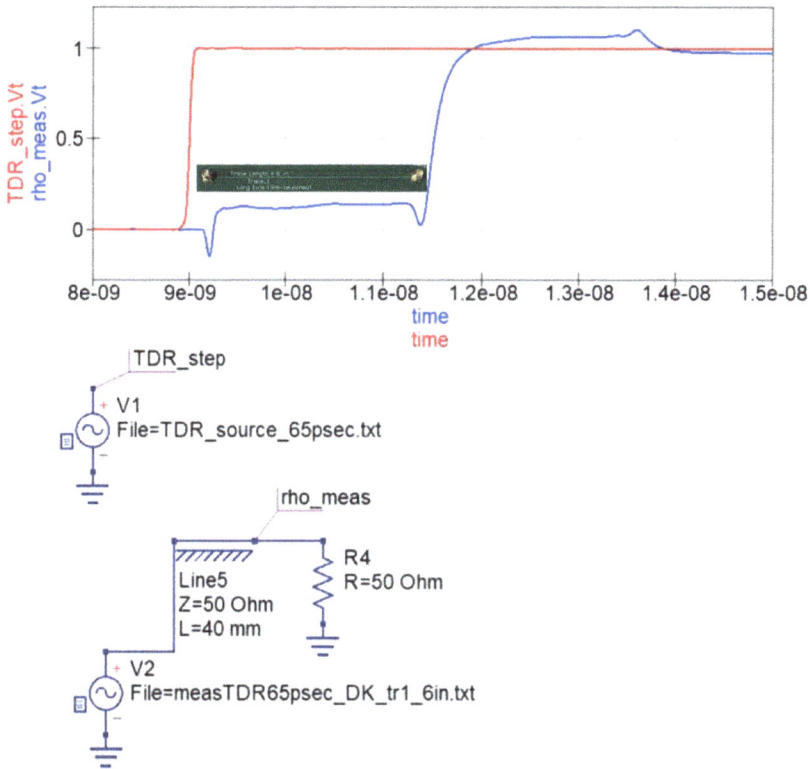

Figure 19.13 Example of the measured TDR step response from a simple uniform stripline transmission line in an FR4 substrate with SMA connectors after brought into the QUCS simulation environment.

After the measured response is brought into QUCS, the circuit topology can literally be read off from the simulated TDR response. From the measured reflection coefficient, the starting place model would be a capacitor, a uniform transmission line and another capacitor followed by a short transmission line, corresponding to the barrel of the SMA and then an open.

The capacitors should be the same at the beginning and the end of the line, since the TDR response was measured to be identical from both ends. This is shown in **Figure 19.14**.

487

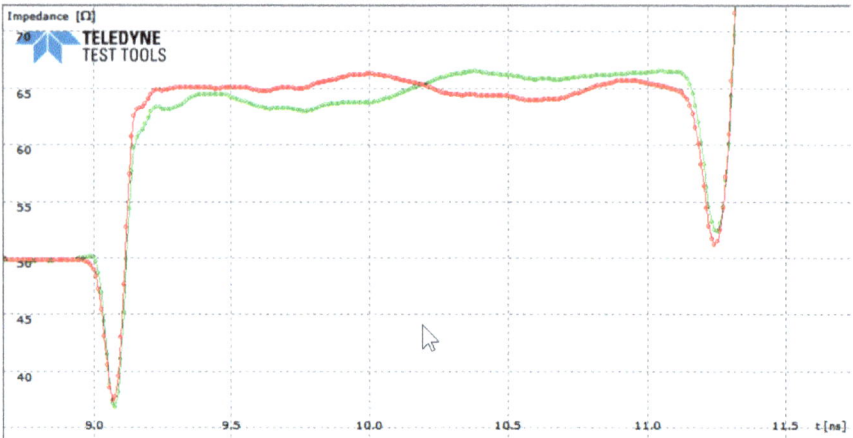

Figure 19.14 Measured TDR response from this stripline transmission line, measured from both ends. Note the launches are identical and there is a little asymmetry in the impedance profile of the line.

Once the model is created, it's a matter of walking from the beginning of the model and adjusting the parameter values to get the best match between the measured and simulated response.

The TDR impedance profiles as measured and simulated with the optimized model elements are shown in **Figure 19.15**. The plot is expanded to align the model elements with their features in the TDR response.

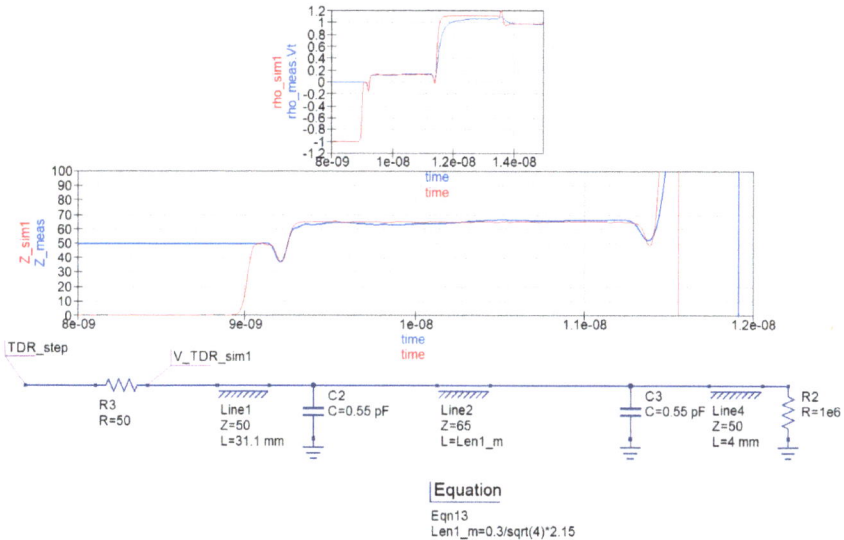

Figure 19.15 An example of the measured and simulated reflection coefficient and impedance profile with the circuit element models aligned to the TDR features. The parameter values are also shown.

The excellent match between the measured and simulated TDR performance can be used to establish high confidence that the circuit elements and parameter values selected are a good model for the interconnect features.

The launch capacitance is 0.55 pF. The very slight difference in simulated and measured impedance profile at the far end discontinuity is due to the fact that the rise time for the measured TDR has increased slightly due to the interconnect losses, while this effect is not included in the simulated model.

19.8 Avoid Mink Holes

In the quest for a more perfect match, it's easy to get caught up in a *mink hole*. A rat hole is a convoluted path that takes you farther and farther away from your goal.

A mink hole is a *rat hole* lined in mink fur. It is a convoluted path that takes you farther and farther away from your goal, but feels really good when you are inside it.

Finding a slight tweak to the model, adding another circuit element, or finding the third decimal place of a parameter for that perfect overlap between the measured and simulated response can be a lot of fun for an engineer. Always be aware of when it is ok to stop.

Is the model good enough to have a figure of merit that describes the discontinuity or complete interconnect?

As a general principle, always start with the simplest model possible and build complexity from there.

It is remarkable how incredibly well very simple models can match the measured performance of very complex-looking structures. Often, the real world can be accurately described in a very simple way.

19.9 Hacking an Interconnect with Multiple Impedances

This technique of hacking can be used to match not just discontinuities but also uniform transmission lines, taking into account the effect of masking. This occurs when the reflections from one section of transmission line distorts the transmitted signal and interferes with the interpretation of the following reflections.

This was illustrated in example circuit B_uniformTrace.

In example H_multipleTlines, the measured interconnect structure is composed of five different sections of uniform transmission lines with different linewidths and different impedances. A close-up of the line is shown in **Figure 19.16**. The connector and launch discontinuity is only on the left side, not the right side. This is reflected in the model for this interconnect.

Figure 19.16 Transmission line with five different impedance sections. Note the connector is only attached on one side. There is no launch discontinuity at the far right end of the transmission line.

The measured and simulated TDR responses based on a simple model of the launch and the five transmission line segments are shown in **Figure 19.17**.

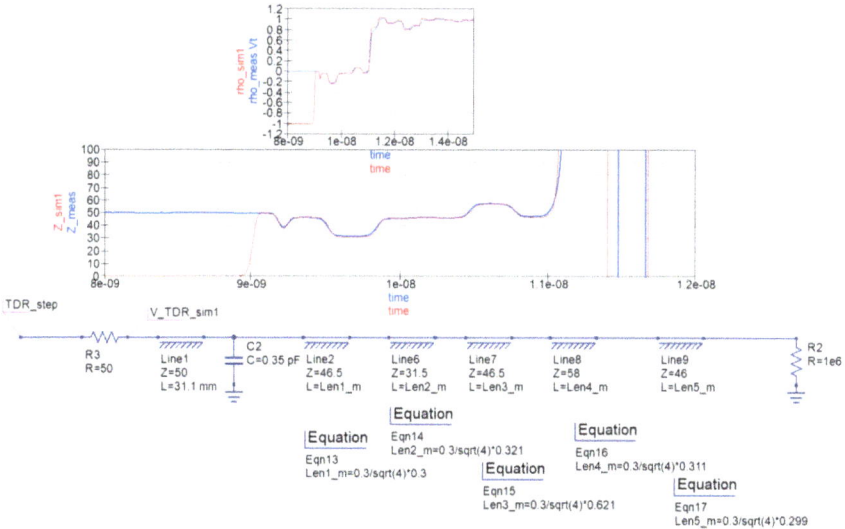

Figure 19.17 Measured and simulated response of the multiple section transmission line structure. Note that all of the small reflections in the reflection coefficient plots after the open end are perfectly matched with this model.

491

The agreement between the measured and modeled response is excellent, even including the multiple reflections apparent in the reflection coefficient out to 15 nsec.

Watch this video and I will walk you through hacking this multiple transmission line device under test.

19.10 A Transmission Line with a Small Capacitive Discontinuity

In this example, there is a small pad in the middle of a uniform transmission line. A close-up of this microstrip test line is shown in **Figure 19.18**. The measured data and circuit simulation of this structure is in schematic file: C_uniform_with_pad.

Figure 19.18 A simple microstrip with a capacitive pad.

The TDR response should be of a simple uniform transmission line with a capacitive discontinuity in the middle. This is confirmed in the measured TDR response brought into QUCS. A simple model with a uniform transmission line on either side of a capacitive discontinuity matches this measured behavior very well. This is shown in **Figure 19.19**.

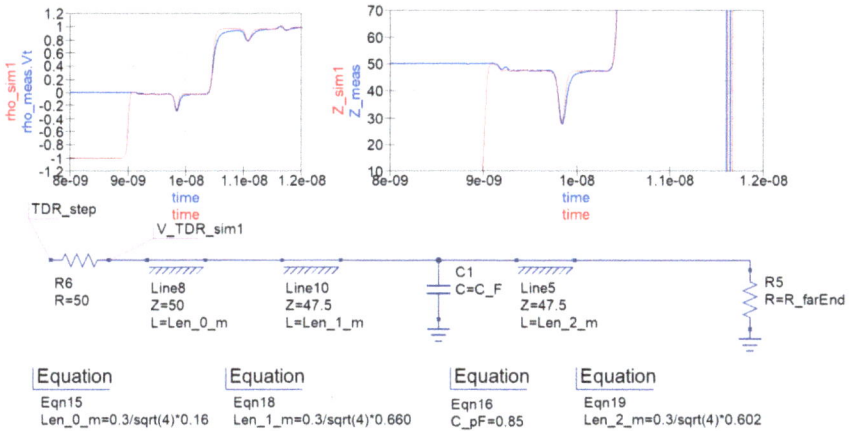

Figure 19.19 The measured and simulated response of the uniform transmission line with a small capacitive discontinuity. This is schematic file C_uniform_with_pad

The capacitance of this pad is measured in this way as 0.85 pF. This example illustrates how it is possible to use a TDR measurement to build equivalent circuit models for PCB structures.

19.11 Hacking a Very Low Impedance Interconnect

In this example, a very wide microstrip transmission line was measured. This is expected to have a very low impedance. A top view of the conductor is shown in **Figure 19.20**.

Figure 19.20 Example of a very low impedance microstrip with two reference pads to dielectric constant measurement.

This can be modeled as a low-impedance, uniform transmission line with two lumped capacitor elements spaced apart. This is a total of five transmission line elements.

The circuit model and simulated response after the transmission lines and capacitor parameters were optimized, as shown in **Figure 19.21**.

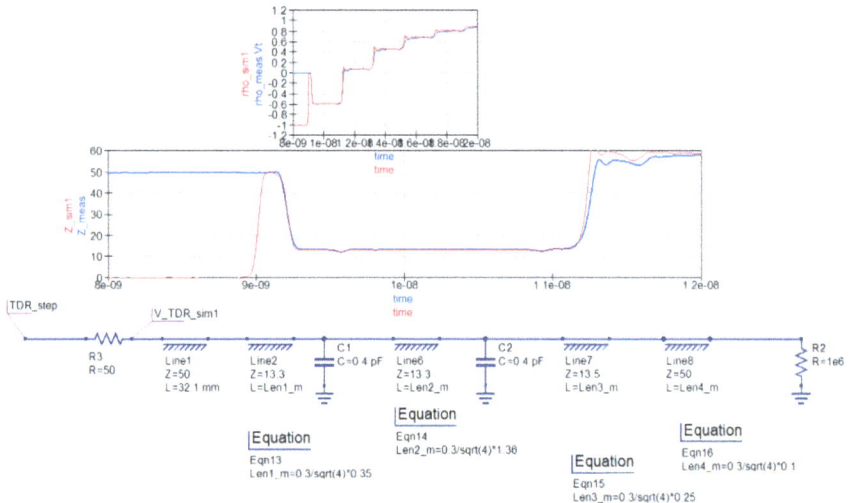

Figure 19.21 An example of the measured and simulated TDR response of a low-impedance transmission line.

The impedance profile after the initial reflection at the far end shows only about a 50-ohm impedance. It is not the high impedance of an open.

This is what is measured, and more importantly, this is what is simulated for the low-impedance transmission line. The masking effect due to the high reflection at the front and each time the signal passes through the 13-ohm to 50-ohm interface, is completely taken into account in the simple model and the simulated TDR response.

It is also remarkable that all the details in the reflection coefficient and its long climb up toward the open at the far end is accounted for in this simple model.

19.12 Extracting Circuit Models for Termination Resistors

This technique of hacking is a generic technique that can be used to extract circuit models for passive components used in interconnect applications, such as terminating resistors.

As an example, a few transmission lines were constructed with a thru-hole and surface -mount resistor termination. These two transmission lines are shown in **Figure 19.22**. The measured data and simulation are in schematic file E_thruHole_R.

Figure 19.22 Examples of two microstrip transmission lines with SMA launched on one end and terminated with 50-ohm resistors at the far end. Top trace has an axial lead thru hole resistor and the bottom trace has a 1206 surface-mount resistor.

The measured and simulated response of the axial lead resistor is shown in **Figure 19.23**. The model that describes this resistor is basically a 0.8-pF capacitor that defines the pad at the launch and a 4.2-nH inductor that defines the leads of the resistor. In this example the resistor is actually 51.5 ohms.

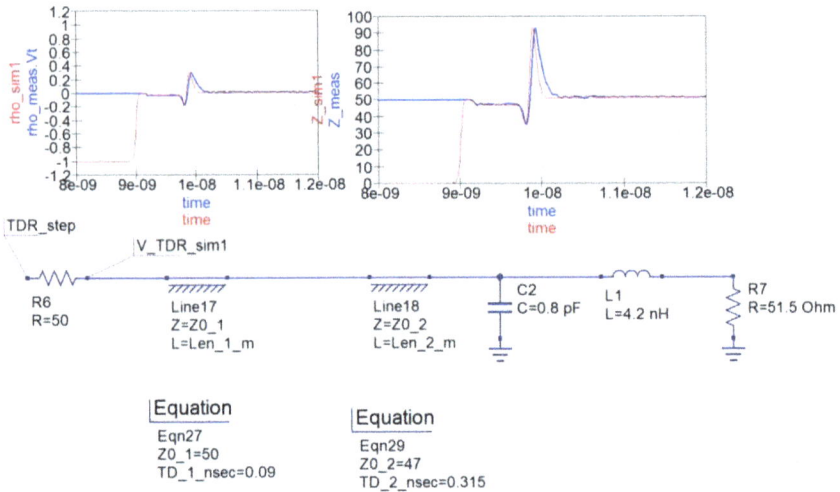

Figure 19.23 The measured and simulated behavior of an axial lead resistor. This is the output of schematic file E_thruHole_R.

Watch this video and I will walk you through hacking a model for the thru-hole axial lead resistor and transmission line.

The same analysis and model can be used to describe the surface mount resistor. Given its smaller size, we expect smaller values for the parameter values in the C and L elements in the model.

The measured and simulated behavior of a transmission line with a 1206 SMD 50-ohm resistor is shown in **Figure 19.24**. This is included in the schematic file F_SMT_R.

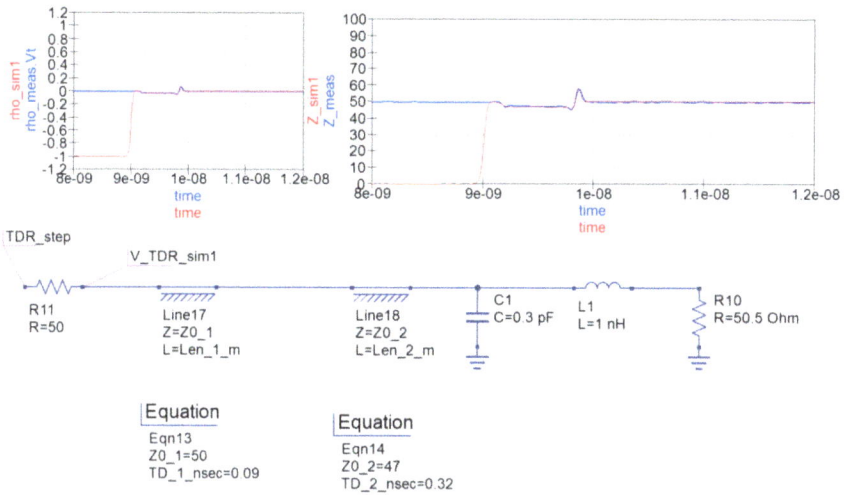

Figure 19.24 Measured and modeled impedance profile for a surface-mount 1206 resistor. This is in schematic file F_SMT_R.

From this model fit, the circuit model for this SMT resistor is a 0.3 pF capacitor and a 1-nH inductor. These are each less than one third of the axial lead resistor parameter values. This is one of the advantages of using SMT devices. They will look more transparent.

This technique of hacking can be applied to frequency domain S-parameter measurements, time domain TDR responses, and can model all aspects of the electrical properties of interconnects, as long as these effects are part of the models' properties available in the simulation engine.

It is a very powerful, general technique that should be one of the tools in every engineer's toolbox.

19.13 Review Questions

1. In a TDR impedance profile, what distinguishes an electrically long and electrically short structure, or a uniform transmission line and a discontinuity?

2. When we see a region in an impedance profile that is flat for some region, what are we implicitly assuming the electrical circuit model for this section of the interconnect to be?

3. What does a figure of merit for an interconnect really refer to?

4. What are three different circuit elements for a discontinuity?

5. What is another name for the process of hacking an interconnect?

6. How do you know what circuit elements to use when building a circuit topology to describe the TDR response of an interconnect?

7. When selecting a simulation tool to hack an interconnect, what is an essential circuit element the simulation tool must understand?

8. Why is it important to use the same source waveform in the simulation as when the TDR response was measured?

9. What is an important limitation in the type of transmission line models available in QUCS? What impact will this have on simulating TDR responses?

10. If a discontinuity is a dip in the impedance profile, what two possible circuit elements could you use to model it? Which one should you use?

11. What is a mink hole? How do they arise and why should they be avoided?

12. When measuring the TDR response from a very low impedance transmission line open at the far end, why does the far end of the transmission line not appear as an open in the TDR impedance profile?

13. What is the difference in capacitor and inductor values of a SMT resistor compared to an axial lead resistor? How does this support the idea to use SMT devices in high-speed applications and avoid axial lead devices?

Appendix A: Answers to the Review Questions

1 Chapter 1

1.1 What is the difference between a microstrip and stripline transmission line geometry?

In a microstrip, there is air above the trace, while in a stripline, there is a top return plane in proximity.

1.2 Why is engineering intuition an important skill?

The design process is basically a creative process. It is based on our intuition. The better our engineering intuition, the better we can apply our creativity to think of new and practical design options.

We do not design using a 3D full-wave field solver. We verify a design with these tools. We create and invent with our intuition.

1.3 How do you answer "it depends" questions?

The most common answer to all engineering questions is "it depends." The way we answer these questions is by putting in the numbers using rules of thumb, approximations, and numerical simulation tools.

Putting in the numbers will allow us to answer the How big? How important? How small? sort of questions.

While it may be possible for an effect to happen, it is often a question of, yes, but will it be a problem, and is the marginal increase in performance worth the cost?

We answer these questions by putting in the numbers.

1.4 What do Maxwell's equations describe and why are they so important?

These four equations describe how electric fields and magnetic fields interact with conductors and dielectrics and their dynamics.

In some region of space, at any position and at any instant in time, the electric and magnetic fields must obey these differential equations. Always. No exceptions have even been observed.

This means if we know the location of the dielectrics and conductors and their material properties, and the initial conditions of the electric and magnetic fields, we can calculate how the fields change in time and at any location.

They also translate an electric field over space into a voltage difference and a tangential component of the magnetic field into a current.

They predict that if the initial condition is a changing electric field, this will self-propagate by creating a changing magnetic field that will create a changing electric field, and so on.

1.5 Why don't we tackle every SI problem by grabbing a pencil and paper and solving Maxwell's equations?

These differential equations are really hard to solve except in the simplest cases. While they give insight into how signals propagate on transmission lines, they are difficult to get answers from for arbitrary geometries.

Even if we use a 3D full-wave field solver, which solves Maxwell's equations, the time required to solve even the simplest problems would take too long to explore design space as well.

To get to the answer faster, we use approximations and

increasingly more accurate simulations, balancing how quickly we want the answer and how accurate we want the answer to be.

1.6 What is a transmission line?

A transmission line is any two conductors with some spatial extent or some length. Every interconnect in every board is a transmission line.

We call one of the conductors the signal path and the other conductor we call the return path. The signal that propagates is the voltage difference between the two conductors.

1.7 After connectivity, what is the only thing interconnects can introduce in an electronic product?

The first step in a board design is connectivity. The main purpose of the traces in a board or interconnects in a product is to connect all the terminals of all the components that need to be connected together.

Once connectivity is established, the only thing the interconnects can do is introduce noise. This is in the form of self-aggression noise and mutual-aggressions noise. The interconnects will distort the pristine signals coming out of the drivers on their way to the receivers.

Self-aggression noise is when the signal interferes with itself as it propagates down an interconnect. This is usually from reflections and losses in the interconnect.

Mutual-aggression noise is cross talk between aggressor lines and victim lines. This is cross talk, coupling of signals by electric and magnetic fields.

1.8 What is the primary tool we will use to measure or characterize a transmission line?

The most valuable tool to use to characterize or measure the properties of a transmission line is the time domain reflectometer (TDR).

This instrument will send a fast rising edge signal on an interconnect and measure the reflections that come back. The only thing that causes a reflection is a change in the instantaneous impedance the signal sees.

Starting with a reference source of 50 ohms, the TDR measures changes in the impedance of the interconnect. When we measure the instantaneous impedance, we interpret this in terms of the characteristic impedance of uniform transmission lines. This is the most important electrical property of an interconnect.

1.9 What are the two most common problems interconnects introduce as rise times get shorter?

The two types of noise generated by interconnects that scale with shorter rise time and first appear in an interconnect are mutual aggression noise in the form of switching noise and self-aggression noise in the form of reflection noise.

Switching noise is the cross talk between the loop inductance of an aggressor and a victim signal-return path loop. This is caused by the dI/dt in the aggressor creating changing magnetic fields in the victim that induces a voltage noise in the victim.

Since it happens only when the current changes or switches, it is called switching noise. This noise from the interconnects can appear at signal bandwidth of about 10 MHz – 50 MHz.

Reflection noise, depending on the length of the interconnects, can arise at signal bandwidths also in this frequency range. It is caused by the signal propagating down the transmission line and reflected between the high impedance of the receiver and the low impedance of the driver.

Below about 10-MHz bandwidth, interconnects are generally transparent. But, above 10 MHz, these two problems, switching noise and reflections, can arise.

1.10 If an interconnect is 12 inches long, like on a PC motherboard, what is the rise time below which the interconnects may not be transparent?

As a rough approximation, reflection noise can arise and be an issue when the rise time in nsec is shorter than the length in inches.

For an interconnect 12 inches long, reflection noise might be an issue when rise times are shorter than 12 nsec.

This does not mean the interconnect will have too much noise at 11-nsec rise time. It means that there is the potential for some noise with rise times less than 12 nsec and you should do your own analysis. There is the possibility that noise may be an issue.

Likewise, if the rise time is longer than 12 nsec, reflection noise will probably not be an issue so move on to other, more important problems.

1.11 If an interconnect is 3 inches long, like on a small microcontroller board, what is the rise time below which the interconnects may not be transparent?

If the rise time is 3 nsec, then when the interconnects are 3 inches long or shorter, they are transparent, and you should not worry about reflection noise being an issue.

This is why on many small microcontroller boards, transmission lines and reflection noise are a nonissue.

1.12 What is the biggest difference between signals in RF applications and high-speed digital applications?

In RF applications, the signals are generally at one frequency, the carrier frequency, with some modulation about this frequency. The signals care about the electrical properties of the interconnects at one frequency. The bandwidth of the signals is narrow.

In digital applications, the signals have frequency components all the way from DC up to a high frequency based on the rise time of the signal. We have to engineer the properties of the interconnects across a wide frequency range.

Because of the very different frequency range over which we engineer interconnects, the design guidelines for interconnects is very different for RF applications than for high-speed digital applications.

1.13 Why can't we just design interconnects for high speed digital applications with the same guidelines as they are designed for RF applications?

In RF applications, we use stubs to tune electrical properties at a specific frequency. We terminate transmission lines with LC loads, with a specific impedance at a specific frequency.

In high-speed digital applications, we need broadband solutions. The only circuit element that has a flat impedance with frequency is a resistor. This means we will leverage resistors for terminations in digital applications.

Stubs are terrible in digital applications because they distort the signals while they are great in RF applications.

506

Chapter 2

2.1 Why is it a good idea to not call the return path, ground?

Ground is really the conductor that provides the reference from which all other voltages are measured. Generally, we are assuming there is no voltage drop between different locations on the same ground conductor and all points are equivalent.

But another application for the ground conductor is to provide the return path for the signals in transmission lines. Even though the same conductor carries the return current for multiple transmission lines, the location of the return currents in the ground plane conductor are spatially separated.

When the ground conductor is used as the return path, the precise location of the ground conductor matters. This is a different role than being used as a reference conductor for measuring voltages.

If we call the ground plane the return path, we will start seeing it as the return path and not just as a conductor to measure voltage against.

2.2 What is generated between the signal and return conductors at the beginning of a transmission line when a signal is launched? What will happen?

When a signal is launched into a transmission line, this means the voltage between the signal and return conductors has increased. This is literally the signal launched into the transmission line.

If the voltage between the signal and return path increases, there will be an electric field generated between the signal and return path. As the voltage changes, the electric field changes. This changing electric field will create a changing magnetic field which creates a changing electric field that self-propagates.

This is really how a signal launched into a transmission line propagates as a signal down the transmission line.

2.3 Is the signal the moving electrons in the conductor or the electric field in the dielectric?

Fundamentally, the signal that propagates down the transmission line is the changing electric and magnetic fields. These changing fields self-propagate.

As they propagate, the electric field creates a voltage between the signal and return conductors that we would measure with a scope, and the magnetic field creates a current in the conductors.

Sometimes, it is easier to analyze problems thinking of the voltages and currents, which are really a result of the fields. Voltages and currents are not a wrong way of thinking of signals in interconnects, they are an approximation. This means sometimes we can get to an acceptable answer faster by considering voltages and currents instead of electric and magnetic fields.

2.4 What is a good rule of thumb for the speed of a signal in an FR4 transmission line?

The speed of light in air is 12 inches per nsec. In a dielectric like FR4, with a Dk = 4, the speed of light slows down to 6 inch/nsec. This is 15 cm/nsec.

There is another useful term, the wiring delay: the time delay per inch that is just the inverse of the speed. In air, the wiring delay is 1/6 nsec/inch = 0.167 nsec/inch or 167 psec/inch.

2.5 What is the difference between the voltage measured by an oscilloscope and the signal on the transmission line?

An oscilloscope cannot tell what direction a signal is traveling, all it can tell is the total voltage between the signal and return conductors. It only measures the scalar voltage, not the vector voltage, which has direction information.

2.6 Using the wiring delay of a transmission line, what is the time delay of a 1-inch long interconnect in a large BGA package? A 3-inch-long interconnect on a small circuit board? A 30-inch-long interconnect in a backplane?

The wiring delay is 167 psec/inch in FR4-type laminate material. For an interconnect 1 inch long, the one-way time delay is 170 psec. Unless you are sure of the Dk value, this wiring delay is just a rule of thumb. Approximating it as 170 psec/inch is just fine.

A 3-inch length interconnect has a delay of 3 in × 170 psec/inch = 510 ps ~ 0.5 nsec.

A 30 inch interconnect has a time delay of 170 psec/inch x 30 inches = 5.10 nsec.

2.7 What does instantaneous impedance refer to? Why is this so important?

The instantaneous impedance is the impedance the signal sees each step along the way. This is the most important electrical property of an interconnect.

If the instantaneous impedance is constant, the signal continues to propagate with no distortion. But when the instantaneous impedance changes, the signal will reflect. This is the most important consequence of why we care about instantaneous impedance. The origin of reflections is the why we care about instantaneous impedance.

2.8 If an interconnect trace gets wider, what happens to the capacitance per length? What happens to the instantaneous impedance?

When the linewidth increases, there is more area of overlap and the capacitance per length increases. This means there is more current required to charge up that region of the transmission line. If more current goes into that region of the transmission line, the instantaneous impedance in that region decreases.

Another way of thinking about it is that if the capacitance increases, there are more electric field lines that change between the two conductors and the dE/dt, the displacement current increases. This means more current, and a lower impedance.

2.9 What is required by an interconnect to have a characteristic impedance?

The characteristic impedance is the one value of instantaneous impedance that characterizes a transmission line. This means the instantaneous impedance is constant down the length of the transmission line. This is the very definition of uniform transmission line.

If the interconnect is not uniform, there is no one value of the instantaneous impedance. The interconnect does not have a single instantaneous impedance that characterizes it. It has no characteristic impedance.

2.10 In the physics model of a transmission line, where did the inductance of the interconnect go?

When we describe the transmission line as a series of capacitors as the footsteps we walk on moving down the transmission line, we place the capacitors some distance apart. This makes it a physics model, in the sense that it includes the physical features. There is no distance between electrical elements in a circuit model.

We added to this physics model the principle that the signal was traveling down the transmission line at a fixed speed, encountering each capacitor every instant in time.

The combination of the speed of the signal and the capacitance per length in the interconnect has information embedded in it about the inductance per length. We are just looking at the properties of the interconnect differently in the physics model.

2.11 What is special about 50 ohms and why is it so common in high-speed digital applications?

There is nothing special about 50 ohms for modern interconnects. Its historical origin is that it offers a minimum in the attenuation of a cable when limited by a maximum outside diameter.

This was important in the early days of radio when every watt of transmitted power to the antenna was important.

Since so many cables were engineered as 50 ohm cables, many test tools were designed to test 50 ohms and many electronics instruments were design for 50 ohms.

For compatibility, if there were no compelling reasons otherwise, all instruments and cables were designs for 50 ohms. It became the de-facto standard.

In modern high-speed digital systems, 50 ohms is a reasonable compromise between manufacturability and performance. However, when specific features of the system are well established, 50 ohms may not be the optimized impedance to use.

In some systems, 66 Ohms is a better impedance or 28 ohms in some systems or 75 ohms in other systems.

For electronic test equipment, 50 ohms has become the standard. The reason all scopes have a 50-ohm input is because it is assumed the cable connected to the scope will be a 50-ohm characteristic impedance cable.

2.12 When 20 mA of current flows into the signal conductor that is 12 inches long, when will the return current come out of the return conductor?

The return current comes out of the return conductor instantaneously with the signal current going in. It flows between the signal and return conductor through the displacement current between the signal and return conductor.

This means the current that comes out of the return conductor has nothing to do with how the end of the transmission line is terminated.

2.13 What is displacement current and where does it flow?

Displacement current is the current that flows through the electric field lines when the field changes. It is a new type of current in addition to conduction current. Whenever there are electric field lines between conductors and those field lines change due to a change in voltage, there will be displacement current flowing between the conductors.

When a signal is launched into a transmission line, the electric field right at the beginning of the transmission line, the displacement current is when enables the current to return through the return path.

2.14 Where in the return path will the return current flow?

As the signal propagates down the transmission line, the current flows between the signal and return path only where the voltage is changing, which is right at the signal edge.

The signal that propagates is at the same time as the voltage wavefront and the current wavefront flowing between the signal and return path.

Chapter 3

3.1 What are the five families of transmission line types?

Every transmission line will have at least five different specific properties:

- Uniform or nonuniform
- Single-ended or differential
- Uncoupled or coupled
- Lossy or lossless
- Balanced or unbalanced

3.2 What is the difference between a uniform and nonuniform transmission line?

A uniform transmission line has one value of instantaneous impedance a signal sees as it propagates. This means there are no reflections or distortions in the signal as it propagates.

In a nonuniform transmission line, the impedance varies and there will be reflections and distortions along the way.

A uniform transmission line is preferred for high-performance interconnects so that the signal is not distorted when it exits the interconnect.

3.3 Why doesn't a nonuniform transmission line have a characteristic impedance?

In a nonuniform transmission line, there is no one instantaneous impedance that characterizes the transmission line. Instead, the instantaneous impedance varies along the line. You would have to separate the nonuniform transmission line into multiple sections that can be approximated by uniform regions. Then each region would have a characteristic impedance.

3.4 What is the difference between a single-ended transmission line and a differential pair transmission line?

A single-ended transmission line is a single transmission line with a signal and return path. The type of signal that can be transported on a single-ended transmission line is a single-ended signal.

A differential pair is comprised of two single-ended transmission lines. When considered as a differential pair, there are two kinds of signals that can be transported on the differential pair, a differential signal and a common signal.

3.5 What does the cross talk between two single-ended lines refer to?

When two single-ended transmission lines are close enough so that the fringe field lines from one signal and return path can spread over and see an adjacent single-ended transmission line, we refer to the impact on the second line as cross talk.

If there were no fringe field lines from the aggressor to the victim line, there would be no cross talk. The amount of cross talk is a measure of the fringe field lines that spread from one line to the other.

We also refer to the amount of fringe field lines that span between the aggressor and victim lines as the coupling. The more fringe electric and magnetic fields, the more the coupling.

3.6 What is the advantage of a tightly coupled differential pair?

The biggest advantage in using tightly coupled differential pairs, meaning keeping the two signal lines that make up the differential pair close together is higher interconnect density. This is important to enable either a smaller board or fewer routing layers.

It is generally NOT correct that a tightly coupled differential pair has less cross talk from other aggressors and should be used to reduce cross talk. In some cases, it can have higher cross talk than loosely coupled differential pairs.

3.7 Which type of differential pair can achieve a better differential impedance, a tightly coupled differential pair or a loosely coupled differential pair?

This is a trick question. You can engineer just as good a differential impedance using a tightly coupled or loosely coupled differential pair. It is a matter of adjusting the linewidth or dielectric thicknesses to compensate for the coupling in the differential pair.

If you use tightly coupled lines, you would use a narrower linewidth than using loosely coupled differential pairs, all else being equal.

515

3.8 What is a loosely coupled differential pair a good balance between?

A loosely coupled differential pair is a good balance between the interconnect density of a tightly coupled differential pair and the lower loss of an uncoupled pair that would enable a wider line.

3.9 What are the two root causes of loss in a lossy transmission lines?

There are two fundamental mechanisms for loss in a transmission line, conductor loss and dielectric loss. Each of these are frequency-dependent and increase with higher frequency.

Conductor loss increases with frequency due to the current redistribution into a smaller cross section from the skin depth effect. This means the series resistance increases with frequency.

The dielectric loss increases with frequency due to the rotation of the dipoles in the material. On one cycle of the sine wave of the signal, the dipoles rotate one way and then the other. This is a current. For the same amount of charge that rotates, the higher the frequency, the shorter the cycle time and the same charge flowing in a shorter time is a higher current.

As the current increases, so does the loss.

3.10 What is the impact on transmitted signals from lossy transmission lines?

Because the loss is frequency-dependent, higher-frequency signal components get attenuated more than lower-frequency signal components. When a fast edge enters the line, its higher-frequency components will be attenuated more than the lower-frequency components. This means the rise time will increase.

The biggest problem with frequency-dependent loss is the rise time degradation of the signals. When the rise time is longer than the unit interval, this long rise time contributes to intersymbol interface, data-dependent jitter, and collapse of the eye.

3.11 If lower dissipation factor materials result in less distortion of high-speed signals, why not always use the lowest dissipation factor laminate available?

So much of interconnect design is about cost-performance trade-off. Generally, if something has higher value, it costs more. A lower dissipation factor laminate with lower-loss, offers more value and generally costs more.

For the lowest-cost system, you don't want to pay for performance you don't use.

3.12 What is the difference between a balanced and unbalanced transmission line?

The balance in a transmission line is about the symmetry between either the signal and return path of a single-ended transmission line, or the two lines that make up a differential pair.

In a single-ended transmission line, a balanced transmission line has the two conductors the same. A twin rod conductor is a balanced transmission line a microstrip is an unbalanced transmission line.

In a differential pair, the microstrip differential pair is balanced, but a broadside-coupled differential pair may not be.

3.13 What properties of a transmission line are affected by whether it is balanced or unbalanced?

The signal propagation on a transmission line is not affected by whether it is balanced or unbalanced, but its sensitivity to radiated emissions is sensitive to when the balance changes. The current distribution in the two conductors that make up a single-ended transmission line changes if the conductors are balanced or unbalanced. This redistribution of currents can contribute to radiated emissions.

3.14 Why are uniform transmission lines preferred for carrying signals?

The advantage of a uniform transmission line is that the signal sees a constant instantaneous impedance and travels undistorted with no reflections.

In a nonuniform transmission line, the instantaneous impedance will vary and there will be reflections along the way. These reflections will distort the signal as it travels down the line, and some of the reflections may rereflect and head in the same forward direction as the signal, further distorting it.

To eliminate this source of noise, always use a uniform transmission line where possible.

3.15 What type of transmission lines does this book focus on?

This book focuses on single-ended and differential, coupled or uncoupled, lossless transmission lines. As single-ended transmission lines, they are microstrip or stripline and unbalanced, but for differential pairs, we assume the two lines are identical and the differential pair is balanced.

3.16 What is an example of a very important type of transmission line that is outside the scope of this book?

When data rates are above 1 Gbps, the losses in the transmission line dominate the important properties of the transmission line and distort the signals.

How the design choices and materials choices influence the losses is really important when it comes to engineering lower-loss transmission lines.

Other books in this series cover this very important topic of lossy transmission lines.

Chapter 4

4.1 What is the fundamental, basic definition of impedance?

Impedance is always V/I. Whether we look in the frequency domain or the time domains for any two-terminal device, impedance is always V/I. IT is the most important electrical quality a signal cares about and ultimately translates how a voltage turns into a current in any circuit.

4.2 What are the five different impedances that can be associated with a transmission line?

While the basic, fundamental definition of impedance is always V/I, there are specific variations of impedance depending on the application.

There is the instantaneous impedance a signal sees each step as it travels down an interconnect.

There is the characteristic impedance of a uniform transmission line, the one value of instantaneous impedance a signal would see.

There is the input impedance in the time domain of a transmission line. This is time-dependent. Initially, it is the characteristic impedance of the transmission line, but if we wait awhile, it becomes the impedance at the load of the transmission line.

There is the input impedance of the transmission line in the frequency domain. This depends on the reflections back and forth from the end of the line and will vary with frequency.

There is the wave or surge impedance. This is the same as the input impedance of the transmission line in the time domain, initially when the signal first sees the transmission line, which is the same as the characteristic impedance of the line.

4.3 What is the instantaneous impedance of a transmission line?

This is the impedance the signal sees each step along the way. This is what you would see if you were the signal traveling down the transmission line.

4.4 The characteristic impedance of a uniform transmission line is 60 ohms. What is the initial input impedance a signal will see entering the line? What is the instantaneous impedance the signal will see as it propagates down the line.

If the characteristic impedance of the transmission line were 60 ohms, this means it is a uniform transmission line. The initial or surge impedance the signal would see upon entering the transmission line would be 60 ohms and the instantaneous impedance each step along the way would be 60 ohm.

4.5 Why is it ambiguous when we just refer to "the impedance" of a transmission line?

With five different impedances, if we just refer to "the impedance" of the transmission line, we don't know which one we are referring to.

I may be thinking the input impedance of the transmission line in the frequency domain, which might start as a high impedance at low frequency, but will show wild fluctuations, and will depend very much on the length of the line.

You may think instantaneous impedance, which is constant down the line and independent of the length of the line. We will be referring to two very different quantities and not know the intent of the other. This makes for confusion.

4.6 When we do refer to "the impedance" of a transmission line, in the context of a digital signal propagating on the line, to which impedance are we probably referring?

Generally, we get lazy and just use the term, "the impedance" of a transmission line. When referring to a digital signal on a transmission line, generally, the term impedance probably refers to the instantaneous impedance or the characteristic impedance. But it is less ambiguous to keep that preface and refer to the characteristic impedance of the transmission line, where possible.

4.7 What lumped circuit element does the front of a transmission line look like initially?

When the signal enters the transmission line initially, it sees an instantaneous impedance that looks like a resistor. As long as we look for a time short compared to the round-trip time of the transmission line, the behavior of the transmission line is indistinguishable from that of a resistor.

4.8 A 50-ohm transmission line is open at the far end. What impedance will a digital ohmmeter read if connected between the signal and return conductors? How does this relate to the 50 ohms of the transmission line?

If the transmission line is open at the far end and we connect a DMM to the front of the transmission line between the signal and return conductor, it will read as open. It is completely unrelated to the fact the transmission line is a 50-ohm impedance.

The 50 ohms refers to the instantaneous impedance of the transmission line. But the ohmmeter will read the input impedance about 1 second after the signal has entered the line. All the reflections have happened and we will see just the termination at the end of the line.

Of course, if the transmission line were really, really long, like from the earth to the moon, and it takes 3 seconds for a signal to travel down and back, in the first second the DMM will measure a resistance of 50 ohms. As long as we look for a time short compared to the round-trip time, the input impedance we will measure will be that of a resistor, equal to the characteristic impedance of the line.

4.9 Why is the source impedance of the driver important to know in order to calculate the voltage launched into a transmission line.

When a driver connects to a transmission line, the initial voltage launched into the transmission line is a result of the voltage driver between the source resistance and the input impedance of the transmission line.

If the source resistance of the driver is high compared with 50 ohms, the voltage launched into the transmission line, which propagates down the line, is a small fraction of the source voltage.

If the source resistance is very small compared to 50 ohms, the voltage launched into the transmission line is close to the source voltage.

4.10 For how long a period of time can you think of a transmission line in a circuit as a resistor? And where would the resistor be placed?

As long as we look for a time that is shorter than the round-trip time, the input impedance of a transmission line will look like a resistor. This resistor would be between the signal and return path.

We can literally replace the front of the transmission line with a resistor between the signal and return conductor.

4.11 A driver has a Thevenin source resistance of 10 ohms and a Thevenin source voltage of 3.3V. What is the voltage initially launched into a 60-ohm transmission line? If its time delay is 1 nsec, for how long will the input of the transmission line look like a 60-ohm resistor?

At the front of the transmission line, the signal sees a 60-ohm resistor initially. This make a voltage divider with the 10-ohm source resistance. The voltage divider is $60/(60 + 10) = 6/7$

If the Thevenin source voltage is 3.3 V, the voltage launched into the transmission line is 3.3 V x 6/7 = 2.83 V. This is the voltage that will be launched and will propagate down the transmission line.

It will get to the end of the line in 1 nsec, and maybe reflect back. It may get back to the beginning of the line in another 1 nsec, or 2 nsec for its round-trip time. During the round-trip time, the front of the transmission line will look like a 60-ohm resistor.

4.12 In the frequency domain, what is the assumption about the sine wave signal entering the transmission line?

In the frequency domain, the sine wave signals that are used to excite any interconnect are at steady state. This means the sine wave that is being used has always been and will always be. There is no transient of the sine wave turning on. Once the sine wave is launched, all the reflections have happened and they have died out and are at steady state.

4.13 Why is there no reflection when a signal, traveling on a 75-ohm impedance transmission line encounters a 75-ohm resistor between the signal and return path?

All a propagating signal cares about is the instantaneous impedance it sees. As it propagates on the transmission line, it sees a 75-ohm instantaneous impedance between the signal and return conductor. As soon as it steps on the 75-ohm resistor, it sees the 75-ohm impedance of the resistor. There is no change in the instantaneous impedance, so no reflection.

4.14 At low frequency, what does the input impedance of a transmission line, open at the far end, look like? What if it were shorted at the far end? What will it look like electrically?

In the frequency domain, the input of a transmission line, open at the far end, looks like a capacitor at low frequency. This means that its impedance starts high and drops with higher frequency. It keeps dropping until the frequency where a quarter wavelength fits in the transmission line and then the input impedance goes up again.

In the frequency domain, the sine wave voltage across the front of the transmission line has always been and will always be, so when we look at any one frequency, all the reflections have happened

524

and died down and we are seeing the steady-state behavior. This means that the entire transmission line, including what happens at the end, is included in the input impedance.

When the far end is shorted, the front of the line looks like an inductor. This means at low frequency, it is a low impedance and increases with increasing frequency.

4.15 If the length of the transmission line, shorted at the far end, is exactly half a wavelength, what will the input impedance of a transmission line look like in the frequency domain?

When the length of the transmission line is exactly one half a wavelength, the input impedance at the front of the transmission line will look exactly the same as the load.

When the length of the transmission line is exactly one half a wavelength and the far end is open, the wave will enter the transmission line, travel to the end, reflect with no phase change, and travel back to the front. In traveling down and back there is one cycle phase change. The wave that reflects from the front of the line and from the back of the line are in phase and the source sees the sum of the waves looking like an open.

If the far end is a short, the phase change from the reflection is one half a cycle. This means that between the signal that reflects from the front and back of the transmission line that makes its way back to the source are out of phase, cancel out, and there is no voltage left and so the input looks like a short.

4.16 If you have the measured impedance vs frequency of a transmission line open at the far end, you will see a minimum impedance at a frequency where the length of the transmission line is exactly one quarter cycle. At what frequency would you look to read the characteristic impedance of the transmission line directly from the impedance curve?

At a frequency where the transmission line length is one eighth a wavelength, the input impedance of the line, whether an open or short at the far end, will look like the characteristic impedance of the transmission line.

The frequency where you see the first dip or peak is the frequency where the length of the line is one half a wavelength. Half of this frequency is the frequency where the length is one eighth a wavelength. This is the frequency to look for to read the impedance directly from the front screen.

4.17 What is different about the input impedance of a transmission line in the time domain than in the frequency domain?

These are completely different domains, with completely different interpretations of the impedance.

In the frequency domain, there is no transient response. The input impedance is the steady-state response. In the time domain, it is all transient response. The input impedance is all about how the input impedance of the transmission line changes over time.

Chapter 5

5.1 What is the problem that results when the instantaneous impedance a signal sees changes?

When the instantaneous impedance changes, the signal will reflect and the transmitted signal will be distorted.

If this reflection is large enough, the signal transmitted at the receiver will be distorted.

If there are multiple changes in the instantaneous impedance, there will be multiple reflections and these distortions can get back to the receiver.

5.2 What is the first goal in interconnect design to reduce reflection noise? What is the second goal?

The first goal is to keep the instantaneous impedance the signal sees constant. This will prevent all reflections. This means engineer the interconnects as uniform transmission lines.

No matter how perfect the transmission lines, there will always be ends. The second goal is to manage the reflections at the ends of the lines, and this is with a termination strategy.

5.3 What is the consequence if there were no reflections generated at an interference between two different characteristic impedance transmission lines?

If the signal passes from one impedance environment to another, the ratio of the voltage to the current of the signal would change. The voltage and current cannot be the same in the two regions because the ratios are different in the two regions.

Right at the boundary, if there were no reflections created, then if the voltage did not change, we would have a different current going into and out of the interface. This would have eventually built up an infinite amount of charge and the universe would blow up.

If the current were constant across the boundary, then the voltage would have to change and there would be an infinite electric field across the boundary.

If no reflections were to occur, the boundary conditions across the interface — a continuous voltage and continuous current — cannot be met. This can only be met by creating a reflected voltage.

5.4 What are the two boundary conditions at any interface that keep the universe from blowing up?

The voltage has to be continuous ,— this means the voltage across one side of the interface must be the same as the voltage on the other side.

The current has to be continuous across the boundary. This means whatever current going in has to be the current coming up out of the interface.

5.5 When a signal is propagating from the left to right direction, incident on an interface between two transmission lines, what are the two signals that exist in the left side of an interface? What are the signals that exist on the right side of the interface?

The net or total voltage of current must be the same across the boundary. On the left side, the total voltage is the sum of the incident and reflected voltages. On the right side of the boundary, there is only one signal, the transmitted signal.

5.6 What are the two boundary conditions that voltage and current must meet?

On the left side of the interface, the continuity of voltage means the incident voltage + the reflected voltage must = the transmitted voltage.

Likewise, the net current must be continuous as well. The incident current + the reflected current must = the transmitted current.

5.7 What is the difference between the signal propagating in a transmission line and the voltage measured by a scope?

A scope can only measure the net voltage between the signal and return conductor it cannot tell what direction the signal is traveling. This means the scope cannot tell you an important property of the signal, its direction of propagation. You have to know something about what is being measured to know about the signal.

If the scope input is 1 meg and you are using a 50-ohm cable, then the voltage measured at the scope is the sum of the incident and reflected signal. However, you would never know if there were multiple reflections going on in the interconnect between the scope and the source.

The signal is the voltage wave that is propagating up and down the transmission line and has a direction of propagation associated with it.

5.8 A 2 V signal is incident from 50 ohm to 75 ohms. What voltage reflects? What is the voltage a scope would measure at the interface, right before and right after the reflection is created?

When a signal hits the impedance transition, the reflection coefficient is $(75 - 50)/(75 + 50) = 0.2$.

When the 2 V signal is incident, $2\text{ V} \times 0.2 = 0.4\text{ V}$ will reflect from the interface.

If a scope were measuring the voltage between the signal and return conductors, it would measure the 2 V incident + 0.4 V reflected, or 2.4 V as the voltage at the interface. We would never know from the scope that there were really two waves at the interface.

Right before the incident signal hit the interface, we would measure 0V. When the incident signal hits the interface, we would see the combination of the incident and reflected wave, which is 2.4 V.

5.9 A 2 V signal is incident to the interface from the other side, from the 75-ohm region propagating to the 50-ohm impedance region. What is the reflection coefficient? What is the voltage on either side of the interface just after the reflection?

Now we are coming from the other side of the interface, from the 75-ohm side. The reflection coefficient when we hit the 50-ohm region from the 75-ohm region is $(50 - 75)/(50 + 75) = -20\%$.

When a 1V signal enters the 75-ohm region and hits the 50-ohm region, -20% of it reflects. This means the voltage at the interface is $1V - 0.2 V = 0.8 V$. There would only be 0.8 V as the voltage at the interface and the voltage that propagates into the other side of the 50 ohms and continues in this interconnect region.

5.10 When a 2V signal travels from 50 ohm to 75 ohms, the voltage at the interface is different than if the signal travels from 75 ohm to 50 ohms. Why?

From the last example, we can see that if a 2 V signal is incident from the 50 Ohm region to the 75-ohm region, the voltage at the interface will be 2.4 V. The 2.4 V signal will propagate from the 50-ohm into the 75-ohm region.

From the other side, a 2 V signal propagating from the 75-ohm region into the 50-ohm region, will see a 1.6 V signal at the interface that propagates into the 50-ohm region and continues propagating in the 50-ohm region.

There is a very different behavior in the signal depending on the direction of propagation. This is fundamentally due to the observation that signals are dynamic and how they propagate, and the order of the impedances they encounter are important factors influencing how the signals behave.

At the interface, the reflection coefficient is different from the two sides and will depend on which direction the signal is propagating.

5.11 A 1V signal travels from a 100-ohm impedance to an open. What is the final voltage at the open and the current through the open?

At the open, the reflection coefficient is 1. This means that the 1V incident will reflect and the reflected signal will be 1V. At the open, we would measure a 2 V voltage with a scope.

The voltage transmitted into the open, which is appearing on both sides of the interface, will also be 2 V. We can see this from the transmission coefficient = (2 × infinite)(infinite + 100) = 2.

With a 1V incident signal, the transmitted signal is 1 V × 2 = 2 V.

5.12 A 1V signal travels from a 100-ohm impedance to a short. What is the final voltage at the short and the current through the short?

The final voltage on the short is the combination of the incident voltage and the reflected voltage. The reflection coefficient is -1. With a 1 V signal incident, the reflected voltage is – 1V so the net voltage is 1 V + -1 V = 0 V. After all, it is a short.

The current through the short is the incident current, I = V/R = 1 V/ 100 ohms = 10 mA plus the reflected current which is also 10 mA.

The sum of these currents passes through the short, or 20 mA.

5.13 When a 1V signal encounters an open, what is the reflection coefficient? What is the transmission coefficient? What is the voltage transmitted into the open?

When a signal encounters an open, its source impedance doesn't matter. The reflection coefficient is always 1 and the transmission coefficient is always 2. The voltage that is transmitted into the open is the same as the voltage measured across the open, which is 2 x the incident voltage, or 2 V in this example.

5.14 What is the initial input impedance a source will see looking into a 50-ohm transmission line? A 25-ohm transmission line?

The initial impedance the signal sees at the beginning of the transmission line is the instantaneous impedance which, for a uniform transmission line, is the characteristic impedance.

For a 50-ohm line, the signal will see s 50-ohm resistor as the front of the line for a time short compared to the round-trip time of the interconnect.

Likewise, if the line is a 25-ohm characteristic impedance line, the input impedance will look like a 25-ohm resistor for a short period of time.

5.15 When a 2 V source with a 50-ohm source resistance drives a 50-ohm transmission line, what is the initial current launched into the transmission line? What is the current into the transmission line after all the reflections have died out, if the far end termination is open, shorted, or 50 ohms?

With a 2 V source voltage and a 50-ohm Thevenin resistance, when connected to a 50-ohm transmission line, we have made a 2:1 voltage divider. Half the 2 V of the source will appear across the input 50 ohms of the transmission line and this 1V signal will end up propagating down the transmission line.

The 1V signal, across the 50 ohms of the transmission line will drive a 1 V/50 Ohms = 20 mA of current into the transmission initially.

If the far end is terminated in 50 ohms, this 20 mA of current will continue flowing into the transmission line at steady state.

If the far end is terminated in an open, then eventually, after all the reflections have died down, there will be no current into the transmission line.

If the far end is terminated in a short, then the current will be limited by the source resistance eventually after all the reflections have died down. This is 2 V/50 ohm = 40-mA current.

5.16 When a 2V source with a 10-ohm source resistance drives a 50-ohm transmission line, what is the initial current launched into the transmission line? What is the current into the transmission line after all the reflections have died out, if the far end termination is open, shorted, or 50 ohms?

The voltage launched into the transmission line due to the source impedance and transmission line impedance voltage divider is 2 V × (50 ohms/(50 + 10 ohms) = 5/3 V = 1.67 V.

533

This is the voltage across the 50 ohms of the transmission line, so a current of I = 1.67 V/50 ohms = 33 mA flows between the signal and return paths.

If the far end is terminated in 50 ohms, this 33 mA will flow down into the transmission line forever.

If the far end is open, the current into the transmission line will eventually be 0.

If the far end is shorted, the current into the transmission line will be limited by the source resistance. The steady state current after all the reflections is 2 V/10 Ohms = 200 mA.

5.17 When we send a 1V signal into a transmission line open at the far end and measure a 2 V signal at the open, where did the extra 1V come from?

The voltage across the open, which a scope would measure, for example, is larger than the source voltage. This comes from the reflected signal, to keep the universe from blowing up, and from matching the boundary conditions.

While it is counterintuitive that we see a 2 V signal when the source is only a 1V source, there is no rule that says voltage is conserved.

At each interface, when the signal is incident, the power flow through the interface is conserved, but voltage is not.

5.18 What's wrong with analyzing a resistor driving a transmission line using the concept of a reflection coefficient?

When the source is a Thevenin voltage source with a Thevenin resistance, it is tempting to think that the model is a resistive impedance at the interface seeing the transmission line on the other side.

This is not what the signal will see. We cannot think about the signal coming from a resistor impedance into a transmission line impedance. The signal will only flow from a transmission line into an impedance, not from an impedance into a transmission line.

It is not correct to think of reflection or transmission from a resistor as an impedance into a transmission line. Instead, we think of the transmission line as a resistor and then the first resistor makes the circuit look like a voltage divider.

This is an issue with how we think about the circuit and applying simple circuit analysis, either lumped circuit analysis or transmission line boundary analysis. In any sort of SPICE simulation model, the behavior of the signal when a resistor encounters a transmission line is fully taken into account.

It's just that we cannot do this simulation in our head. We have to use one of two different approximated views of the circuit.

Chapter 6

6.1 What fundamentally causes the ringing when viewing the output from a fast driver by a scope?

While we call the signature of the signal at the scope ringing, it is not really ringing, it is due to the reflection noise. When the reflected signal heads back to the source from the high impedance of the scope, it sees a lower impedance at the source. This means the reflection coefficient is negative. It is negative every time the signal encounters this interface.

This means that every time the reflection from the scope hits the source again, it changes its sign. Initially, the first reflection turns the positive signal into a negative. When this signal comes back from the 1 meg of the scope, it is changed into a positive signal, then a negative, then a positive. This is the origin of the ringing.

6.2 What does a bounce diagram describe?

A bounce diagram is a way of mapping out the dynamic nature of the signal as it reflects from each discontinuity. It is a shorthand way of keeping track of each reflected signal given the reflection coefficient at each interface.

As long as the rise time of the signal is very short compared to the time delay of the reflections, it can also keep track of the voltage at the receiver over some time period.

6.3 What are two common features of a real transmission line system a bounce diagram does not do a good job of calculating?

A bounce diagram can do nothing to describe how the signal behaves when the rise time is comparable or longer than the time delay of the interconnect. There will be smearing out of the signal over the rise time as the reflections bounce back and forth during the rising edge. this behavior is too complicated to include manually in a bounce diagram.

The second feature the bounce diagram cannot include is when there are multiple reflections from 3 or 4 transmission line segments. There is a limit to how complex a circuit can be analyzed manually in a bounce diagram. Keeping track of all the reflections that change over time is not too complicated with more than 3 or 4 different segments.

6.4 If the source resistance is lower than the transmission line impedance is the voltage launched into the transmission line higher or lower than half the source voltage?

When the source resistance is lower than the transmission line input impedance, then more voltage gets launched into the transmission line. The lower the source impedance, the larger the fraction of source voltage gets into the transmission line.

6.5 When simulating the reflections from an interface between two transmission lines, why did we terminate each end in its characteristic impedance?

So far, we have wanted to focus on what happens at just one interface without all the multiple reflections from all the interfaces.

By keeping the source impedance the same impedance as the transmission line impedance, we eliminated the reflection back from the source of the reflected signal from the interface. There was just one reflection to worry about.

Likewise, by adding a termination at the far end, we eliminated any reflection back to the interface from a transmitted signal. This allowed us to focus on just the reflection at an interface.

6.6 When the source impedance that drives a transmission line is much higher than the impedance of the line, and the receiver is a high impedance, what is the shape of the signal at the receiver?

When the source impedance is higher than the line impedance, only a small fraction of the source voltage gets launched into the line. And when the receiver impedance is high, there will be a reflection back to the source. This will result in steps rising up steadily until the full source voltage is reached at the receiver.

537

6.7 When the source impedance that drives a transmission line is much lower than the impedance of the line, and the receiver is a high impedance, what is the shape of the signal at the receiver?

When the source impedance is very small compared to the impedance of the line and the receiver is high impedance, the reflections back to the source will always change sign and result in ringing at the source.

A high-impedance source impedance means staircase steps increasing the voltage at the receiver. A low-impedance source means a ringing-like waveform at the receiver.

6.8 When there are three transmission line segments, each with different impedances, why is it nearly impossible to solve using a bounce diagram, but trivial to set up in a TDR simulator?

With three different transmission line segments in series, we can track the reflections at each interface, but the bounce diagram will have to keep track of all the multiple bounces back and forth between each interface. As the bounces happen, we are assuming the time delay of each segment is long compared to the rise time, so we can resolve each reflection.

In the time for the signal to go from the entrance to the exit, there could be at least nine different reflections to track. To list out the signal that appears at the receiver over a round trip-time, there could be more than 20 different reflections back and forth between every interface, some of them making their way to the receiver, and some to the transmitter.

Other than just the simplest problem, it is incredibly tedious to diagram all the reflections at all the interfaces.

In contrast, keeping track of all the reflections, even when the rise time is comparable to the time delay of any segment, is trivial to do with any SPICE-like circuit simulation. It is simply a matter of drawing the circuit topology, adding a Thevenin source, and simulating the voltage at each node and the receiver.

If once you do a bounce diagram once to see the dynamics of the reflections, it is not really necessary to do one again.

6.9 What are two examples of SPICE compatible simulation tools which can be used to simulate a transmission line circuit?

There are many SPICE-like circuit simulation tools. A good, open-source version is QUCS. The version offered through National Instruments is Multisim. Mentor Graphics offers HyperLynx. Keysight offers ADS. LTSpice is offered for free from Linear technologies. Another open-source version of SPICE is NGSPICE.

Chapter 7

7.1 What are the typical features of the signal on the compensation port of a scope?

Every source has three important figures of merit: the Thevenin voltage, the Thevenin resistance, and the rise time. If the signal is repetitive, it may also have a frequency.

In the case of the compensation source on most scopes, the voltage is 1V peak to peak. The source resistance is about 800 ohms, and the rise time is about 5 nsec. The period of the square wave is typically 1 kHz.

7.2 Does the measured rise time of the signal into the scope increase with cable length?

The measured rise time we see on the scope screen does seem to increase with cable length. It appears that a longer cable results in a longer rise time signal.

But this is just an artifact from the way we do the measurement. It does not mean that the rise time of the source is changing with cable length, just the measurements in some cases.

7.3 If you want to efficiently fix a problem, what do you need to know about the problem?

To fix a problem and have confidence you have fixed the problem, you need to know the root cause of the problem. Then you can fix the problem at the root cause.

7.4 If the root cause of the longer rise time from the compensation signal is capacitance in the cable, what two features would we want to change about the cable to reduce the rise time?

If the longer rise time were due to the capacitance of the cable, then we would want to reduce the capacitance of the cable to reduce the rise time of the signal we measure. This means use a shorter cable, and a cable with a lower capacitance per unit length.

Of course, this would mean that we would want to use the absolute shortest cable if we were measuring short rise times.

A lower capacitance per length cable also has a higher characteristic impedance. If it were the capacitance in the cable that slowed the edge down, then we would want to use a high-impedance cable.

Based on this assumed root cause, even if we were to pay extra for a really high impedance cable, it would only make the problem even worse.

The root cause of the longer rise times is not capacitance in the cable. Following this path would not fix this problem.

7.5 What does DUT stand for?

DUT is a commonly used term to refer to the device under test. It is the interconnect or source, or board or component that we are measuring or characterizing.

7.6 What are three figures of merit for every transient voltage source? If we know these three terms, we know the important features of the source.

The three important figures of merit that characterize any transient voltage source are the Thevenin voltage, the Thevenin resistance, and the rise time. We can define the rise time as the 10% to 90% or the 20% to 80%.

7.7 What is the real root cause of the long rise time from the compensation signal?

The real root cause of why the rise time appears to increase with longer cable length is from the combination of high source resistance compared to the cable impedance, and high termination resistance into the scope at the end of the cable.

These two combinations mean a small voltage gets launched in the transmission line, reflects from the far end, and reflects again from the high impedance of the source. All these reflections build up to the full voltage at the receiver after multiple reflections.

541

The rise time increasing with the cable length is because the time delay between reflections increase with longer cable length so it takes longer for the reflections to build up.

7.8 What are three possible tests we can do to verify this explanation?

If this is true, that the longer cable rise time is due to reflections in the cable, then we should see the individual steps if we zoom in on the received signal.

In addition, the time between the steps should be a round-trip time for the signal to reflect from the high impedance, head down and hit the high impedance of the source, reflect, and then hit the receiver again. This is a round-trip time between step increases. For a 3-foot-long cable, this is a time delay between steps of 10 nsec. A 9-foot-long cable would have about 30 nsec for the length of each step.

Finally, if it is really due to the reflections including the reflection from the high impedance of the scope, then using a 50-ohm termination in the scope, to match the impedance of the 50-ohm cable should stop the reflections and we should see the intrinsic rise time of the signal with 50 ohms in the scope input.

7.9 If we want to measure the intrinsic rise time coming from a source, how do we do that?

Always use the 50-ohm input impedance of the scope, with 50-ohm cables connected to the scope. The reason there is a 50-ohm input in the scope is because we assume you will be using a 50-ohm coax cable as the cable. This will prevent any reflections from the scope input so that whatever signal is launched into the transmission line is what is measured by the scope.

7.10 In a 12-foot-long, low cost coax cable, what is the increase in rise time we might expect to see? How does this compare to the rise time initially measured in the scope with a 1 Meg input impedance?

It is absolutely true that the rise time in transmitting through a low-cost cable will increase due to the attenuation. Higher-frequencies will attenuate more than low-frequency components because of conductor loss and dielectric loss. Each of these are frequency dependent.

As the signal travels down the transmission line, the high frequencies are turned into heat and are taken out of the signal. This leaves only the low-frequency components which means the rise time increases.

But this effect is small at the time frame measured from the compensation source.

In a 12-foot-long low-cost coax cable, if a 1-psec rise time signal were to enter the transmission line, the rise time would only be about 0.5 nsec. At most, we would expect to see about 0.5-nsec increase in the rise time of the signal due to the losses in the cable. This absolutely happens, it's just much less than the rise time degradation we see in the actual measurements.

7.11 Why is 50 ohms one of the options as the input impedance of a scope?

There are only two resistances available for the input impedance of a scope, 50 ohms and 1 meg. The 1 meg is to provide a high impedance so as not to load signals down.

The 50 ohms is specifically to terminate a cable, assuming you are using a 50-ohm characteristic impedance cable. This means, you should not use a cable with any characteristic impedance other than 50 ohms. If you do, expect to see reflections and distortions.

7.12 If the maximum power dissipation the 50-ohm resistor inside a scope can handle is 0.5 watt, what is the max RMS voltage that should ever be applied to a scope set for 50-ohms input?

The maximum power is 0.5 watts. This is V^2/R. If the R is 50 ohms, then the voltage is sqrt(0.5×50) = 5V.

This is the rms voltage. If the rms voltage exceeds 5V, then the power consumption in the 50-ohm resistor will exceed 0.5 watts and it may possibly damage the resistor.

7.13 When the output impedance of the source is 50 Ohms, what is the voltage measured by the scope set on 50 Ohms input? On 1 Meg input?

When the Thevenin output resistance is 50 ohms, and the source is connected to a 50-ohm cable, half of the Thevenin voltage will be launched into the transmission line.

If the scope is set for 50 ohms, half the Thevenin voltage is what would be measured by the scope.

If the scope is set for 1-meg input, the signal into the scope will double and the scope will measure the full Thevenin voltage of the source. The reflected wave will terminate into the 50-ohm source resistance and no more reflections will occur.

7.14 When the output impedance of the source is very low, what is the signature of the signal measured by a scope with its input on 50 ohms? 1 meg?

When the source impedance is very small, most of the voltage of the source will be launched into the transmission line.

When the scope is terminated with 50 ohms, we will see this signal out of the source with the source rise time and no reflections.

With a 1 meg termination in the scope, there will be multiple reflections back and forth that will appear as ringing.

7.15 Why do some function generators output a 2 V peak to peak signal when you set them for a 1 V peak to peak signal?

Many function generators have a 50-ohm source resistance and assume that their outputs will be connected into a 50-ohm cable. The voltage for which you set their output is the voltage that will be launched into the 50-ohm cable.

What you do with that signal after it gets to your system is up to you.

If 1V is launched into the 50-ohm transmission line and the source impedance is 50 ohms, this means the Thevenin source voltage must be 2 V.

If you connect the 50-ohm cable to the scope with 50 ohms in the scope, you will measure the 1 V signal launched into the cable, which is the voltage set on the function generator.

If you connect the 50-ohm cable to the scope with 1 meg ohm input impedance, you will measure the voltage launched into the cable plus the reflected voltage from the high impedance, which will be twice the voltage set on the function generator. This is the Thevenin source voltage, 2 V.

7.16 When the source impedance is very low, the scope sees a voltage pattern that looks like ringing. What is the time interval between adjacent peaks related to? For a 3-foot coax cable, what is this period and what is this ringing frequency?

The ringing we see in the scope is due to the successive reflections from the high impedance at the scope and the low impedance at the source.

The initial signal gets to the scope and we measure the peak. One time delay later, the reflected signal has hit the low-impedance source.

Another time delay later and this reflection hits the scope. The voltage is pulled down. One round-trip time delay and another reflection has gotten to the scope and brought the signal level high.

This means the time between peaks is $4 \times$ TD. If the cable is 3 ft long, the time delay is about 5 nsec. This means there will be 20 nsec between peaks in the ringing.

7.17 What is an advantage of a 10x probe?

A 10x probe will attenuate the signal by 10x and act as a relatively high-impedance load to the source. The chief advantage is that it will not load the source very much.

The 10x cable itself is special. It is designed to damp out reflections so that even though it is a coax, when the scope is set for 1 meg, there are no reflections in the cable.

7.18 What are two disadvantages of a 10x probe?

One problem with the 10x probe is that it attenuates signals. This means that you will lose 90% of the signal. Most scopes have a most sensitive scale of 1 mV/div. This corresponds to a trip voltage of 10 mV/div as the most sensitive scale.

If you need to measure signals less than 20 mV, don't use a 10x probe.

The second problem is the limited bandwidth. At best, a 10x probe will have a bandwidth of about 200 MHz. It will be degraded when the tip loop inductance is large.

7.19 When should you always be using a direct coax cable connection between the DUT and the scope?

If you care about small signal levels or higher bandwidth, you should not use a 10x probe. But as a general-purpose probe it should be the first go-to probe to use.

Chapter 8

8.1 What is the difference between the real world and the ideal world of electrical models?

The real world is the world of real interconnect structures with conductors and dielectric materials. It is the world of real measurements on real physical components. Only real components can be measured.

The ideal world is the world of ideal components that have very precise mathematical descriptions. They can be complicated. They can have very sophisticated behaviors, but they are all described by the precision of mathematics and do not deviate from these descriptions.

Only ideal components can be used in a simulator. Only real components can be measured.

8.2 What are the five ideal circuit element components found in all SPICE simulators?

All SPICE simulators understand an ideal resistor, capacitor, inductor, mutual inductor, and transmission line. These are the ideal circuit elements with very precise definitions and behaviors.

8.3 What are three different measurement instruments that can measure properties of transmission lines?

There are three general instruments that can be used to measure passive elements, the TDR, the VNA, and the impedance analyzer.

To measure a passive interconnect element, these instruments generate their own stimulus signals and measure the response of the interconnect.

8.4 What is special about the mathematical world?

The mathematical world is the world where all the mathematical functions are defined. This is an abstract world. The only criterion is that the rules of mathematics are obeyed.

The mathematical rules do not have to have any relationship or connection to the physical world. They just have to be self-consistent in their own world.

8.5 What is special about the ideal world?

The ideal world is the world of precisely defined models. These models have a very specific behavior and can be used in calculations or simulations.

The goal is to define the elements that make up the models in such a way so they can provide some connection to the behavior of elements in the real world. They are more closely to the real world components, the higher their value.

It is remarkable how close the predicted behavior, like their impedance, of combinations of ideal elements can match the behavior of real circuit elements.

8.6 What is special about the real world?

The real world is the world in which we live. It is the world in which we have real components with real voltage and currents or electric and magnetic fields.

The real world is the world in which we perform measurements and see all the fine details of how the real world behaves.

8.7 What is the behavior of an ideal capacitor, as different from a real capacitor?

The impedance of an ideal capacitor in the frequency domain, for example, will start at a high impedance at low frequency and drop off linearly on a log-log scale.

The impedance of an ideal capacitor will continue to decrease at higher and higher frequency, with no limit.

A real capacitor component will show an impedance profile very similar to an ideal capacitor's at low frequency. It is just that as frequency increases, we reach a point where a real capacitor's impedance starts to deviate from an ideal capacitor's impedance behavior.

8.8 What can be done to the model of a real capacitor to increase its accuracy? Is it still an ideal model?

The accuracy of an ideal model is a measure of how well the predicted behavior of the ideal model matches the measurements of a real component.

To make an ideal capacitor model higher accuracy to match the behavior of a real capacitor component, we could add a series ideal inductor and ideal resistor.

This combination of ideal circuit elements would have a predicted impedance match the behavior of a real capacitor much closer and to higher bandwidth.

It is still an ideal model, it's just configured to provide a better match to the behavior of a real component.

8.9 What are the two properties of an ideal lossless transmission line?

An ideal transmission line will have a characteristic impedance and a time delay. The characteristic impedance will be the instantaneous impedance the signal sees.

8.10 What features would be added to an ideal lossy transmission line to include the effects of loss?

In the real world, there are two important loss mechanisms: conductor loss and dielectric loss. These two properties can be added to the ideal lossless transmission line model to make an ideal lossy transmission line model.

The conductor loss can increase with the square root of frequency and the dielectric loss with frequency. These are ideal properties.

Real lossy transmission lines match this behavior very well.

8.11 When we describe a real transmission line as a 47-ohm impedance and time delay of 1.3 nsec, what assumption are we really making?

When we interpret the TDR impedance profile of a real transmission line and say it is 47 ohms and a time delay of 1.3 nsec, we are really saying that if we build an ideal lossless transmission line model with a characteristic impedance of 47 ohms and time delay of 1.3 nsec, its simulated TDR response would match the measured TDR response really well.

We could literally replace the real transmission line with this simple ideal model and accurately predict any measurement behavior.

8.12 Why is the T-element a good ideal model for a real transmission line?

A real transmission line behaves like an ideal transmission line really well. In TDR measurements, the losses are a second-order factor influencing performance. All the reflection issues and properties are fully taken into account by the ideal lossless transmission line model.

The bottom line is, it works. The predicted TDR response of an ideal lossless transmission line matches the measured TDR response of a real transmission line really well. It is the way the world seems to work.

8.13 When we describe a real transmission line as an n-section LC model, what is the assumption we are making about n? what does this make the L and C values?

When we approximate a real transmissions line as an n-section LC model, we are assuming that n is large enough to match the measured behavior up to the bandwidth we care about.

If we have a long line and want high bandwidth performance, this could mean a really large number for n. This means long computation time and no guarantee of a stable solution.

The total L and total C values should be constant as n is increased. This means the inductance and capacitance of each element will get smaller with larger n.

While an n-section LC model might be a good approximation for the real transmission line, for the effort involved in using this model, it may not have much ROI compared to using a single ideal transmission line element, which is just one element of a high bandwidth model.

8.14 A transmission line has a Z0 = 45 ohms and a TD = 1.5 nsec. How much total capacitance does it have and total loop inductance?

The total capacitance in a transmission line is just TD/Z0. This means for this transmission line, the total capacitance is 1.5 nsec/45 ohms = 33 pF.

The total loop inductance is TD × Z0. This means this transmission line has a total loop inductance of 1.5 nH × 45 ohms = 67 nH.

8.15 A real transmission line is open at the far end. What ideal circuit element will its input impedance look like in the frequency domain? What if it were shorted at the far end?

A real transmission line open at the far end looks like an ideal capacitor at low frequency. However, it also looks like an ideal transmission line open at the far end.

If it were shorted at the far end, it would look like an ideal inductor at low frequency or like an ideal transmission line shorted at the far end.

8.16 Up to what frequency will a single ideal L or C match the impedance of an ideal transmission line? What frequency is this for a time delay of 1 nsec? 10 nsec?

Generally, a single LC model will match the behavior of an ideal transmission line up to about the self-resonant frequency of the LC. This is basically the pole frequency of the 2-pole LC circuit model.

This is also about the one-quarter wave resonant frequency, which is about one-quarter × 1/TD. For the case of time delay of 1 nsec, the self-resonant frequency of the transmission line is about ¼ × 1/1 nsec = 0.25 × 1 GHz = 250 MHz. This is about the bandwidth of a single-section LC model.

If the time delay is 10 nsec, the bandwidth of the 1-section LC model would be about 10× lower frequency or 25 MHz.

8.17 What might be the bandwidth of an ideal T element compared to a real, low loss transmission line?

The bandwidth of an ideal lossless T element could easily be above 10 GHz, comparing the TDR response or the input impedance of the ideal element compared to the real transmission line.

If we compare the insertion loss or the transfer function, the real transmission line will show a dropoff more than the ideal transmission line. How much more depends on the amount of loss in the real transmission line. This will be the first term that shows a bandwidth limit.

8.18 What is the capacitance per length of all 50-ohm transmission lines in FR4? Suppose a 50 ohm line is 12 inches long? How much total capacitance is in the line? What is the underlying assumption we are making?

All 50-ohm transmission lines in FR4 have about 3.4 pF/inches of capacitance per length. A transmission line 12 inches long would have 12 inches × 3.4 pF/inches = 41 pF.

We are just assuming the transmission line is uniform with a characteristic impedance of 50 ohms and it is composed of a dielectric with a dielectric constant of 4. That's it.

This says all 50 ohm lines in FR4 have the same capacitance per inch.

8.19 What is the loop inductance per length of all 50-ohm transmission lines in FR4? Suppose a 50 ohm line is 12 inches long? How much total loop inductance is in the line? What is the underlying assumption we are making?

The loop inductance per length of a 50-ohm line in FR4 is 8.4 nH/inch. A trace 12 inches long has a total loop inductance of 12 inch × 8.4 nH/inch = 101 nH.

We are assuming this is a uniform transmission line with a 50 Ohm impedance and composed of FR4 dielectric. These are the only assumptions. This means all 50-ohm lines in FR4, regardless of their geometry, have the same loop inductance per inch.

8.20 Why is an n-section LC model not a preferred circuit model for an ideal transmission line?

The return on investment is not very good. The ideal transmission line model is smaller, more computationally efficient, and has a higher bandwidth than an n-section lumped circuit model.

There is no situation where an n-section LC model is preferred over an ideal T element model.

8.21 Why does the dip frequency of the single LC circuit not match the dip frequency for the ideal transmission line when using the total L and total C?

An ideal transmission line and an LC circuit are really different circuits. They are not the same. While the LC model is an approximation to a transmission line it is still only an approximation. There is no reason they have to be the same.

In the LC model, we assume the total L and total C are each concentrated in single lumped points. All their values are in one point element.

In an ideal transmission line, the loop inductance of the transmission line is distributed down the length of the line, and is intertwined with the capacitance. There will be some deviation between how the distributed inductance and capacitance interact down the line. This is a different behavior than a lumped LC model.

8.22 What happens to the pole frequency of an n-section lumped circuit model as we increase the number of LC elements? What is the pole frequency of an ideal transmission line?

With more LC sections, the value of each LC will decrease. This means the pole frequency for a low-pass filter with more LC elements will increase.

An ideal transmission line has an infinite bandwidth and an infinite pole frequency. This means the LC model will always be an approximation to the transmission line. The more LC elements, the higher the bandwidth the approximation is.

Chapter 9

9.1 What does a TDR actually measure?

Ultimately, the TDR measures the voltage at an internal point where the incident step edge passes by. We interpret this voltage as first the incident and reflected component. From these we calculate the reflection coefficient and from this we back out a first-order estimate of the impedance profile of the interconnect.

9.2 What feature of the interconnect corresponds to the time between the start of the signal coming out of the TDR and receiving the reflected signal back in the TDR?

Between the signal coming out of the cable of the TDR and into the DUT and receiving the reflection from the other end of the DUT is the round-trip time delay of the DUT. From this, we calculate the one-way time delay.

9.3 What is the shape of the waveform coming out of the TDR and entering the interconnect?

The signal generated by the TDR is a very fast step voltage wave, the leading edge of a square wave. The rise time is about 35 psec.

9.4 If only one node inside the TDR is being measured, how does it separate what is the incident voltage and what is the reflected voltage?

The total voltage at the internal point is what is being measured. If there were no reflections, this voltage measured would be the incident signal.

Any measured voltage different from this constant voltage is reflected. From the difference, we separate out the reflected voltage.

Then we look at the time at which we measure any different voltages that are due to reflections spatially separated from each other.

9.5 What is a discontinuity?

A discontinuity is any change in impedance from a structure that is electrically short. We see a reflection and cannot see a flat top or flat bottom.

If there is a flat top or flat bottom, we can describe this structure as a uniform transmission line structure rather than a discontinuity.

9.6 What is the impact of a discontinuity at the beginning of a transmission line on the reflected signal and the interpreted impedance?

If there is a reflection at the beginning of the line from a discontinuity, we see this initially. The TDR response we see for the first few round-trip times is distorted by the initial reflection.

But if we wait a few round-trip times of the discontinuity, the impedance on the other side will be a good measure of the impedance profile of the DUT.

9.7 If a discontinuity has a peak in impedance, how else do we refer to this discontinuity?

If we see a peak or a dip, we call this a discontinuity. It may be a uniform transmission line segment, but we cannot resolve it as a uniform structure. It is electrically short and a different impedance than the rest of the interconnect.

9.8 If a discontinuity has a dip in impedance, how else do we refer to this discontinuity?

A discontinuity that is a dip shows a lower impedance than the rest of the impedance profile. We refer to this as a capacitive discontinuity.

9.9 If a resistor is connected to the end of a DUT, what impedance will the TDR measure if we wait long enough?

Regardless of the impedance profile of the DUT, if we wait long enough, a few round-trip time delays of the DUT, we will see the impedance of the resistance of the resistor at the end of the line.

This could be an open, a short, or anywhere in between.

9.10 If you measure a resistor or a long transmission line and look for a time short compared to the round-trip time, how could you distinguish which measurement is the resistor and which is the transmission line?

In principle, you can't. The reflected signal will look identical if it is a long uniform transmission line or a resistor.

In practice, you might see a different launch discontinuity, or some small instantaneous impedance variations in the transmission line as the signal propagates down the length. The reflection from the resistor will be very constant.

9.11 What is the difference between an electrically short and electrically long interconnect?

An electrically long transmission line will be resolved with a flat top or bottom. This identifies this structure as a uniform transmission line.

If the reflected profile has just a peak or a dip, you cannot resolve the uniform transmission line's instantaneous impedance. We describe this impedance change as electrically short.

The feature of the interconnect to make it electrically short, in practice, is that its time delay is shorter than the rise time of the signal. An electrically long interconnect has a time delay longer than the rise time of the signal.

9.12 If the rise time of the TDR is 50 psec, how long does a discontinuity have to be in order to see its flat top or flat bottom?

To be electrically long, the transmission line's TD should be longer than 50 psec. This means TD > 50 psec.

In an FR4-type material, the time delay is 170 psec/inch ×Length.

The condition for electrically long is:

170 psec/inch × Length > 50 psec

This means Length > 0.3 inches to be electrically long.

9.13 What feature would be adjusted in a TDR to resolve a shorter interconnect as a uniform line and not a discontinuity?

The only way to resolve the uniform impedance profile of an electrically short discontinuity is to reduce the rise time of the TDR signal so that the flat top or flat bottom can be resolved.

9.14 If two structures are 0.5 inches apart, how short a rise time is needed to resolve them as two different structures?

To resolve two discontinuities as separate, we need to be able to see two separated peaks. This means the time delay of the distance between the edges of the structures must be $> \frac{1}{2}$ RT.

The time delay difference is the 170 psec/inch × distance $> \frac{1}{2}$ RT. If the distance is 0.5 inches, then we need:

$$170ec/inch \times 0.5\ inch > \frac{1}{2}\ RT\ or\ RT < 170\ psec.$$

9.15 What is the assumption we make when we read the impedance of an interconnect directly from the front screen?

The reflected voltage is measured by the TDR from which we calculate the reflection coefficient.

The assumption we make when we interpret the reflection coefficient as an impedance profile is that at each interface, we have the transition from a 50-ohm transmission line to some other impedance.

As long as the actual initial impedance at each interface is close to 50 ohms, this is a good approximation. When there are large deviations in impedance from 50 ohms, these large impedances will mask the impedance on the other side.

9.16 When does masking play a role?

Masking will play a role when the impedance of a region of the DUT is far off from 50 ohms. It would have to be roughly 20 ohms away from 50 ohms to cause noticeable masking problems.

To unmask the change in impedance, you would have to hack the impedance on the other side.

9.17 If there are multiple reflections, how can we extract a more accurate model of the interconnect from the measured TDR response?

To get a better model of the impedance profile if there is masking, we would have to build a model of the impedance profile and fit the parameter values. This process is called hacking.

This is a very good method that will enable extracting complex structures, not just a single transmission line on the other side.

Chapter 10

10.1 What are the two most important measurements to take with a TDR before doing any other measurements?

Before doing any other DUT measurements, it's always important to measure the TDR response from the open end of the cable from the TDR so we know where the end of the TDR is and the beginning of the DUT.

It's a good habit to also do a second measurement to measure a 50-ohm calibration resistor. This will help verify the quality of the calibration.

10.2 Why is it important to always use a torque wrench?

If you do not use a torque wrench connecting the TDR cable to the DUT, there will always be some uncertainty about the contact resistance and the size of the discontinuity at the launch. More than anything, these two features would not be reproducible.

Using a torque wrench reduces the contact resistance and makes the discontinuity at the launch reproducible.

10.3 What is the shape of the reflection from a connection that is not tight?

Generally, if the pin of the SMA connector is not well seated in the cup of the other connector, there will be an inductive discontinuity. In addition, there will be a contact resistance at the interface. This will contribute to a series resistance in the TDR impedance profile.

If you see a step increase in impedance at a connection, chances are it is a contact resistance of the connector. Try tightening the connector or getting a new cable. The gold on the tip of the connector may have worn off, in which case the contact resistance will only increase next time.

10.4 Why is the launch into a coax cable usually inductive?

Inside a coax cable, the braid of the cable needs to be pulled back in order to slip over the connector that covers the dielectric of the cable. It is very difficult to make a tight connection of the shield to the dielectric.

Depending on the quality of this termination of the cable to the connector, there may be a gap between the shield and the dielectric, which means the return is pulled farther away from the signal, increasing the impedance in this short region.

Generally, the termination of a cable to a connector looks inductive because the signal and return spacing increases.

10.5 What is the impedance noise floor of a typical TDR?

Depending on the TDR, the noise floor of how small an impedance change is real or just measurement noise is about +/- 0.2 ohms. This is the smallest change we can measure as a real impedance variation, whether for an electrically long or short structure.

In some TDR instruments, the smallest impedance change is larger, on the order of 0.5 ohms. This is why it is important to measure a calibrated 50-ohm resistor. This will give a direct indication of the noise on the impedance measurement.

10.6 What is the typical impedance variation in an airline?

An airline is a structure that is a precision machine cylinder with a solid copper wire as the center conductor. This cross section can be very uniform and is another indication of the variation in instantaneous impedance that can be measured.

Its typical instantaneous impedance variation is also on the order of 0.2-ohms.

10.7 What is the typical impedance variation in a PCB trace?

Due to the manufacturing variation in linewidth and dielectric thickness and local Dk value, there is real impedance variation in a transmission line fabricated in a circuit board.

This variation can change from 0.5 ohms to as much as 2 ohms. This impedance variation is a real impedance variation; it is not due to noise in the measurement process.

It is difficult to predict what the variation in the impedance profile should be because it depends so much on the quality of the manufacturing for that specific circuit board, and the local variation in Dk due to the presence of glass yarn and resin.

10.8 What is the most important consistency test to determine if a measured impedance variation is real?

Sometimes it is hard to decide is a measured impedance variation real or due to some other effect. The best confirmation that an impedance profile is real is to measure the impedance from both ends of the transmission line. If the impedance variation is real then the impedance profile form one end should be the mirror image as measured from the other end.

10.9 If there is a large series resistance distributed down a transmission line, what will the impedance profile look like?

If the impedance is increasing down the length of the transmission line due to series resistance, then the impedance profile would look uphill from both ends. There will be the same increase in impedance from both sides of the interconnect.

10.10 How would you verify an impedance variation is due to a high series resistance or a real spatial variation in impedance?

Measure the impedance profile from both ends of the interconnect. A real impedance variation that increases from one side of the transmission line to the other will appear as a decreasing impedance from the other end of the line.

If the increase in impedance is due to a series resistance, then it will look increasing from both ends. This experiment would distinguish which root cause is at the heart of the increasing impedance.

10.11 Why do identical connectors look different when measured from different ends of an interconnect?

An electrically short discontinuity will show an impedance peak or dip whose value depends on the rise time of the signal. A longer rise time means a smaller change in displayed impedance.

As the TDR signal travels down the transmission line, the rise time will increase a little bit due to the losses in the transmission line. This means the rise time will be different when it encounters the connector at the beginning of the line and the end of the line.

When it encounters the beginning of the line, there would be a large reflection. When the longer rise time encounters the identical connector at the far end, its rise time is longer and the reflection will be smaller.

Measured from one of the line, the near connector will look worse than the far connector, as viewed from either end. The near-connector measured from either end will look the same. This is an indication that we see the difference from opposite ends due to the rise time degradation.

10.12 How would you confirm the two ends of the transmission line have identical connectors?

The best way of checking the similarity of the two connectors is to measure the TDR profile from both ends. If the connectors are the same, we will see the same reflection profile when the rise time of the signal is the same at the beginning of the line.

10.13 If the far end of a transmission line is open, what will be the reflection coefficient we see if we wait long enough?

If we wait for all the reflections to die down, so masking is not an issue, we will always see the impedance of the far end of the line. If the far end is open, we will see a reflection coefficient of 1 if we wait long enough.

10.14 Why is a copper pour on a signal layer a bad practice?

When we add a copper pour to a layer with signal lines on it, we will increase the chance of getting enhanced cross talk between adjacent transmission lines. There will be resonances in the floating copper and these resonances will couple enhanced noise to the other transmission line signals.

This is why adding a copper pour is dangerous. It solves no problem and can add additional noise.

Chapter 11

11.1 What is the Dk of a dielectric a measure of?

The dielectric constant of an insulator is a measure of how much the material will slow down the speed of a signal compared to the speed of light.

In addition to this behavior, a dielectric material will also increase the capacitance of two conductors separated by the dielectric material.

11.2 What combination of materials contribute to the Dk of a laminate layer?

In a single layer of dielectric in a circuit board, the material is made up of a layer of glass yarn and then embedded by resin.

This means any single layer of dielectric material is a composite of glass fiber and resin material. Their relative proportion will determine the specific dielectric constant of the composite.

11.3 Why would the measured Dk of a stripline structure not be the Dk of each layer?

In a stripline, there are really two layers between the signal lines and the planes. The signal propagating down the stripline structure will see both dielectric layers. It is not possible to separate out the contribution of the dielectric constant of each layer from the composite measurement.

It is possible for the Dk of the top layer to be very different from the bottom layer Dk. All that can be extracted is an effective Dk value.

11.4 What are three ways of measuring the Dk of a transmission line?

Ultimately, the Dk of a laminate will affect the time delay of the signal in traveling from one end to the other. This suggests a few methods of measuring the time delay for a fixed length.

We could just measure the time delay of a length of transmission line. There is the slight uncertainty of where is the beginning and end of the line.

We could build two different length lines and measure the time delay difference of the two different length lines.

Alternatively, we can build a line with two small discontinuities spaced a precision distance apart. The round-trip time delay between these discontinuities and the spacing will be a measure of the speed of the signal in this dielectric. This and the speed of light in air will result in the effective Dk value.

11.5 Why is the measured Dk based on the time delay from the beginning to the end of the line inaccurate?

When we use the time delay from the beginning to the end of the line, there is always some uncertainty in where the end of the connector and the beginning of the line might be. This adds some uncertainty in the time delay for the length of the line.

11.6 What is the advantage of measuring the Dk using two different length transmission lines?

When two different length lines are used, each with the same connectors on the ends, the only difference in time delay is the length difference. They each have exactly the same delays for the connectors.

Using the two different length lines means the delay difference is only about the length difference.

11.7 What is the assumption we are making when using two different transmission lines to measure the Dk?

When two different length transmission lines are used to measure the Dk, we are assuming the connectors for each line are the same. This way the only difference between the time delays of the two lines is due to the length difference.

But what if there is really a small difference in the connectors? Then there will be an artifact introduced in the time delay that is not due to just a longer length segment.

11.8 What is the advantage of using one line with two small strategically placed discontinuities?

When a single transmission line with two small discontinuities are used to measure the time delay between them to get the speed of the signal, we are not subject to any connector variation that may happen between two different lines. This should be a more robust solution.

11.9 What is a typical value measured for the composite Dk of a stripline composite?

It is difficult to say what a typical value of the composite DK might be in a stripline structure. It depends so much on the specific layers.

In the one specific case illustrated in this chapter, the composite Dk value of two layers is 4.39.

11.10 How does this compare with the bulk Dk of the microstrip layer?

In the specific example shown, the bulk Dk of the stripline layer was seen to be 4.49. This is compared to 4.39 for the stripline layers. They are very close.

There is no connection between these two layers. They are different dielectric materials and in different boards. There is no reason they should be the same or even close together. They happen to be similar.

This is why it is important to measure the Dk of each layer in a circuit board. They will vary depending on what the fab shop happened to grab from their inventory of laminate layers. Sometimes, if they have a lot of one specific layer, they may use these one week and a different laminate layer the next week.

11.11 Why is the Dk measured for a microstrip not the bulk Dk?

When the time delay of a microstrip transmission line is measured, it is the time delay of the interconnect. The signal will see the combination of the bulk dielectric material below the signal line and the dielectric constant of air above.

This means that the effective Dk measured in a microstrip is always less than the bulk value by some depending on the geometry.

11.12 What do you have to know about the microstrip to convert the effective Dk into the bulk Dk?

To convert the effective Dk to the bulk Dk, we need to know what the cross-section geometry is of the microstrip. As the cross section changes, the relative contribution of the dielectric below and the air above will also change.

To convert between the effective and bulk Dk, we need to know the linewidth and dielectric thickness of the layers as a minimum.

11.13 What do we call the property of the material if the Dk varies with frequency?

When the Dk varies with frequency, the speed of a signal will depend on frequency. We call this property dispersion. This is the property of any material to have a speed of a signal that depends on frequency.

11.14 If the dispersion of a material is large, what must also be true of the electrical properties of the material?

There is a connection built into the behavior of materials, that the dispersion is coupled to the losses in a dielectric. The more the dispersion, the more the loss in the materials as measured by the dissipation of the material.

This also means that a low-loss laminate will have very little dispersion and the Dk will be more constant with frequency. There is a reason real materials behave this way, related to the root cause of the Dk and Df as connected by the rotation of dipoles in the material.

11.15 What does a causal model mean?

This is a very subtle quality. It is a property all functions that describe real physical performance must meet. It means that an output cannot arrive before an input. The response has to happen after the stimulus.

This translates to a specific relationship between the dielectric loss and the dispersion in the dielectric constant of the material.

Any model that describes the frequency dependence of the dielectric constant must be causal.

11.16 How does the Dk vary with frequency in most materials?

In most materials, to the first order, the DK is constant with frequency. To second order, the Dk will decrease with the log of frequency. At higher frequency, the Dk gets smaller, but by a very small amount.

11.17 If the small variation in Dk with frequency is important in your application, what should you do?

The TDR measurements are not a good way of characterizing the frequency dependence of the Dk of a material. If this is an important property, methods other than the TDR should be used to characterize the Dk, such as a frequency domain measurement with a VNA.

11.18 When we use a TDR to measure interconnect properties, what are we assuming about the frequency dependence of the figures of merit?

The TDR is not a good way of measuring the frequency dependence of the Dk of a material. To first order, it is probing the high-frequency properties of the Dk. It measures the Dk at the rise time of the signal, which is in the 10-GHz frequency range.

When we use the TDR to extract a Dk value, we are assuming this one value of DK is roughly at 10 GHz.

Chapter 12

12.1 What is the difference between analysis and design?

Design is about creating something from nothing. This is a synthesis process. Analysis is about evaluating performance of a completed design. This is after the design is created and its performance is being evaluated, either as a virtual prototype or as a real system that has been measured.

12.2 Why are we able to calculate the characteristic impedance of a coax exactly?

Maxwell's equations, which are differential equations, can only be solved exactly for a few geometries, either spherically or cylindrically.

In two dimensions, a coax is a cylindrical geometry and Maxwell's equations can be solved exactly for this geometry. Likewise, two rods and rod over ground are cylindrically symmetrical and can be solved exactly.

For any other geometry, there is no closed-form analytical formula for the characteristic impedance of other cross sections.

12.3 Why is TEM mode propagation important when considering using a 2D field solve to calculate the characteristic impedance of a transmission line?

The assumption of the TEM mode of propagation is that the electrical fields are varying only in the x-y direction, not in the z direction — the direction of propagation.

This means the electric and magnetic fields have a constant amplitude in the direction of propagation. We only need to know their value in a cross section. They will be the same everywhere down the line. This is why we can use a 2D field solver to calculate the characteristic impedance instead of a 3D field solver.

12.4 How high a frequency can we go and have TEM field lines in a microstrip if the dielectric thickness is 20 mils? 5 mils?

The limit to the TEM mode depends on the thickness of the dielectric. The highest frequency at which the TEM mode propagates is roughly 600 GHz/h[mils].

If the dielectric thickness is 20 mils, the TEM mode will propagate up to frequencies of 600 GHz/20 = 30 GHz.

Above 30 GHz, a microstrip may show dispersion due to a non-TEM mode of propagation.

If the dielectric thickness of the top layer is 5 mils, the frequency up to which the TEM mode will propagate is 600 GHz/5 mils = 120 GHz.

Generally, signals will not see frequency components up to 120 GHz for data rates below 112 Gbps.

12.5 What is the hidden assumption in most 2D field solvers about the return plane?

Most 2D field solvers assume the return plane is infinitely wide. This means that the return current has spread out however far it needs to based on the current redistribution from the skin depth effect.

12.6 At a minimum, how wide should the return plane be to not affect the characteristic impedance of the line to less than 1%?

As a general rule of thumb, for a 50-ohm transmission line, the return path should be about 3-linewidths on either side to enclose all the return current.

If the linewidth were 5 mils, this would be a 30-mil-wide return path under the signal line.

12.7 What is rule #9 and why is it important?

Rule #9 is never do a measurement or simulation without first anticipating the result. It is the first and most important consistency test that we can do to confirm the quality of the result.

We can never prove we are right; all we can do is verify that the result is consistent with all the other tests we can think of doing.

Testing to see if the result is consistent with our expectations is the most important test we can perform.

12.8 What is an example of an error you could make using a 2D field solver that you could catch if the result was way off from what you expected?

A 2D field solver will calculate the characteristic impedance for a specific set of geometry and material properties. If, for example, we had an 8-mil-thick dielectric and we meant to type in 16-mil trace, we would have expected to see a calculated characteristic impedance of about 50 ohms.

If instead, we had typed in 61 mils as the linewidth, we would have seen an impedance closer to 10 ohms, and known immediately that something was not correct. The calculated value of impedance did not come close to what we expected.

This is how rule #9 could help find a simple error in how we entered terms.

Chapter 13

13.1 What are the two most commonly used transmission line topologies found in all circuit boards?

The two common circuit board transmission line geometries are microstrip on surface traces, and stripline for buried traces.

13.2 For a 50-ohm FR4 microstrip, if the line width is 10 mils, what should the dielectric thickness be approximately?

As a rough rule of thumb, in a microstrip, the line width to dielectric thickness should be 2:1. A linewidth of 10 mils should use about a 5-mil-thick dielectric.

13.3 As the dielectric thickness increases, what should happen to the characteristic impedance?

When dielectric thickness increases, the capacitance per length decreases and the characteristic impedance should increase.

13.4 What is the difference between a first order and second order term? What are examples of a first order term and a second order term in the design of a microstrip?

A first-order term means it has a strong influence on the characteristic impedance on the line. A second-order term means it has only a small impact on affecting the characteristic impedance of the line.

For example, line width will be a first order term. A small percentage change in the line width of a trace will have a comparable change on the characteristic impedance.

But a small change in the conductor thickness will have a miniscule impact on the characteristic impedance. It is a second-order term.

13.5 What is a good rule of thumb for the ratio of the line width to dielectric thickness for a 50-ohm microstrip line?

In a microstrip, the linewidth to dielectric thickness for 50 ohms in FR4 is 2 to 1. A 10-mil-wide line width would need a 5-mil-thick dielectric layer.

13.6 If the laminate layer is 5 mils thick, what line width would give about 50-ohms microstrip?

For a laminate thickness of 5 mils, the linewidth would be twice the thickness, or 10 mils wide.

13.7 What design terms influence the characteristic impedance of a microstrip the most?

The three most important first-order terms that influence the characteristic impedance of the microstrip are the linewidth, the dielectric thickness, and the dielectric constant of the laminate.

13.8 What is the limitation to using on-line calculators for the characteristic impedance of transmission lines compared to a 2D field solver?

All online calculators are equation-based. This means they are an approximation. If you don't know the quality of the approximation used, there is no guarantee it is any good. Even the approximation recommended by the IPC is off by more than 20% for impedances below 30 ohms.

If accuracy is important, a 2D field solver should be used to calculate the characteristic impedance of a stackup.

13.9 What are three examples of second order terms in the design of a microstrip that influence the characteristic impedance only slightly.

There are many second-order terms, such as the conductor thickness, the presence of a solder mask, and the presence of a conductor nearby.

13.10 How many dielectric thicknesses do the fringe fields extend laterally from the edge of a trace?

As a rough rule of thumb, the fringe field extends laterally from the signal line about 6 dielectric thicknesses. This dimension includes virtually all the fringe field lines from the edge of the transmission line.

13.11 If you do not consider the various second order design features in the design of a microstrip, how far off can the fabricated impedance be from what was estimated?

While each second-order design feature might impact the characteristic impedance by only a few percent, taking all of them together and combining them in the worst direction so they all add up could change the characteristic impedance by as much as 15% or more.

Chapter 14

14.1 If the dielectric thickness is 20 mils in each layer of a symmetric stripline, roughly what is the line width for a 50 Ohm line?

For a symmetrical stripline, the ratio of linewidth to dielectric thickness is about 0.75 to 1. If the dielectric thickness in each layer is 20 mils, the linewidth would be about 20 mils × 0.75 = 15 mils wide.

14.2 What is one of the assumptions about the geometry in the IPC approximation for stripline?

In the IPC approximation for stripline, some of the assumptions are that the return planes above and below are infinite in extent, the dielectric thicknesses in each layer are the same, and the dielectric constant, in each layer are the same.

14.3 What is the advantage of using the narrowest line width the vendor can fabricate?

Using the narrowest linewidth the fabricator can provide at no extra charge means you can route with the highest interconnect density. This makes routing easier and can enable either a smaller board size or fewer routing layers, both of which could result in a lower-cost board.

14.4 How far away should the top plane be from the trace compared to the bottom plane in order to not have much return current in the top plane?

If the top plane is 4 × away from the signal line as the signal line is from the bottom plane, the top plane will not have much impact on the impedance of the trace and there will not be much return current in the top plane.

The top plane could be a split plane and not have much impact on the signal line.

14.5 What are the most important terms influencing the characteristic impedance of a stripline?

There are really three terms that influence the characteristic impedance of a stripline: the linewidth, the dielectric thickness of each layer, and the Dk of each layer. These are the first-order terms.

Chapter 15

15.1 What are the two applications for differential pairs?

Differential pairs are used in both low-speed analog signaling and in high-speed digital signaling. The requirements on the differential pair design are different in these two applications.

15.2 Why are the requirements different for these two applications?

When differential pairs are used for analog signals, the most important requirement is to reduce the noise pick-up they might see. Whatever noise one line sees, the other should also see it.

This means routing the traces close together with wide traces and with a continuous return plane underneath them.

The differential impedance is only of secondary importance.

For high-speed digital signals, the differential impedance is very important, as is low loss in very high data rate applications and symmetry between the two lines.

15.3 What is the difference between earth ground and chassis ground?

Earth ground is a connection to a copper pipe stuck into the ground outside a building. It is a safety feature. We often refer to a conductor or system that is not connected to earth ground as floating with respect to earth ground.

The chassis ground is the connection to the outside metal of the enclosure of an electronic product. The chassis may connect to earth ground, or it may float above earth ground.

15.4 What is a better name for digital or analog ground?

When associated with the digital or analog ground, these conductors are really used as return paths. A better term for analog ground is return path, rather than ground.

15.5 What is the difference in the signals on differential pairs in analog signaling vs digital signaling?

Analog signals on a differential pair generally have low bandwidths; less than 10 MHz is typical. Their voltage levels may be a few volts, but generally are very sensitive to noise. A few mV of noise may be considerable. This is why engineering the interconnects for low-noise pick-up is so important.

Digital signals generally will have much higher bandwidths, starting at 1 GHz and extending well above 10 GHz. These signals are in the volt range and noise is not nearly as important a concern. What is important is achieving a target differential impedance, keeping loss acceptable, and keeping the two lines symmetrical.

15.6 What are two important benefits of differential signaling?

The benefit of differential signaling is really in the transmitters and receivers. When a differential signal switches, the total current from the driver is usually constant. This means there is less switching noise.

When a differential signal encounters a discontinuity in the return path, the return currents may overlap and cancel out, so the differential signal is more robust to return path discontinuities.

15.7 What is a pure differential signal? A pure common signal?

A pure differential signal means there is no common signal component. Since the common signal is the average, this means the average voltage of the p and n lines in a pure differential signal is 0 V. A pure differential signal would have 0V average, like a 0 to 1V signal on one line and a 0 to -1V signal on another line.

A pure common signal means there is no differential signal component. Since the differential component is the difference in voltage between the p and n line, a pure common signal has no difference between the p and n lines. They have exactly the same voltage, such as in a 0 to 1 V signal on one line and a 0 to 1 V signal on the other line.

15.8 What is the difference between a balanced and unbalanced differential pair?

The balance refers to the geometry of the two lines that make up the differential pair. A balanced differential pair means the two lines are the same, like a microstrip or stripline pair.

An unbalanced differential pair would be a single ended microstrip that uses another plane as a common return for both conductors as the p and n lines. It is rare to have an unbalanced differential pair.

15.9 What is differential impedance or common impedance?

The differential impedance is the instantaneous impedance the differential signal sees each step along the interconnect. The common impedance is the instantaneous impedance the common signal sees each step along the interconnect.

15.10 When the p and n lines are far apart and each has a single-ended impedance of 50 ohms, what is their differential impedance and their common impedance?

A differential pair that has its traces very far apart can be just as good a differential pair as one with traces very close together, as long as they are engineered for the same impedance.

When the traces are far apart and their single-ended impedance is 50 ohm, then their odd-mode impedance and even mode impedances are also 50 ohms.

The differential impedance will be twice the odd-mode impedance, or 100 ohms, and the common impedance will be half the even-mode impedance, or 25 ohms.

15.11 What happens to the input impedance of a single-ended line when an adjacent single-ended line tied to ground is brought in from far away, but never closer than 2 × w?

The single-ended impedance of a line is unaffected by the proximity of an adjacent line until they get very close together. This is contrary to popular belief, but can easily be demonstrated in a simple simulation.

15.12 What happens to the input impedance of a signal line when an adjacent signal line driven opposite is brought in from far away?

When the second line is driven opposite and brought in proximity, the impedance of the second line will decrease. This is, by definition, the odd-mode impedance. The odd-mode impedance will decrease with increased coupling. This means the differential impedance will decrease as well.

15.13 What happens to the input impedance of a signal line when an adjacent signal line driven the same is brought in from far away?

When the pair of lines is driven the same — in other words, a common signal is applied — the impedance of one line, the even-mode impedance, will increase as coupling increases.

This means the common impedance will also increase as the two lines are brought closer together.

15.14 What is the odd-mode of a differential pair?

The odd-mode impedance of a differential pair is the impedance of one line when the pair is driven by a differential signal in the odd mode.

15.15 What is the even-mode of a differential pair?

The even-mode impedance of a differential pair is the impedance of one line when the pair is driven by a common signal in the even mode.

15.16 What is the difference between the odd-mode impedance and differential impedance of a differential pair?

The odd-mode impedance and the differential impedance are both impedance terms of a differential pair, but they refer to different aspects. They both apply to a differential pair when a differential signal is applied.

The odd-mode impedance refers to the impedance of one line.

The differential impedance refers to the impedance the differential signal will see interacting with the interconnect.

15.17 What is the difference between the common impedance and the differential impedance?

The common impedance is the instantaneous impedance the common signal sees when it propagates on the differential pair. The common signal will only be sensed when a common signal is applied.

The differential impedance is the impedance the differential signal sees. It will only sense this impedance when a differential signal is applied.

15.18 When the traces in a differential pair are far part, what is the single-ended impedance, the odd mode impedance and the even mode impedance of a 50-ohm impedance transmission line?

When traces are far apart, the transmission lines are uncoupled and the single-ended impedance is 50 ohms, the odd-mode impedance is 50 ohms, and the even-mode impedance is 50 ohms.

The impedance of one line is the same and independent of how the other trace is driven when they are far apart.

15.19 Why is it confusing referring to the impedance of a line as just 45 ohms?

This is ambiguous. Are you referring to the single-ended impedance? The input impedance in the frequency domain? The input impedance in the time domain, or even the odd-mode impedance of a line? There are too many choices without adding the preface.

Chapter 16

16.1 How does a TDR measure the single-ended impedance of a transmission line?

A TDR sends a step voltage wave and measures the reflections. From the reflections we back out what impedance change must have caused the reflection.

To measure the single-ended impedance of a transmission line, the other line should be either grounded or kept floating. Then the measured impedance is the single-ended impedance.

16.2 How does the TDR measure the odd mode impedance of a differential pair?

To measure the odd-mode impedance, we measure the impedance of one line while at the same time driving the other line in the odd mode by applying a differential signal to the pair.

16.3 How does the TDR measure the differential impedance of a differential pair?

Fundamentally, differential impedance is measured by measuring the odd-mode impedance of each line in the pair and adding up the two odd-mode impedances.

16.4 As the coupling in a differential pair increases, what happens to the difference between the single-ended impedance and the odd mode impedance?

As coupling increases, the single-ended impedance will not change very much, but the odd mode-impedance will decrease. The higher coupling will enable more current to flow through the electric field lines from the coupling when a differential signal is applied.

16.5 If you want to maintain a constant differential impedance and the coupling increases, what can you do to the line width?

If the coupling increases, the differential impedance will decrease. To keep the differential impedance constant as we increase coupling, we would have to decrease the linewidth.

16.6 If the odd mode impedance of a line in a differential pair is 45 ohms, what is the differential impedance?

The differential impedance of a pair is twice the odd-mode impedance of either line. This means if the odd-mode impedance is 45 ohm, the differential impedance is 90 ohms.

Chapter 17

17.1 What is the difference between a tightly coupled and uncoupled differential pair?

Coupling is about the relative spacing between the signal lines. Tightly coupled means the traces are close together. This generally is a spacing as close as can be fabricated, which is the linewidth. Uncoupled is when the spacing is far enough away so that one trace does not affect the other. This is about 3 linewidths apart.

17.2 If all you know is the differential impedance or the odd-mode impedance of a differential pair, how do you estimate the coupling between the two lines that make up the differential impedance?

You cannot. This is a trick question. You can design a differential pair with any differential impedance or odd-mode impedance you want, with any range of coupling. These are independent terms.

17.3 Which is better, a tightly coupled or an uncoupled differential pair?

Unfortunately, the answer is it depends. But you have to evaluate what it does depend on.

If loss is not important, tightly coupled is usually better because this means higher interconnect density and fewer layers or a smaller board.

On the other hand, in stripline, this means a thicker dielectric than you might need if you used a loosely coupled differential pair.

If loose is important, then a loosely coupled differential pair might be preferred because this would mean a wider line and lower loss.

17.4 If the line width is adjusted in a differential pair to keep the differential impedance constant as coupling changes, what will a differential signal see propagating down such a differential pair?

If the instantaneous differential impedance is constant to the signal, it will not care what coupling the pair has. All it will see is the constant differential impedance.

17.5 What is the Johnny Cash principle and why is this important designing a constant differential impedance pair?

When coupling changes and you want to keep the differential impedance constant, you have to adjust the linewidth to keep the differential impedance constant.

The design curve of the signal linewidth needed as the coupling spacing changes is the line you must walk to keep the differential impedance constant. Walking this line to keep the differential impedance constant is called the Johnny Cash principle.

17.6 What is an advantage of using a tightly coupled differential pair?

It will result in higher interconnect density, which could mean f ewer routing layers and a smaller board.

17.7 What is an advantage of using an uncoupled differential pair?

This would enable thinner dielectrics either in microstrip or stripline and wider lines, which means less conductor loss.

17.8 As coupling increases and the differential impedance decreases, what happens to the common impedance?

As coupling increases, the even-mode impedance will increase and the common impedance will increase.

17.9 If the coupling in a differential pair increases and the line width is decreased to keep the differential impedance constant, what happens to the common impedance?

It's always easier to analyze these sorts of problems by looking at the odd- or even-mode impedance, then see what happens to the differential or common impedance.

As the coupling increases and the traces are moved closer together, the odd-mode impedance decreases and the even-mode impedance increases.

As the linewidth decreases, the odd-mode and even-mode impedance will both increase. This means that the even-mode impedance will increase even more, from the closer spacing and the narrower linewidth. This means that while the differential impedance will be constant, the common impedance will be much higher.

17.10 When are losses in a differential pair important?

Generally, at data rates below 1 Gbps, the losses in the interconnects are not important. But above 1 Gbps, the losses from dielectric and conductor losses can dominate the performance of a channel.

17.11 What are the two loss mechanisms in a transmission line?

There are two general root causes for losses, conductor loss and dielectric loss. These are both frequency-dependent.

17.12 For lowest conductor loss at a target impedance, what line width should be used?

For lowest conductor loss, always use as wide a linewidth as practical. This means looser coupling, thicker dielectric, and as low a dielectric constant as practical.

17.13 What are three ways of engineering the widest possible line width at a fixed impedance?

The three most important terms to adjust to enable a wider line are to use a thicker dielectric, use loose coupling, and use a lower dielectric constant.

17.14 What is the advantage of a loosely coupled differential pair?

A loosely coupled differential pair means we can use a thinner dielectric and a wider line. This is valuable when trying to reduce conductor loss.

17.15 What determines the speed of a differential signal or a common signal?

Ultimately, the speed of a signal is due to the effective dielectric constant the signal sees. In an uncoupled differential pair, the electric field distribution between the signal return conductors are the same for the differential signal or the common signal. The field lines between the two signal lines do not interact.

This means the speed of the differential signal and common signal are the same.

But when the pair is tightly coupled, the electric field lines are very different for the differential signal or the common signal. The deferential signal will have many more electric field lines in the air and see a lower effective dielectric constant than the common signal.

Of course, in stripline, the differential and common signals will always see exactly the same dielectric distribution and have the same speed.

17.16 In a differential microstrip, which travels faster, a differential signal or a common signal?

In a tightly coupled differential pair, the differential signal will have more electric field lines in the air and see a lower effective dielectric constant than the common signal.

The differential signal will travel faster than the common signal.

17.17 In a differential stripline, which travels faster, a differential signal or a common signal?

In a stripline, while the electric field distribution may be different for a differential signal and common signal, they will see the same Dk and have the same effective DK and travel at the same speed.

17.18 For line widths of about 10 mils and 50 ohm single-ended impedance transmission lines, how close can the lines get before the differential impedance drops by more than about 1%?

This small change is about the threshold for uncoupled differential pairs. This happens when the spacing is closer than 3-linewidths. With a 10-mil-wide line, this is a spacing of 30 mils. When the spacing is closer than 30 mils, the differential impedance will begin to decrease by more than 1%.

17.19 What is the disadvantage of a tightly coupled differential stripline pair?

A tightly coupled differential pair will mean a narrower linewidth than could be achieved with a loosely coupled differential pair.

It will also require a thicker dielectric than a loosely coupled pair.

17.20 Why is it more effective to decrease the line width for tightly coupled differential stripline pairs than increase the dielectric thickness?

Decreasing the linewidth is something that can be done on the same layer. It is far easier to adjust the linewidth than to select different layers with different dielectric thicknesses.

17.21 How strong an impact is there on the differential impedance in a stripline geometry from the resin rich region between the traces? Is this a design issue to worry about?

The resin-rich region between the two lines in a differential pair will increase the differential impedance due to its lower Dk. It is only by using a 2D field solver that we can put in the numbers and determine by just how much the impedance will increase.

When we do, it is a tiny amount, in most cases negligible. There are more important features to worry about,

Chapter 18

18.1 What is displacement current and when does it flow?

Displacement current is the current that flows through changing electric field lines. This is how the return current flows between the signal and return paths through the insulating dielectric.

It will only flow when the voltage between the signal lines change, which is when the electric field lines between the conductors change.

18.2 As the return plane is moved farther way what happens to the single-ended and common impedance in a microstrip or stripline differential pair?

As the return plane is moved farther away, the single-ended impedance will always increase. This is the case in a microstrip or stripline.

Likewise the common impedance will increase as well.

18.3 What is the difference in the electric field lines between the p-line and the n-line in a differential pair when driven by a differential signal and a common signal?

When driven by a differential signal, there will be electric field lines between the p- and n-lines. There is a large voltage between them, as well as a large electric field between them.

When driven by a common signal, the voltage between the p and n-lines will be the same. This means there is no electric field between them.

18.4 What is "coupling" between the p-line and other conductors a measure of?

Coupling is a measure of the fringe electric fields between conductors. The higher the coupling, the more the electric fields between them, when they have different voltages.

When they have a common signal, coupling refers to the impact on the proximity of the other conductors.

18.5 What are two different explanations as to why there is no return current distribution in the return plane of a differential pair when the plane is very far away from the signal lines and driven with a differential signal?

One way of seeing that there is no return current in the return plane when the plane is moved farther away is that the return currents of the p- and n-lines when driven with a differential signal, will cancel out. The presence of the return plane is not needed.

When the return plane is far away, there is not electric field between it and either of the two lines that make up the differential pair. This means there is no displacement current flowing between them, and no return current in the return plane.

18.6 In a differential microstrip, the signal lines are 10 mils wide and the spaced 20 mils apart. Roughly how far away would the plane have to be so that its presence did not influence the differential impedance?

As a rough rule of thumb, if the spacing to the return plane is farther than the span between the two lines. When the linewidth is 10 mils wide and spaced by 20 mils, the span is 10 mils + 20 mils + 10 mils = 40 mils.

This means when the plane is more than 40 mils away, there will be no return current in the return plane and it can be removed with no impact on the differential impedance.

18.7 In a differential pair driven by a differential signal, what is the direction of propagation of the voltage wavefront in each line of the pair? What is the direction of propagation of the current wavefront into each line of the pair? What is different in each line of the pair?

This is a trick question. When a differential signal drives a differential pair, the voltage wavefront and the current wavefront both travel in the same direction.

What is different in the p-line and the n-line is the direction of circulation of the current wavefront in each line. In the p-line, the current circulates in the clockwise direction. In the n-line, the current circulates in the counterclockwise direction.

18.8 What is the fundamental principle that drives the current redistribution in a transmission line as frequency increases?

As frequency increases, the current always tries to take the path of lowest loop inductance. This drives the current to redistribute to the outer surface in the signal traces and to the upper surface in the return path. This effect is called the skin depth effect.

18.9 When the plane is very far away in a differential pair, what other impedance is the differential impedance the same as?

If the plane is very far away, the differential impedance is the same as the single-ended impedance of the pair. The field distribution looks identical whether applying a differential signal or a single-ended signal.

18.10 When a single-ended signal crosses a gap in the return path, what happens to the instantaneous impedance in the gap region and why?

The impedance of the single-ended signal will increase when the signal crosses the gap in the return path. The electric fields to the return path decrease because the conductor is farther away. This means the displacement current decreases and the current decreases. If the current decreases, the impedance will increase in this region.

18.11 When a differential signal crosses a gap in the return path, what happens to the instantaneous differential impedance in the gap region and why?

The differential impedance a differential signal sees will increase when it crosses a gap in the return path. But since there will be some coupling between the two signal lines, and some displacement current between them, the differential impedance change will not be as large as the odd-mode impedance change.

18.12 In an unshielded twisted pair, what do you expect to see as the single-ended impedance of one line with the other line grounded, the odd mode impedance of one line and the differential impedance of one line?

When the other line in a twisted pair is grounded, the single-ended impedance will be related to the presence of the other line. They will act as a single-ended transmission line.

The differential impedance will be the same as the single-ended impedance.

When the pair is driven with a differential signal, the odd-mode impedance of one line will be half the differential impedance.

18.13 Where is the return current in a twisted pair in CAT5 cable?

When driven by a differential signal the return current of one line is carried by the other line. Any return plane is so far away all return current for the differential signal has overlapped and cancels out.

18.14 CAT6 cable has a shield surrounding all four twisted wire pairs. The spacing between wires in each twisted pair is about 50 mils. The spacing from the twisted pair to the shield is about 200 mils. Where do you expect the return current to flow in a CAT6 cable when a pair is driven by a differential signal?

In a CAT6 cable, the return current in the shield cancels out between the p- and n-lines. The presence of the shield does not affect the differential impedance or the differential signal.

The purpose of the shield is to provide a return path for any common signal that might be on the cable.

Chapter 19

19.1 In a TDR impedance profile, what distinguishes an electrically long and electrically short structure, or a uniform transmission line and a discontinuity?

If the TDR profile shows a flat bottom or flat top, we can see the structure as a uniform transmission line and read its characteristic impedance off the screen. This is electrically long.

When the response looks like a peak or dip, the impedance value also depends on the rise time and we call the structure electrically short.

19.2 When we see a region in an impedance profile that is flat for some region, what are we implicitly assuming the electrical circuit model for this section of the interconnect to be?

We assume the model of the structure is an ideal uniform lossless transmission line. It has just a characteristic impedance and a time delay. When the structure is electrically long, we can read these two figures of merit off the front screen of the TDR.

19.3 What does a figure of merit for an interconnect really refer to?

A figure of merit is a parameter that defines a specific feature of the interconnect. It is based on the parameter value of an equivalent ideal model element.

For an ideal capacitor element, its figure of merit is its capacitance.

For an ideal lossless transmission line, its figures of merit are the characteristic impedance and time delay. These are a few parameters that tell us most of what we want to know about the properties of the interconnect in a format we can use to evaluate this interconnect in an application.

19.4 What are three different circuit elements for a discontinuity?

A discontinuity can be described with an ideal uniform transmission line, an ideal capacitor or an ideal inductor.

19.5 What is another name for the process of hacking an interconnect?

This process of building a circuit topology and fitting parameter values based on matching the simulation to the measurement is also called measurement-based modeling.

19.6 How do you know what circuit elements to use when building a circuit topology to describe the TDR response of an interconnect?

The process of selecting the circuit topology based on the TDR response is very simple. If the TDR response is flat, use a transmission line element. If there is a peak, use an inductor element. If it is a dip, use a capacitor circuit element. These are the common elements to include in a circuit topology.

19.7 When selecting a simulation tool to hack an interconnect, what is an essential circuit element the simulation tool must understand?

The simulation tool must have an ideal lossless transmission line element as a minimum. This is adequate for most examples.

19.8 Why is it important to use the same source waveform in the simulation as when the TDR response was measured?

The TDR response of a discontinuity depends on the rise time of the signal. This means to match the simulation with the measurement, the simulated step signal has to be the same rise time as used in the measurement.

19.9 What is an important limitation in the type of transmission line models available in QUCS? What impact will this have on simulating TDR responses?

For time domain transient simulations, the only element available in QUCS is a lossless transmission line. It cannot simulate any losses, such as the series resistance, which causes the TDR response to increase down the line.

19.10 If a discontinuity is a dip in the impedance profile, what two possible circuit elements could you use to model it? Which one should you use?

If there is a dip in the impedance profile, this could be a capacitor or a low-impedance transmission line. The specific model you use depends on how the answer you are looking for.

If you know it is a transmission line segment and know either the characteristic impedance or time delay, you can hack the other term. If you need an answer in terms of the capacitance, then a capacitor is the element to use.

19.11 What is a mink hole? How do they arise and why should they be avoided?

A mink hole is a path that takes us far away from our goal but feels really good while we are following it. It is easy to get caught up in making small changes until we get an ever-more-perfect match between the simulation and measurement. We always have to ask, Are we done? There may be more important problems to go after.

19.12 When measuring the TDR response from a very low impedance transmission line, open at the far end, why does the far end of the transmission line not appear as an open in the TDR impedance profile?

The TDR response on the other side of a low-impedance interconnect open at the far-end should look an open. But the masking from the low-impedance section will make its far end look less than an open.

This masking is an effect you cannot interpret from the front screen. It is only possible to determine what is real and what is an artifact from masking by hacking the interconnect.

19.13 What is the difference in capacitor and inductor values of a SMT resistor compared to an axial lead resistor? How does this support the idea to use SMT devices in high speed applications and avoid axial lead devices?

A discrete resistor will look like an inductor and a capacitor. The pads look capacitive and the resistor body looks inductive.

Using hacking, we can extract a simple C-L model of a resistor element.

Generally, an axial lead resistor will have an inductance and capacitance at least 5x what a surface-mount resistor has. This is why at high frequency, always use a surface-mount component. Otherwise, its parasitics will distort the signals at the higher frequency and not be a very effective termination.

References

Bogatin, Eric, *Signal and Power Integrity- Simplified*, 3rd edition, Prentice Hall, 2018

Hall, S. and Heck, H, *Advanced Signal Integrity for High-Speed Digital Designs* (Wiley - IEEE) 2011

Hall, S., Hall, G, and McCall, J, *High-Speed Digital System Design: A Handbook of Interconnect Theory and Design Practices*, IEEE-Wiley, 2008

Paul, Clayton, *Analysis of Multiconductor Transmission Lines*, 2nd Edition, Wiley-IEEE Press, 2007

Paul, Clayton, *Transmission Lines in Digital and Analog Electronic Systems: Signal Integrity and Crosstalk*, 1st edition, Wiley, 2010

Resso, Mike and Bogatin, Eric, *Signal Integrity Characterization Techniques*, Addie Rose Press, 2019

Thierauf, Stephen, *High-Speed Circuit Board Signal Integrity, Second Edition*, Artech House, 2017.

About the Author

Eric Bogatin is currently a Signal Integrity Evangelist with Teledyne LeCroy and the Dean of the Teledyne LeCroy Signal Integrity Academy, at www.beTheSignal.com . Additionally, he is an adjunct professor at the University of Colorado - Boulder in the ECEE dept, and technical editor of the Signal Integrity Journal.

Eric received his BS in physics from MIT in 1976 and MS and PhD in physics from the University of Arizona in Tucson in 1980. He has held senior engineering and management positions at Bell Labs, Raychem, Sun Microsystems, Ansoft and Interconnect Devices. He has written seven technical books in the field and presented classes and lectures on signal integrity worldwide.

Eric is a prolific author, writing engineering textbooks, project experiment books for makers and science fiction novels. In his free time, he is also an amateur astronomer and a member of the Longmont Astronomical Society and the Boulder Astronomy and Space Society.

Contact him at eric@EricBogatin.com

www.ingramcontent.com/pod-product-compliance
Lightning Source LLC
Chambersburg PA
CBHW072250210326
41458CB00073B/943